Annals of Mathematics Studies
Number 66

ADVANCES IN THE THEORY OF RIEMANN SURFACES

Proceedings of the 1969 Stony Brook Conference

EDITED BY

LARS V. AHLFORS
ROBERT C. GUNNING
LIPMAN BERS
IRWIN KRA
HERSHEL M. FARKAS
HARRY E. RAUCH

PRINCETON UNIVERSITY PRESS
AND
UNIVERSITY OF TOKYO PRESS

PRINCETON, NEW JERSEY
1971

LC Card: 72-121729
ISBN: 0-691-08081-x
AMS 1968: 3001

Published in Japan exclusively by the
University of Tokyo Press;
in other parts of the world by
Princeton University Press

Printed in the United States of America

PREFACE

The Stony Brook Conference (June 1969) was a gathering of mathematicians working in certain inter-related fields: quasiconformal mappings, moduli, theta-functions, Kleinian groups. All of these fields belong directly or indirectly to the theory of Riemann surfaces. The first such gathering was the conference at Tulane University in May, 1965, sponsored by the Air Force Office of Scientific Research. The Tulane proceedings were published informally and are not generally available. This may have been a mistake, and it was decided not to repeat the mistake after Stony Brook.

The present volume contains all but two of the papers read at the conference, as well as a few papers and short notes submitted afterwards. We hope that it reflects faithfully the present state of research in the fields covered, and that it may provide an access to these fields for future investigations.

In the name of all participants, we thank the State University of New York at Stony Brook for its hospitality, the Office of Naval Research for the financial support which made the conference possible, and Princeton University Press for assuring rapid publication of the results.

The Editors

CONTENTS

ADVANCES IN THE THEORY OF RIEMANN SURFACES

SOME REMARKS ON KLEINIAN GROUPS

Willaim Abikoff

In this note, we will construct four Kleinian groups. The first is finitely generated and possesses limit points which are not in the boundary of any component of the ordinary set of the group. The construction yields a counter-example to the following assertion of Lehner [1]: If G is a discontinuous group with limit set Σ and ordinary set Ω, then $\Sigma = \cup \operatorname{Bd}(\Omega_i)$ where the Ω_i are the components of Ω. The remaining examples are of infinitely generated groups whose limit set has positive area. Each example will show that a statement, either proved or believed to be true for finitely generated groups, is not true for infinitely generated groups.

The First Group

We construct a group G_1 having a limit point which does not lie in the boundary of any component of the ordinary set of the group.

Let $G_1{}'$ be any finitely generated Fuchsian group of the first kind whose elements all possess isometric circles. The Ford fundamental region then has two components R_b and R_∞ - the first inside the principle circle C and the second a neighborhood of ∞. Choose a Möbius transformation T with the following properties:

a) the fixed points of T are x and x', $x \in \operatorname{Int} R_b$ and $x' \in \operatorname{Int} R_\infty$,

b) the isometric circle of T, $I(T) \subset \operatorname{Int} R_b$,

c) the isometric circle of T^{-1}, $I(T^{-1}) \subset \operatorname{Int} R_\infty$ and ∞ is in the exterior of $I(T^{-1})$.

The free product G_1 of $G_1{}'$ and the group generated by T is discontinuous, Furthermore, x is in the limit set of G_1 but is not in the boundary of any component of the ordinary set of G_1. To see this, suppose x is in the boundary of Ω_1, a component of $\Omega(G_1)$. Then there is a point $z \in \Omega_1$, $|z-x| = \alpha > 0$. But for sufficiently large n, $T^n(C)$, which is a subset of Σ, will separate z from x. Hence the component containing z cannot have x as a boundary point.

The Second Group

It is a conjecture of several years standing that the limit set of a finitely generated Kleinian group has zero area. We first recall that an Osgood curve is a Jordan curve of positive area; a construction is described in Osgood [2]. We will construct an infinitely generated Kleinian group whose limit set and Ford fundamental region both contain two Osgood curves. The following statement of Ahlfors [3] is thus shown not be extended to infinitely generated groups: if G is a finitely generated Kleinian group, then the portion of the limit set of G lying in the exterior of all isometric circles has area zero.

Let C and C′ be disjoint Osgood curves with disjoint interiors, and such that one may be obtained from the other by translation. By Schoenflies' version of the Jordan curve theorem, there exist homeomorphisms of the plane f and f′ which map U, the open disc, onto the interiors of C and C′ respectively and map the unit circle onto C and C′ respectively. Let $\{U_i\}$ be a filling up of the unit disc by disjoint discs such that for every point on the unit circle, there exists a subsequence (U_i') of (U_i) which converges to that point. Let x_i and x_i' be the images of the center of U_i under f and f′ respectively. There exist discs C_i and C_i' contained in $f(U_i)$ and $f'(U_i)$ about x_i and x_i'. Further, every point on the curves C and C′ is a limit of some subsequence of the C_i or the C_i'. Pick a Möbius transformation T_i whose isometric circle is C_i and whose inverse has isometric circle C_i'.

Let G_2 be the group generated by the T_i. G_2 is discontinuous in the common exterior of C and C′, since that set is contained in the Ford fundamental region of G_2. Furthermore, the isometric circles of the elements of G_2 accumulate at every point of C and C′; hence $C \cup C' \subset \Sigma$, where

Σ is the limit set of G_2. We have shown that there exists a Kleinian group whose limit set has positive area.

The Third Group

Our first construction showed that $N(G) = \Sigma(G) - \cup \operatorname{Bd}(\Omega_i)$ need not be void even for finitely generated groups. We now show that for infinitely generated groups $N(G)$ may have positive area. Let C and C' be as in the previous example. We consider four sequences (J_i), (J_i'), (K_i) and (K_i') of Jordan curves having the following properties:

a) all curves are disjoint from each other and from C and C',

b) $J_i \subset \operatorname{Int} J_{i+1} \subset \operatorname{Int} C \subset \operatorname{Int} K_{i+1} \subset \operatorname{Int} K_i$,
$J_i' \subset \operatorname{Int} J_{i+1}' \subset \operatorname{Int} C' \subset \operatorname{Int} K_{i+1}' \subset \operatorname{Int} K_i'$

c) the sequences (J_i) and (K_i) converge uniformly to C and the sequences (J_i') and (K_i') converge uniformly to C' (i.e., (J_i) converges uniformly to C if for each positive ϵ, there is an i sufficiently large so that

$$\max(|x - J_i|, |z - C|) < \epsilon \text{ for each } x \text{ on } C \text{ and } z \text{ on } J_i).$$

Using the techniques of the previous construction, we get a group G_3 having each curve in the four sequences in the limit set of the group, and by the same argument as in the first construction, $C \cup C' \subset N(G_3)$. Since both C and C' have positive area, we have constructed a group for which $N(G)$ has positive area.

The Fourth Group

Maskit [4] has shown that every finitely generated Kleinian group G leaving a Jordan curve invariant, that is every quasi-Fuchsian group, is the quasiconformal deformation of a Fuchsian group. We now exhibit an infinitely generated quasi-Fuchsian group which is not the quasiconformal deformation of a Fuchsian group. We construct a group G_4 whose limit set Σ is an Osgood curve C. Thus G_4 leaves C invariant. If G_4 were the quasiconformal deformation of a Fuchsian group, C would have zero area

(Ahlfors [5], p. 33) which it does not. In the previous two constructions the existence of Osgood curves was sufficient for our purposes, but we now need some specific properties of Osgood's example.

Osgood constructs not a Jordan curve but a Jordan arc of positive area. With the addition of two Euclidean line segments, the arc becomes a Jordan curve J. J is then the closure of a countable number of Euclidean line segments, I_k, which are pairwise disjoint. The measure of J is supported on $J - \cup I_k$, which is a measurable set. The idea of the example is to construct a group whose limit set is a curve passing through each point of $J - \cup I_k$. The curve J cannot be the limit set of a Kleinian group since as a function of a parameter t, a subarc of J is differentiable. In this case J would have to be a circle (Appell-Goursat [6], p. 82).

We first construct a reflection group G having the following properties:

a) the centers of the generating circles lie in $\cup I_k$,
b) the generating circles have disjoint interiors,
c) a generating circle intersects exactly two other generating circles, each in a single point.
d) except for its endpoints, each I_k is covered by the closed discs corresponding to the generating circles whose centers lie on I_k,
e) as we approach an endpoint of I_k, the radii of the generating circles tend to zero,
f) no circle whose center lies on I_k intersects a circle whose center lies on $I_{k'}$, for $k \neq k'$.

We let G_4 be the subgroup of G consisting of the orientation preserving motions. It is of index two in G and therefore has the same limit set as G. The interior of each generating circle contains one (in fact many) limit points of G_4 hence $\Sigma(G_4) \supset J - \cup I_k$ and $\Sigma(G_4)$ has positive area. G_4 is discontinuous, so we must only show that $\Sigma(G_4)$ is a Jordan curve. Let K be an arbitrary generating circle whose center lies on I_k. Then by considering reflections in K and the generating circles tangent to K, we note that $K \cap I_k \subset \Sigma(G_4)$. Further, no other points of $\Sigma(G_4)$ lie on K. $K \cap I_k$

consists of exactly two points a and b. It suffices to show that the part of $\Sigma(G_4)$ contained in the closed disc bounded by K is a simple arc. The argument is tedious but straightforward. One looks at the a's and b's corresponding to all other generating circles, reflects them into the inside of K and uses the images to define densely the arc between a and b. The arc carries an induced parametrization. Simplicity of the arc is guaranteed by an argument similar to the nexted interval theorem.

REFERENCES

[1] J. Lehner, Discontinuous Groups and Automorphic Functions, *Amer. Math. Soc.*, 1964 p. 105.

[2] W. F. Osgood, A Jordan curve of positive area, *Trans. Amer. Soc.*, 4 (1903), p. 107-112.

[3] L. V. Ahlfors, Some Remarks on Kleinian Groups, *Proceedings of the Conference on Quasiconformal Mappings, Moduli and Discontinuous Groups*, Tulane University, 1965.

[4] B. Maskit, On Boundaries of Teichmüller Spaces and on Kleinian Groups, II, *Ann. of Math.*, (to appear).

[5] L. V. Ahlfors, *Lectures on Quasiconformal Mappings*, Von Nostrand, 1966.

[6] P. Appell and É. Goursat, *Theorie des fonctions algébriques et de leurs intégrales*, Vol. 2, Gauthiers-Villars, 1930.

Bell Telephone Laboratories
and Polytechnic Institute of Brooklyn

VANISHING PROPERTIES OF THETA FUNCTIONS FOR ABELIAN COVERS OF RIEMANN SURFACES

(unramified case)

Robert D. M. Accola [1]

1. INTRODUCTION. The vanishing properties of hyperelliptic theta functions have been known since the last century [3]. Recently, Farkas [1] discovered special vanishing properties for theta functions associated with surfaces which admit fixed-point free automorphisms of period two. The author has discovered other vanishing properties for special surfaces admitting abelian automorphism groups of low order. The purpose of this report is to give a partial exposition of a theory that will subsume most of the above cases in a general theory. Due to limitations of time and space, a full exposition must be postponed.

Let W_1 be a closed Riemann surface of genus p_1, $p_1 \geq 2$, admitting a finite abelian group of automorphisms, G. The space of orbits of G, W/G ($= W_0$), is naturally a Riemann surface so that the quotient map, $\underline{b}$, is analytic. In this report we shall develop the theory in the case where no element of G other than the identity has a fixed point; that is, the map $\underline{b} : W_1 \to W_0$ is without ramification. We shall, however, state some theorems in a more general context especially when proofs are omitted.

[1] The research for this report has been carried on during the last several years during which the author received support from several sources. 1) Research partially sponsored by the Air Force Office of Scientific Research, Office of Aerospace Research, United States Air Force, under AFOSR Grant No. AF-AFOSR-1199-67. 2) National Science Foundation Grant GP-7651. 3) Institute for Advanced Study Grant-In-Aid.

II. REMARKS ON GENERAL COVERINGS. Let $\underline{b} : W_1 \to W_0$ be an arbitrary n-sheeted ramified covering of closed Riemann surfaces of genera p_1 and p_0 respectively. Let M_1 be the field of meromorphic functions on W_1 and let M_0 be the lifts, via $\underline{b}$. of the field of meromorphic functions on W_0. Then M_0 is a subfield of M_1 of index n. We now define an important abelian group, A, as follows:

Definition: $A = \{f \epsilon M_1^* \mid f^n \epsilon M_0^*\}/M_0^*$. [2]

Now let M_A be the maximal abelian extension of M_0 in M_1 .

LEMMA 1: A is isomorphic to the (dual of the) Galois group of M_A over M_0.

PROOF: (omitted). A proof in the case where $M_1 = M_A$ will follow in Section V.

Now, fix a point in W_1, z_1, and let $z_0 = \underline{b}(z_1)$. Fix canonical homology bases in W_1 and W_0 and choose bases for the analytic differentials dual to these homology bases. Thus maps u_1 and u_0 from W_1 and W_0 into their Jacobians, $J(W_1)$ and $J(W_0)$, are defined:

$$\begin{array}{ccc} W_1 & \xrightarrow{\quad u_1 \quad} & J(W_1) \\ \Big\downarrow \underline{b} & & \\ W_0 & \xrightarrow{\quad u_0 \quad} & J(W_0) \ . \end{array}$$

The maps u_1 and u_0 are extended to divisors in the usual way. A map $\underline{a}$ [3] is now defined from divisors on W_0 to those of W_1 as follows: for $x_0 \epsilon W_0$, $\underline{a}x_0$ is the inverse image of x_0 under $\underline{b}$ with branch points counted according to multiplicity. Thus $\underline{a}x_0$ always has degree n. $\underline{a}$ is extended by

[2] If K is a field, K* will stand for the multiplicative groups of non-zero elements in K.

[3] The symbol $\underline{a}$ will be used consistently to denote homomorphisms from abelian groups associated with W_0 into the corresponding abelian groups associated with W_1. The particular group will be clear from the context.

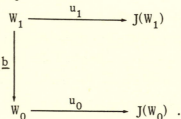

linearity to arbitrary divisors on W_0. Now we define a map from $J(W_0) \to J(W_1)$, again denoted by $\underline{a}$ as follows: if D_0 is a divisor on W_0 of degree zero, then $\underline{a} u_0(D_0) = u_1(\underline{a} D_0)$. $\underline{a}$ is easily seen to be a homomorphism. Let M_{UA} be a maximal unramified abelian extension of M_0 in M_1; thus $M_0 \subset M_{UA} \subset M_A \subset M_1$.

LEMMA 2: The kernel of $\underline{a}: J(W_0) \to J(W_1)$ is isomorphic to the Galois group of M_{UA} over M_0.

PROOF: (omitted). A proof in the case when $M_{UA} = M_1$ will follow in Section VI.

With the homology bases and the dual bases of analytic differentials chosen, let $(\pi i E, B_0)^{p_0 \times 2p_0}$ and $(\pi i E, B_1)^{p_1 \times 2p_1}$ be the corresponding period matrices where E is the appropriate identity matrix. Finally let $\theta[\chi_0](u; B_0)$ and $\theta[\chi_1](u; B_1)$ be the corresponding first order theta functions with arbitrary characteristic.

LEMMA 3: For any characteristic χ_1 there is an exponential function E(u) so that $E(u)\theta[\chi_1](\underline{a}u; B_1)$, as a multiplicative function on $J(W_0)$, is an nth order theta function.

PROOF: (omitted). The proof is an immediate adaptation of the simplest parts of transformation theory.

III. Resume of the Riemann Vanishing Theorem.

The proofs of the results in this report depend on Riemann's solution to the Jacobi inversion problem. We summarize here those portions of the theory that will be needed later.[4]

Let W be a closed Riemann surface of genus p, $p \geq 1$, let a canonical homology basis be chosen, let a dual basis of analytic differentials be chosen, let a base point be chosen, and let u be the map of W into J(W).

---

[4] The material in this section is covered in Krazer [3]. For a complete and more modern treatment see Lewittes [4].

Riemann's theorem asserts the existence of a point K in J(W) so that if we choose any $e \in J(W)$, then there is an integral divisor D on W of degree p so that

$$u(D) + K \equiv e (\bmod J(W)) .$$

If $\theta(e) \neq 0$, then D is unique. If $\theta(e) = 0$, then the above equation can be solved with an integral divisor of degree $p-1$. Moreover, in this latter case, the order of vanishing of $\theta(u)$ at e equals i(D), the index of speciality of D. (By the Riemann-Roch theorem, i(D) equals the number of linearly independent meromorphic functions which are multiples of $-D$ since the degree of D is $p-1$.) Moreover, $\theta(u(D)+K) = 0$ whenever D is an integral divisor of degree at most $p-1$. Finally, if D is a canonical divisor, then

$$u(D) \equiv -2K \ (\bmod J(W)) .$$

IV. ARIBITRARY UNRAMIFIED COVERINGS. For this section assume $\underline{b}: W_1 \to W_0$ is unramified but otherwise arbitrary. Continue the previous notation and let θ_j be the theta-divisor in $J(W_j)$ for $j = 0, 1$.

LEMMA 4: There exists a half-period, e_1, in $J(W_1)$ so that

$$\underline{a}\,\theta_0 \subset \theta_1 + e_1 . \tag{1}$$

(e_1 depends only on the canonical homology bases chosen.)

PROOF: Let D_0 be an integral canonical divisor on W_0 of degree $2p_0 - 2$. Then $D_0 - (2p_0 - 2) z_0$ is a divisor of degree zero whose image under u_0 is $u_0(D_0)$ since $u_0(z_0) \equiv 0$. $\underline{a}D_0$ is canonical in W_1 so we have

$$\begin{aligned} \underline{a}(-2K_0) &\equiv \underline{a}\,u_0(D_0) \\ &\equiv \underline{a}\,u_0(D_0 - (2p_0 - 2) z_0) \\ &\equiv u_1(\underline{a}D_0 - (2p_0 - 2)\underline{a}z_0) \\ &\equiv -2K_1 - (2p_0 - 2)\, u_1(\underline{a}z_0) . \end{aligned}$$

Dividing by two gives

(2) $$\underline{a}K_0 \equiv K_1 + (p_0 - 1)\,u_1(\underline{a}\,z_0) + e_1$$

where $2e_1 \equiv 0$.

If $g_0 \in \theta_0$, then there is a divisor D_0 of degree $p_0 - 1$ on W_0 so that

$$u_0(D_0) + K_0 \equiv g_0 \quad .$$

Thus

$$\underline{a}\,u_0(D_0 - (p_0 - 1)z_0) + \underline{a}K_0 \equiv \underline{a}g_0$$

or

$$u_1(\underline{a}D_0) - (p_0 - 1)\,u_1(\underline{a}z_0) + \underline{a}K_0 \equiv \underline{a}g_0 \quad .$$

By formula (2) it follows that

$$\underline{a}g_0 \equiv u_1(\underline{a}D_0) + K_1 + e_1 \quad .$$

Since the degree of $\underline{a}D_0$ is $n(p_0 - 1) = p_1 - 1$ we have

$$u_1(\underline{a}D_0) + K_1 \in \theta_1 \quad .$$

Thus

$$\underline{a}g_0 \in \theta_1 + e_1 \quad .$$

q.e.d.

V. ABELIAN COVERS.[5] In this section assume $\underline{b} : W_1 \to W_0$ is a possibly ramified abelian cover; i.e., $W_0 = W_1/G$ where G is an abelian group of automorphisms of W_1 whose elements may have fixed points. Let R be the set of characters of G; i.e., the set of homomorphisms of G into the multiplicative group of complex numbers of modulus one. Since G is a finite group, R is isomorphic to G although there is no canonical isomor-

[5] The author wishes to express his thanks to Professor M. S. Narasimhan for many valuable discussions concerning the material of this report, especially this section.

phism. The group A is, however, canonically isomorphic to R. For suppose f is meromorphic on W_1 and f^n is a function lifted from W_0. Then the divisor of f is invariant under G. Thus if $T \in G$, then $f \circ T = \chi_f(T) f$ where χ_f is easily seen to be a character. If f and g yield the same character, then f/g gives the identity character and so lies in M_0. Thus the map $f \to \chi_f$ is an isomorphism of A into R. That this map is onto is seen by examining, for each $\chi \in R$, the cyclic extension of M_0 given by the fixed field for kernel χ. This completes a proof of Lemma 1 when $M_1 = M_A$. In the above situation where we have a function f whose divisor is invariant under G, we shall say that f is *associated* with the character χ_f.

Thus the field extension M_1 over M_0 can have as a vector space basis functions corresponding to each character of R. If f is an arbitrary function, then let $f_\chi = n^{-1} \Sigma_{T \in G} \chi(T^{-1}) f \circ T$. Then $f = \Sigma_{\chi \in R} f_\chi$ and the f_χ's which are not zero are linearly independent since they are associated with different characters.

VI. Unramified Abelian Covers. In this section assume $\underline{b}: W_1 \to W_0$ is an unramified abelian cover. In this context we give a proof of Lemma 2; that is, we now show that the kernel of $\underline{a} : J(W_0) \to J(W_1)$ is isomorphic to A.

For the proof define a map from A into ker $\underline{a}$ as follows. If $f \in M_1^*$ and $f^n \in M_0^*$, then the divisor of f, (f), is invariant and so $(f) = \underline{a} D_0$ where D_0 is a divisor on W_0 of degree zero. Let $\omega_f = u_0(D_0)$. Then $\underline{a}\omega_f \equiv \underline{a} u_0(D_0) \equiv u_1(\underline{a} D_0) \equiv 0$, and so $\omega_f \in$ ker $\underline{a}$. The map $f \to \omega_f$ is easily seen to be well-defined on A and one-to-one. To show the map is onto suppose $\underline{a}\omega \equiv 0$. Then $\omega = u_0(D_0)$ where D_0 is a divisor of degree zero on W_0. Since $\underline{a} D_0 \equiv 0$, suppose $(f) = \underline{a} D_0$. Since $\underline{a} D_0$ is invariant under G, $f \circ T = \chi_f(T) f$ for some character χ_f. Consequently, $f^n \in M_0$. This completes the proof of Lemma 2 in the unramified abelian case.

Let us index ker $\underline{a}$ by R rather than A; that is, ker $\underline{a} = \{\omega_\chi \mid \chi \in R\}$ where ω_χ is defined as follows. If $u_0(D_0) = \omega_\chi$ and $(f) = \underline{a} D_0$, then f is associated with χ. The method of indexing, $\chi \to \omega_\chi$ is an isomorphism of groups.

The reader is reminded that throughout this discussion canonical homology bases on W_0 and W_1 are fixed and all results are related to these bases. Thus the bases of analytic differentials are determined as are the theta functions and the half-period e_1 of Lemma 4. Base points are also fixed and so maps u_0 and u_1 are defined, but these are in a sense auxiliary devices and the base points do not enter into the statements of the final results.

At this point let us introduce a convenient abuse of notation. If $\theta(u;B)$ is a theta function and $\tau = \pi i h + Bg$, where g and h are real p-vectors, then there is an exponential function E(u) so that

$$\theta(u+\tau;B) = E(u)\theta[{}^{g}_{h}](u;B) .$$

In this context write $\theta[\tau](u)$ for the usual $\theta[{}^{g}_{h}](u;B)$. This notation will be extremely convenient and will lead to no confusion provided the canonical homology bases remain fixed.

A main result of this report is the following. By Lemma 3, there is an exponential, E(u), so that $E(u)\theta[e_1](\underline{a}u;B_1)$ is an nth order theta function with respect to $(\pi i E, B_0)^{p_0 \times 2p_0}$. Then there is a constant $\ell \neq 0$ so that

$$E(u)\theta[e_1](\underline{a}u;B_1) = \ell \prod_{\chi \epsilon R} \theta[\omega_\chi](u;B_0) . \tag{3}$$

That this is a plausible result can be seen from Lemma 4. If $\omega_\chi \epsilon \ker \underline{a}$, then

$$\underline{a}(\theta_0 - \omega_\chi) \subset \theta_1 + e_1 .$$

Thus if $g_0 \epsilon \theta_0 - \omega_\chi$, then $\underline{a}g_0 \epsilon \theta_1 + e_1$; that is, if $\theta[\omega_\chi](g_0;B_0)$ vanishes, then so does $\theta[e_1](\underline{a}g_0;B_1)$. Thus the two sides of formula (3) are n^{th} order theta functions and whenever the right-hand side vanishes, so does the left-hand side. Both sides have rational characters, half-integer characters in fact. That this is true for the left-hand side follows from the generalization of the usual transformation theory mentioned after the statement of Lemma 3. For the right-hand side it is true because the sum of the elements in an abelian group is always an element of order two. Formula (3) would be proven if we knew that ℓ was not zero, but this does not seem immediate. The final theorem of this report among other things establishes formula (3) with an $\ell \neq 0$.

THEOREM: Fix $g_0 \in J(W_0)$. Suppose that $\theta[e_1](u; B_1)$ vanishes to order N at $\underline{a}g_0$. Suppose for each $\chi \in R$, $\theta[\omega_\chi](u; B_0)$ vanishes to order N_χ at g_0. Then $N = \Sigma_{\chi \in R} N_\chi$.

PROOF: Let χ_0 be the identity character. Denote the characters for which N_χ is positive by $\chi_1, \chi_2, \ldots, \chi_s$. χ_0 may or may not be among these characters. For convenience write N_j for N_{χ_j} and ω_j for ω_{χ_j} for $j = 1, 2, \ldots, s$. Then $\theta[\omega_j](u; B_0)$ vanishes to order N_j at g_0. Thus there is an integral divisor D_j of degree $p_0 - 1$ on W_0 so that

$$g_0 + \omega_j \equiv u_0(D_j) + K_0 . \tag{4}$$

Let L_j be the linear space of meromorphic functions which are multiples of $-D_j$. Then $N_j = \dim L_j$. Now apply $\underline{a}$ to formula (4). Since $\underline{a}\omega_j \equiv 0$ we have

$$\underline{a}g_0 = \underline{a}u_0(D_j - (p_0 - 1)z_0) + \underline{a}K_0 \quad .$$

By formula (2) we have

$$\underline{a}g_0 \equiv u_1(\underline{a}D_j) - (p_0 - 1)u_1(\underline{a}z_0) + K_1 + (p_0 - 1)u_1(\underline{a}z_0) + e_1$$

or
$$\underline{a}g_0 \equiv u_1(\underline{a}D_j) + K_1 + e_1 \quad .$$

Thus all the divisors $\underline{a}D_j$ are linearly equivalent on W_1, $j = 1, 2, \ldots, s$. Suppose h_j is a function on W_1 whose divisor is $-\underline{a}D_1 + \underline{a}D_j$. If $\underline{a}L_j$ is the linear space of lifts to W_1 of functions in L_j and $h_j\underline{a}L_j$ represents those lifted functions multiplied by h_j, then the families, $\underline{a}L_1$, $h_2\underline{a}L_2, \ldots,$ $h_s\underline{a}L_s$ represent $\Sigma_{j=1}^{s} N_j$ linearly independent multiples of $-\underline{a}D_1$. Thus $N \geq \Sigma_{j=1}^{s} N_j$. If $N = 0$ we have equality.

For the opposite inequality suppose $\theta[e_1](u; B_1)$ vanishes to order $N > 0$ at $\underline{a}g_0$. Then there is an integral divisor D_0 of degree p_0 on W_0 and an integral divisor D_1 of degree $p_1 - 1$ on W_1 so that

(5) $$g_0 \equiv u_0(D_0) + K_0 \quad (\text{mod } J(W_0))$$

(6) $$\underline{a}g_0 \equiv u_0(D_1) + K_1 + e_1 \quad (\text{mod } J(W_1)) \ .$$

Applying $\underline{a}$ to formula (5) gives

$$\begin{aligned} \underline{a}g_0 &\equiv \underline{a}u_0(D_0 - p_0 z_0) + \underline{a}K_0 \\ &\equiv u_1(\underline{a}D_0) - p_0 u_1(\underline{a}z_0) + K_1 + (p_0 - 1)u_1(\underline{a}z_0) + e_1 \\ &\equiv u_1(\underline{a}D_0) - u_1(\underline{a}z_0) + K_1 + e_1 \ . \end{aligned}$$

By formula (6)

$$u_1(D_1) \equiv u_1(\underline{a}D_0 - \underline{a}z_0)$$

or

$$D_1 \equiv \underline{a}(D_0 - z_0) \ .$$

Let f be a multiple of $-\underline{a}(D_0 - z_0)$. For each $\chi \in R$ let $f_\chi = n^{-1} \Sigma_{T \in G} \chi(T^{-1}) f \circ T$. Since $f \circ T$ is a multiple of the invariant divisor $-\underline{a}(D_0 - z_0)$ so is f_χ. But the divisor of f_χ, (f_χ), is invariant. Consequently, if $f_\chi \neq 0$, then there exists an integral divisor D_χ of degree $p_0 - 1$ on W_0 so that

(7) $$(f_\chi) = \underline{a}(D_\chi - D_0 + z_0) \ .$$

For each $\chi \in R$, let N'_χ be the dimension of the space of multiples on W_1 of $-\underline{a}(D_0 - z_0)$ which correspond to χ. Let N''_χ be the dimension of the space of multiples on W_0 of $-D_\chi$. Note that $N = \Sigma_{\chi \in R} N'_\chi$. If

$$(g_\chi) = \underline{a}(D'_\chi - D_0 + z_0)$$

where g is another multiple of $-\underline{a}(D_0 - z_0)$ corresponding to χ, then $(f_\chi/g_\chi) = \underline{a}(D'_\chi - D_\chi)$. Since f_χ/g_χ corresponds to the identity character, D'_χ is linearly equivalent to D_χ. Consequently, the multiples of $-\underline{a}(D_0 - z_0)$ corresponding to χ arise from divisors of degree $p_0 - 1$ on W_0 linearly equivalent to D_χ. It follows that $N'_\chi = N''_\chi$.

To complete the proof it now suffices to show that the χ's for which $N'_\chi \neq 0$ occur among the characters $\chi_1, \chi_2, \ldots, \chi_s$ mentioned at the beginning of the proof. Suppose then $N'_\chi \neq 0$ and $f_\chi \in L_\chi$, $f_\chi \neq 0$. Then using the notation of formula (7) we have

$$u_0(D_\chi - D_0 + z_0) \equiv \omega_\chi$$

or
$$u_0(D_0) + \omega_\chi + K_0 \equiv u_0(D_\chi) + K_0$$

or
$$g_0 + \omega_\chi \equiv u_0(D_\chi) + K_0 \ .$$

Since D_χ is of degree $p_0 - 1$ and is integral, $\theta[\omega_\chi](u\,;B_0)$ vanishes at g_0 to order $N''_\chi > 0$. Thus any χ for which $N'_\chi \neq 0$ belongs to the χ_j's, $j = 1, 2, \ldots, s$. Thus $N \leq \Sigma_{j=1}^{s} N_j$.

q.e.d.

Formula (3) now follows since we can find values of g_0 for which $\theta[\omega_\chi](g_0\,;B_0) \neq 0$ for all χ. Thus $\theta[e_1](\underline{a}u, B_1)$ cannot be identically zero and so $\ell \neq 0$.

VII. APPLICATIONS. If G is a cyclic group, then e_1 of Lemma 4 can be determined. For $n = 2$ this has been done by Farkas, Rauch and Fay. Formula (3) for $n = 2$ seems to have been known to Riemann.[6] For $n = 2$ the formula is equivalent to the relations between θ and η constants derived by Farkas [2], and these in turn, as shown by Farkas, yield the Schottky-Jung relations. Also, the vanishing properties discovered by Farkas [1] follow immediately from the formula and theorem. Thus the

6 See Riemann [6], Nachträge, p. 108.

theorem can be viewed as a generalization of the work of Farkas and Rauch. One can hope that applying the methods of Farkas to formula (3) for other cyclic covers $(n > 2)$ will yield further relations between more general types of theta constants.

Another type of application of the theorem is as follows. It is known that on the surface of genus p, p odd, the vanishing of the theta function to order $(p+1)/2$ at some half-period is equivalent to the surface being hyperelliptic.[7] This is an immediate consequence of Riemann's solution of the Jacobi inversion problem and Clifford's theorem. We now use this face to prove the following. Suppose W_0 is a Riemann surface of genus p_0, $p_0 \geq 4$ and p_0 even. Suppose that the theta function vanishes at two distinct half-period to order $p_0/2$. Then W_0 is hyperelliptic.[8]

An outline of the proof is as follows.[9] Suppose that for a given canonical holology bases on W_0, $\theta[\sigma](u\,;B_0)$ and $\theta[\tau](u\,;B_0)$ vanish to order $p_0/2$ at $u = 0$. The half-period $(\sigma\tau)$ defines a smooth two sheeted cover, W_1, of W_0. By suitable choice of canonical homology bases on W_0 and W_1 we can arrange that $(\sigma\tau) = \left(\begin{smallmatrix} 0 & 0 & 0 & \cdots & 0 \\ 1 & 0 & 0 & \cdots & 0 \end{smallmatrix}\right)$ and $(e_1) = \left(\begin{smallmatrix} 0 & 0 & 0 & \cdots & 0 \\ 1 & 0 & 0 & \cdots & 0 \end{smallmatrix}\right)$ where e_1 is the half-period of Lemma 4. By suitable relabeling, if necessary, it follows that $[\sigma] = \left[\begin{smallmatrix} 0 & \varepsilon \\ 0 & \varepsilon' \end{smallmatrix}\right]$ and $[\tau] = \left[\begin{smallmatrix} 0 & \varepsilon \\ 1 & \varepsilon' \end{smallmatrix}\right]$ where $\left[\begin{smallmatrix} \varepsilon \\ \varepsilon' \end{smallmatrix}\right]$ is a $p_0 - 1$ theta character whose parity is that of $p_0/2$. Then formula (3) applied to this situation yields

$$(8) \qquad E(u)\theta\left[\begin{smallmatrix} 0 & 0 & \cdots & 0 \\ 1 & 0 & \cdots & 0 \end{smallmatrix}\right](\underline{a}u\,;B_1) = \ell\,\theta\left[\begin{smallmatrix} 0 & 0 & \cdots & 0 \\ 0 & 0 & \cdots & 0 \end{smallmatrix}\right](u\,;B_0)\,\theta\left[\begin{smallmatrix} 0 & 0 & \cdots & 0 \\ 1 & 0 & \cdots & 0 \end{smallmatrix}\right](u\,;B_0)\ .$$

---

[7] For $p = 3$, the result is due to Riemann [6], Nachträge, p. 54.

[8] For $p \geq 8$, p even, Martens has proved that the vanishing of the theta function at *one* half-period to order $p/2$ suffices to insure hyperellipticity [5]. He also covers the $p = 6$ case. For $p = 4$ the result seems to be due to Weber [7].

[9] For the remainder of this report all theta characteristics will be half-integer characteristics.

Setting u equal to the half-period whose character is $\binom{0\ \varepsilon}{0\ \varepsilon'}$ shows that $\theta\left[\begin{smallmatrix}0 & \varepsilon & \varepsilon\\ 1 & \varepsilon' & \varepsilon'\end{smallmatrix}\right](u_1 ; B_1)$ vanishes to order p_0 at $u_1 = 0$. Since the genus of W_1 is $p_1 = 2p_0 - 1$ we have $p_0 = (p_1 + 1)/2$ and so W_1 is hyperelliptic. But then W_0 must also be hyperelliptic.

REFERENCES

[1] Farkas, H. M., Automorphisms of compact Riemann surfaces and the vanishing of theta constants. *Bulletin of the American Mathematical Society*, Vol. 73 (1967), pp. 231-232.

[2] ______, On the Schottky relation and its generalization to arbitrary genus.

[3] Krazer, A., *Lehrbuch der Thetafuncktionen*. B. G. Teubner, Leipzig, 1903.

[4] Lewittes, J., Riemann surfaces and the theta function. *Acta Mathematica*. Vol. 111 (1964), pp. 37-61.

[5] Martens, H. H., Varieties of special divisors on a curve. II. *Journal für die reine und angewandt Mathematik*. Vol. 233 (1968), pp. 89-100.

[7] Riemann, B., *Gesammelte mathematische Werke*. Dover, 1953.

[7] Weber, H., Ueber Gewisse in der Theorie der Abel'sehen Funktionen auftretende Ausnahmfalle. *Mathematische Annalen*. Vol. 13 (1878), pp. 35-48.

REMARKS ON THE LIMIT POINT SET OF A FINITELY GENERATED KLEINIAN GROUP

Lars V. Ahlfors

1. It is a conjecture of several years' standing that the limit point set of a finitely generated Kleinian group has areál measure zero at least under some restrictive assumptions. This question has proved to be very elusive and all attempts to find the answer have been abortive. The method I am going to describe is no exception. It leads deceptively close to a solution, but my efforts to push it through have again been futile.

I am nevertheless using this opportunity to publish some of my ideas on the subject that I believe to be independently interesting.

2. We use conventional notations: Γ is a Kleinian group, Λ the limit point set, Ω the set of discontinuity. dm denotes two-dimensional euclidean measure, and $d\omega$ is spherical area. The spherical (chordal) distance is denoted by $[z_1, z_2]$.

Let S be the Riemann sphere and let $A \in \Gamma$ act on $S^3 = S \times S \times S$ by $A(z_1, z_2, z_3) = (Az_1, Az_2, Az_3)$. We denote by S_0^3 the subset of S^3 on which all three coordinates are distinct. Then S_0^3 is invariant under the Möbius group, and there are no fixed points. We observe that the measure

$$dM = (|z_1 - z_2|\, |z_2 - z_3|\, |z_3 - z_1|)^{-2}\, dm(z_1) dm(z_2) dm(z_3) \tag{1}$$

is invariant and can be identified with the Haar measure. It can also be written in the form

$$dM = ([z_1, z_2][z_2, z_3][z_3, z_1])^{-2} d\omega(z_1) d\omega(z_2) d\omega(z_3) \ . \tag{2}$$

* This research was partially supported by the Air Force Office of Scientific Research under Contract No. F44620 69 C 0088.

THEOREM 1. Γ is properly discontinuous on S_0^3 in the sense that every point in S_0^3 has a neighborhood U whose images AU, $A \in \Gamma$, are disjoint.

Proof: We write $z = (z_1, z_2, z_3)$, $\zeta = (\zeta_1, \zeta_2, \zeta_3)$ and $d(z, \zeta) = \Sigma_1^3 [z_k, \zeta_k]$. For fixed $z \in S_0^3$, $d(z, Az) \to 0$ only if $A \to I$. Therefore $d_\Gamma(z) = \inf_{A \in \Gamma - \{I\}} d(z, Az)$ is positive. It is lower semicontinuous and has therefore a positive minimum on any compact subset of S_0^3. Given $\zeta \in S_0^3$ let d_0 be the shortest distance from ζ to $S^3 - S_0^3$. Denote by K the compact set in S_0^3 whose points have distance $\geq d_0/2$ from $S^3 - S_0^3$ and let δ be the minimum of $d_\Gamma(z)$ on K. Let U be a neighborhood of ζ whose diameter is $< \min(\delta, d_0/2)$. If $z \in U \cap AU$ for some $A \in \Gamma - \{I\}$, then $d(z, A^{-1}z) < \delta$ while $z \in K$. This contradicts the definition of δ. Hence $U \cap AU = \emptyset$.

3. We shall use the notation

$$[A'(z)] = \left(\frac{d\omega(Az)}{d\omega(z)}\right)^{1/2} = \frac{|A'(z)|(1+|z|^2)}{1+|Az|^2} \tag{3}$$

for the spherical magnification ratio. It is identically one if A is a rotation of the sphere. In all other cases $[A'(z)] > 1$ on an open cap C(A) that covers less than a hemisphere. We call it the *isometric cap* of A. C(A) and $C(A^{-1})$ are congruent. A maps C(A) on the complement of $C(A^{-1})^-$ and the complement of $C(A)^-$ on $C(A^{-1})$.

To avoid trivial complications we are going to assume that no A is a rotation. Then all the C(A) are defined. One proves exactly as in the case of Ford's isometric circles that the C(A) accumulate only toward Λ.

4. Our immediate aim is to find a fundamental set for the action on S^3. For this purpose we prove the following convergence theorem:

THEOREM 2. Let α, β, γ be positive numbers that satisfy $\alpha+\beta > \gamma$, $\beta+\gamma > \alpha$, $\gamma+\alpha > \beta$ and $\alpha + \beta + \gamma > 2$. Then the series

$$(4) \qquad \sum_{A \in \Gamma} [A'(z_1)]^{\alpha} [A'(z_2)]^{\beta} [A'(z_3)]^{\gamma}$$

converges almost everywhere on S^3, everywhere on Ω^3, and uniformly on every compact subset of Ω^3.

Proof: Suppose that a, b, c < 2 and $a + b + c < 4$. It is routine to show that

$$\int_S [z_1, z_2]^{-a} [z_3, z_1]^{-c} d\omega(z_1) = O(1 + [z_2, z_3]^{2-a-c})$$

and it follows easily that

$$\int_{S^3} [z_1, z_2]^{-a} [z_2, z_3]^{-b} [z_3, z_1]^{-c} d\omega(z_1) d\omega(z_2) d\omega(z_3) < \infty .$$

If $U \subset S^3$ is such that all the AU are disjoint we have consequently

$$\int_U \sum_{A \in \Gamma} \{[Az_1, Az_2]^{-a} [Az_2, Az_3]^{-b} [Az_3, Az_1]^{-c} d\omega(Az_1) d\omega(Az_2) d\omega(Az_3)\} < \infty .$$

By use of the identity

$$[Az, A\zeta] = [z, \zeta][A'(z)]^{1/2} [A'(\zeta)]^{1/2}$$

this can be rewritten as

$$\int_U [z_1, z_2]^{-a} [z_2, z_3]^{-b} [z_3, z_1]^{-c} \sum_A \{[A'(z_1)]^{2-\frac{a+c}{2}} [A'(z_2)]^{2-\frac{b+a}{2}} [A'(z_3)]^{2-\frac{c+b}{2}}\} \cdot$$

$$\cdot\, d\omega(z_1) d\omega(z_2) d\omega(z_3) < \infty \quad .$$

Now choose a, b, c so that the exponents are α, β, γ. The conditions are fulfilled, and we conclude that the series (4) converges almost everywhere in U. Because of Theorem 1 it follows that the convergence takes place almost everywhere in S_0^3, and S_0^3 differs from S^3 only by a null-set.

The stronger result for $(z_1, z_2, z_3) \in \Omega^3$ follows in familiar manner by subharmonicity. In our case the formula

$$\Delta \log [A'(z)] = 4(1+|z|^2)^{-2}(1-[A'(z)]^2)$$

shows that $\log[A'(z)]$ is subharmonic when $[A'(z)] \leqq 1$, that is to say outside of $C(A)$. The same is then true of every positive power $[A'(z)]^\alpha$. A point $(z_1, z_2, z_3) \in \Omega^3$ has a neighborhood $U = U_1 \times U_2 \times U_3$ such that each U_k meets only finitely many $C(A)$. The rest of the reasoning is obvious.

REMARK. If $m(\Lambda) = 0$ the first part of the theorem follows from the second. If $m(\Lambda) > 0$ it is important that we have convergence almost everywhere even if one or more of the z_k lie on Λ, and in particular, almost everywhere on Λ^3.

5. Let $S_{\alpha\beta\gamma}$ be the subset of S^3 on which $[A'(z_1)]^\alpha [A'(z_2)]^\beta [A'(z_3)]^\gamma < 1$ for all $A \in \Gamma - \{I\}$.

THEOREM 3. Almost every $(z_1, z_2, z_3) \in S^3$ is equivalent under Γ to exactly one point in $S_{\alpha\beta\gamma}$.

Proof: We use the convergence of (4). Except on a countable number of hypersurfaces the series has a term that is strictly larger than all the others. If this term corresponds to $A_0 \in \Gamma$ we have clearly

$$[A'(A_0z_1)]^\alpha [A'(A_0z_2)]^\beta [A'(A_0z_3)]^\gamma < 1 \quad \text{for all } A \in \Gamma - \{I\} .$$

In other words $(A_0z_1, A_0z_2, A_0z_3) \in S_{\alpha\beta\gamma}$. The uniqueness is trivial and the theorem is proved.

6. To avoid special considerations for the point ∞ we have given preference to the spherical derivative $[A'(z)]$. On the other hand the ordinary derivative $A'(z)$ has the advantage of having not only an absolute value but also an argument, while suffering from the drawback that it becomes infinite at $A^{-1}\infty$.

In the following we shall assume that ∞ is not a limit point. Then Λ is bounded, and for $z \in \Lambda$ the derivatives $|A'(z)|$ and $[A'(z)]$ differ only by a bounded factor. As a result

$$\sum_{A \in \Gamma} |A'(z_1)|^{\alpha} \, |A'(z_2)|^{\beta} \, |A'(z_3)|^{\gamma}$$

converges almost everywhere on Λ^3 and we can define a fundamental set $E_{\alpha\beta\gamma}$ for Λ^3 by the condition $|A'(z_1)|^{\alpha} \, |A'(z_2)|^{\beta} \, |A'(z_3)|^{\gamma} < 1$ for all $A \in \Gamma - \{I\}$. Naturally, this makes sense only if $m(\Lambda) > 0$ as we shall suppose from now on.

7. A complex-valued measurable function μ on Λ^3 is called a *Beltrami differential* if it is bounded and satisfies

$$\mu(Az_1, Az_2, Az_3)\overline{A'(z_1)}\,\overline{A'(z_2)}\,\overline{A'(z_3)} = \mu(z_1, z_2, z_3)A'(z_1)A'(z_2)A'(z_3)$$

for all $A \in \Gamma$. Because $E_{\alpha\beta\gamma}$ is a measurable fundamental set, every $\mu \in L^{\infty}(E_{\alpha\beta\gamma})$ can be extended to a Beltrami differential on Λ^3. This is the only use we shall make of $E_{\alpha\beta\gamma}$ and we shall even replace it by an arbitrary measurable fundamental set E.

Given a Beltrami μ we form the function

$$\phi_{\mu}(\zeta_1, \zeta_2, \zeta_3, \zeta_4) = \int_{\Lambda^3} \frac{\mu(z_1, z_2, z_3)\,dm(z_1)\,dm(z_2)\,dm(z_3)}{\prod_{i=1}^{3} \prod_{j=1}^{4} (z_i - \zeta_j)}$$

which is obviously holomorphic in all variables when $\zeta_j \in \Omega$. One verifies at once that

$$\phi_{\mu}(A\zeta_1, A\zeta_2, A\zeta_3, A\zeta_4)A'(\zeta_1)^{3/2}A'(\zeta_2)^{3/2}A'(\zeta_3)^{3/2}A'(\zeta_4)^{3/2}$$

$$= \phi_{\mu}(\zeta_1, \zeta_2, \zeta_3, \zeta_4)$$

for all $A \in \Gamma$. As a consequence the function

$$F_\mu(\zeta) = \phi_\mu(\zeta_1, \zeta_2, \zeta_3, \zeta_4) \prod_{1 \le i < j \le 4} (\zeta_i - \zeta_j)$$

is invariant when the ζ_i are replaced by $A\zeta_i$.

LEMMA 1. $F_\mu(\zeta)$ is bounded in Ω^4.

Proof: Because μ is bounded we need only show that

$$\prod_{i<j} |\zeta_i - \zeta_j|^{1/3} \int_\Lambda \frac{dm(z)}{|z-\zeta_1|\,|z-\zeta_2|\,|z-\zeta_3|\,|z-\zeta_4|} \tag{5}$$

is bounded. If we replace Λ by S in (5) the integral is invariant under all Möbius transformations. Hence we may choose $(\zeta_1, \zeta_2, \zeta_3, \zeta_4) = (0, 1, \infty, \zeta)$ and have to show that

$$|\zeta - 1|^{1/3} |\zeta|^{1/3} \int_S \frac{dm(z)}{|z|\,|z-1|\,|z-\zeta|} \tag{6}$$

is bounded. The last expression is still invariant under Möbius transformations that permute $0, 1, \infty$. Therefore it is sufficient to consider (6) when $\zeta \to 0$. Since the integral behaves like $\log 1/|\zeta|$ we conclude that $F_\mu(\zeta)$ is indeed bounded.

8. We have constructed bounded holomorphic functions F_μ, presumably in great number, that are automorphic under the action of Γ on Ω^4. If Γ is finitely generated the quotient space Ω/Γ is essentially a finite union of compact surfaces. For this case the result looks paradoxical, but I have been unable to prove that it actually is.

We show here that F_μ can be replaced by a function G_μ which has the same properties with respect to a Fuchsian group. Let Δ be the unbounded component of Ω, and let $\omega(z)$ be a conformal mapping of $|z| < 1$ on the universal covering of Δ. There is a Fuchsian group Γ_0 such that $A_0 \in \Gamma_0$ if and only if $\omega(A_0 z) = A\omega(z)$ for some $A \in \Gamma$. When Γ is finitely generated it is known that Γ_0 has a fundamental region with finite noneuclidean area. The function $G_\mu(z_1, z_2, z_3, z_4) = F_\mu(\omega(z_1), \omega(z_2), \omega(z_3), \omega(z_4))$ is bounded and automorphic with respect to the action of Γ_0.

Let us now make the hypothetical assumption that a function G_μ with these properties is necessarily a constant. There is some support for this assumption, for by a theorem of E. Hopf it is true for functions $G_\mu(z_1, z_2)$ of two variables under the weaker hypothesis that G_μ is harmonic in each variable. Admittedly, there is a big gap from two to four variables that I am not able to bridge.

9. We continue under the hypothetical assumption. If G_μ is constant it is identically zero and we obtain $F_\mu \equiv 0$ in Δ^4. The same is true of ϕ_μ. Multiply ϕ_μ by the third powers of $\zeta_2, \zeta_3, \zeta_4$ and let these variables tend to ∞. It follows that

$$\int_{\Lambda^3} \frac{\mu(z_1, z_2, z_3)\, dm(z_1)\, dm(z_2)\, dm(z_3)}{(z_1-\zeta)(z_2-\zeta)(z_3-\zeta)} = 0 \tag{7}$$

for all $\zeta \epsilon \Delta$ and all choices of μ. By continuity (7) remains true for $\zeta \epsilon \partial\Delta$.

We add a new hypothesis, namely that Γ is a function group. This means that Ω has an invariant component and we may of course choose Δ to be this component. Then $\partial\Delta = \Lambda$, and the boundary of each component of Ω is contained in $\partial\Delta$. It follows by the maximum principles that (7) holds for all ζ.

Because μ satisfies the Beltrami condition (7) can be replaced by

$$\int_E \mu(z_1, z_2, z_3)\theta(z_1, z_2, z_3, \zeta)\, dm(z_1)\, dm(z_2)\, dm(z_3) = 0 \tag{8}$$

where

$$\theta(z_1, z_2, z_3, \zeta) = \sum_{A \epsilon \Gamma} \frac{A'(z_1)^2 A'(z_2)^2 A'(z_3)^2}{(Az_1-\zeta)(Az_2-\zeta)(Az_3-\zeta)}\,.$$

Since (8) holds for all $\mu \in L^{\infty}(E)$ it follows by standard reasoning that $\theta(z_1, z_2, z_3. \zeta) = 0$ almost everywhere.

This however, is definitely not possible. We quote the following lemma due to Denjoy and Beurling.

LEMMA 2. A series of the from $\Sigma_1^{\infty} a_k(z-\alpha_k)^{-1}$ cannot converge to zero almost everywhere unless all the a_k are zero.

The contradiction is obvious. Under a hypothetical assumption that I would not dignify with the name of a conjecture we have proved that a finitely generated function group cannot have a limit point set with positive area.

REFERENCES

[1] L. V. Ahlfors, Finitely generated Kleinian groups, *Amer. J. of Math.*, Vol. 86, No. 2, (1964), pp. 413-429.

[2] A. Beurling, Sur les fonctions quasi analytiques des fractions rationnelles, 8. *Scand. Congr. Math.*, (1934), pp. 199-210.

[3] L. Ford, *Automorphic Functions*, 2. Ed., Chelsea Publishing Company, No New York (1951).

EXTREMAL QUASICONFORMAL MAPPINGS

Lipman Bers

§1. Introduction

This paper originated in an attempt to interpret R. S. Hamilton's important contribution to the theory of quasiconformal mappings. Our main result (Theorem 5) is a generalization of Hamilton's theorem [12]. It deals with the following problem.

Let D be an open set in the Riemann sphere, whose complement contains more than two points, and let G be a Kleinian group which maps D onto itself; the trivial group $G = 1$ is included. We consider quasiconformal automorphisms w of the Riemann sphere, such that wGw^{-1} is again a Kleinian group. Two such mappings are called equivalent if they coincide on the complement of D and if on every component Δ of D they are homotopic to each other, modulo of the ideal boundary curves of Δ. It is easy to see (Theorem 4) that every equivalence class contains one or several extremal elements, that is, elements with smallest dilatation. It is required to describe these elements, and our discription extends the one found by Hamilton for the case when D is a half plane. Under a certain finiteness hypothesis Hamilton's condition implies the existence of a classical Teichmüller mapping, but when D is not connected, a Teichmüller mapping must be defined in a more general way than in the theory of Riemann surfaces. One may also require that D, G and all mappings considered by symmetric about the real axis, and one obtains similar results (Theorems 6 and 7).

Two auxiliary results may be of independent interest. One (Theorem 2) gives a canonical way of replacing a mapping of not too large a dilatation, by an equivalent one which is very smooth in D. The other (Theorem 3) permits one to construct many one-parameter families of mappings equivalent to the identity.

Our results imply all known and some new theorems characterizing extremal quasiconformal mappings of Fuchsian groups and Riemann surfaces (see §7), and some new theorems on deformations of Kleinian groups (see §8).

We make use of some known facts about quasiconformal mappings, about Teichmüller spaces, and about automorphic forms. The necessary information can be found in the papers and books listed in the references.

§2. Beltrami Coefficients

In this section we collect some essentially known information which will be used later.

Let $\mathfrak{U}$ denote the upper half plane. If $\mu \in L_\infty(\mathfrak{U})$ and $\|\mu\| < 1$, let w_μ denote the unique quasiconformal automorphism of $\mathfrak{U}$ which keeps $0, 1, \infty$ fixed, and has the *Beltrami coefficient* μ, that is, satisfies

$$\frac{\partial w}{\partial \bar{z}} = \mu \frac{\partial w}{\partial z} .$$

If $w_\mu | R = w_\nu | R$ we call μ and ν *equivalent* and we write $\mu \sim \nu$.

LEMMA 1. *If* $\{\mu_j\} \subset L_\infty(\mathfrak{U})$, $\|\mu_j\| \leq k < 1$, $\mu_j \sim \mu_1$, $j = 1, 2, \ldots,$ *and* $\lim \mu_j(z) = \mu(z)$ a.e. *in* $\mathfrak{U}$, *then* $\mu \quad \mu_1$.

Proof: It is known [3] that under the hypotheses of the lemma, $w_{\mu_j}(z)$ converges normally to $w_\mu(z)$.

Let $\mathfrak{L}$ denote the lower half plane. For μ as above, let W be an automorphism of $\mathbf{C}$ such that $(W \; \mathfrak{U}) \circ w_\mu^{-1}$ and $W | \mathfrak{L}$ are conformal, and let $\phi^\mu(z)$, $z \in \mathfrak{L}$, denote the Schwarzian derivative of $W | \mathfrak{L}$. Define the mapping

$$\mu \longmapsto \nu = \Pi(\mu)$$

by

$$\nu(z) = -2y^2 \phi^\mu(x - iy), \quad z = x + iy \in \mathfrak{U} .$$

LEMMA 2. *The mapping* Π *is a well defined holomorphic mapping of the open unit ball in* $L_\infty(\mathfrak{U})$ *into* $L_\infty(\mathfrak{U})$. *with* $\Pi(0) = 0$. *We have* $\mu_1 \sim \mu_2$ *if and only if* $\Pi(\mu_1) = \Pi(\mu_2)$.

This is a well-known basic result in the theory of Teichmüller spaces [2,7,8].

LEMMA 3. *Let* $\phi(z)$, $z \in \mathfrak{L}$, *be holomorphic and such that* $\|y^2\phi(z)\| < 1/2$. *Set* $\nu(z) = -2y^2\phi(x-iy)$, $z \in \mathfrak{U}$. *Then* $\Pi(\nu) = \nu$.

This is a restatement of the Ahlfors-Weill lemma [4].

Now we choose, once and for all, a number k_0, $0 < k_0 < 1$, and a number θ_0, $0 < \theta_0 < 1$. such that

$$\|\Pi(\mu)\| < k_0 \quad \text{for} \quad \|\mu\| < \theta_0 \quad .$$

Let Γ be a Fuchsian group hwich leaves $\mathfrak{U}$ fixed. We denote by $L_\infty(\mathfrak{U},\Gamma)$ the set of $\mu \in L_\infty(\mathfrak{U})$ with

$$\mu(\gamma(z))\overline{\gamma'(z)}/\gamma'(z) \ , \qquad \gamma \in \Gamma \ ;$$

this is a closed linear subspace of $L_\infty(\mathfrak{U})$.

LEMMA 4. *If* $\mu \in L_\infty(\mathfrak{U},\Gamma)$ *and* $\|\mu\| < 1$. *then* $\Pi(\mu) \in L_\infty(\mathfrak{U},\Gamma)$.

The proof is a direct calculation.

Let Δ be a subdomain of the Riemann sphere $\hat{C} = C \cup \{\infty\}$ with at least 3 boundary points. Then there is a holomorphic universal covering

$$h : \mathfrak{U} \to \Delta \ .$$

The *Poincaré metric* $\lambda_\Delta(z)\,|dz|$ on Δ is defined by

$$\lambda_\Delta[h(z)]\,|h'(z)| = |z-\bar{z}|^{-1}, \qquad z \in \mathfrak{U} \ :$$

It is the unique complete conformal Riemannian metric on Δ of constant curvature -4, and hence independent of the choice of h. We assume now, for the sake of simplicity, that $\Delta \subset C$.

LEMMA 5. *Let* $\Delta_1, \Delta_2, \ldots,$ *be domains such that* $\Delta_1 \subset \Delta_2 \subset \cdots \subset \Delta$ *and* $\Delta = \Delta_1 \cup \Delta_2 \cup \cdots$. *Then* λ_{Δ_j} *converges normally to* $\lambda_\Delta(z)$, *as* $j \to \infty$.

The proof is routine and may be left to the reader.

Let $\mu \in L_\infty(\Delta)$. We denote by $h^*\mu$ the element $\nu \in L_\infty(\mathfrak{U})$ such that

$$\mu(h(z))\overline{h'(z)}/h'(z) = \nu(z) \quad .$$

LEMMA 6. *Let* Γ *be the covering group of* h. *Then*

$$h^* : L_\infty(\Delta) \to L_\infty(\mathfrak{U}, \Gamma)$$

is an isometric bijection.

The proof is clear.

Let μ and μ_1 be elements of the open unit ball of $L_\infty(\Delta)$. We say that μ and μ_1 are *equivalent*, and we write $\mu \sim \mu_1$ if and only if $h^*\mu \sim h^*\mu_1$. This definition does not depend on h.

We recall now the definition of the union $b(\Delta)$ of the set of *ideal boundary curves* of Δ. Let Γ be as in Lemma 6 so that we may identify Δ with $\mathfrak{U}/\Gamma$. If Γ is of the second kind, that is, if the maximal open subset $\mathfrak{M}_\Gamma$ of $\hat{\mathbf{R}} = \mathbf{R} \cup \{\infty\}$ on which Γ acts discontinuously is not empty, one sets $\tilde{\Delta} = (\mathfrak{U} \cup \mathfrak{M}_\Gamma)/\Gamma$, $b(\Delta) = \tilde{\Delta} - \Delta$. If Γ is of the first kind, that is, if $\mathfrak{M}_\Gamma = \emptyset$, one sets $b(\Delta) = \emptyset$, $\tilde{\Delta} = \Delta$.

LEMMA 7. *Let* $w : \Delta \to \hat{\mathbf{C}}$ *and* $\hat{w} : \Delta \to \hat{\mathbf{C}}$ *be quasiconformal homeomorphisms with Beltrami coefficients* μ *and* $\hat{\mu}$, *respectively. Then* $\mu \sim \hat{\mu}$ *if and only if there exists a conformal mapping* $f : \hat{w}(\Delta) \to w(\Delta)$ *such that* $w^{-1} \circ f \circ \hat{w}$ *(or rather its continuous extension to* $\tilde{\Delta}$, *which always exists) is homotopic to the identity modulo* $b(\Delta)$.

This is a known result in the theory of Teichmüller spaces [2, 7, 8]. It can be restated as follows: two Beltrami coefficients, on the same domain Δ, are equivalent if and only if they belong to mappings which define the same point in the *Teichmüller space* $T(\Delta)$.

LEMMA 8. *Let* w *and* $\hat{w}$ *be as in Lemma 7. If there is a compact set* $K \subset \Delta$ *with* $w \mid \Delta - K = \hat{w} \mid \Delta - K$, *then* $\mu \sim \hat{\mu}$.

This is an immediate corollary of Lemma 7.

LEMMA 8. *Let* $\Delta \subset \hat{C}$ *be a domain bounded by finitely many disjoint Jordan curves. Let* $w : \Delta \to \hat{C}$ *be a quasiconformal homeomorphism with Beltrami coefficient* μ, *and let* $\hat{\mu} \sim \mu$. *Then there exists a quasiconformal homeomorphism* $W : \Delta \to \hat{C}$ *with Beltrami coefficient* $\hat{\mu}$ such that $W(\Delta) = w(\Delta)$ *and* $w^{-1} \circ W$ *has a continuous extension to the closure of* Δ *which leaves every boundary point of* Δ *fixed.*

Proof: Let $\hat{w} : \Delta \to \hat{C}$ be some quasiconformal homeomorphism with Beltrami coefficient $\hat{\mu}$. Apply Lemma 7, set $W = f \circ \hat{w}$, and note that, under the hypotheses of Lemma 9, we may identify $b(\Delta)$ with the set theoretical boundary of Δ.

An element $\mu \in L_\infty(\Delta)$ will be called *canonical* if $\lambda_\Delta^2 \bar{\mu}$ is holomorphic. Such elements form a closed linear subspace $L_\infty^{can}(\Delta) \subset L_\infty(\Delta)$. If Γ is a Fuchsian group which leaves R fixed, we set

$$L_\infty^{can}(\mathfrak{U}, \Gamma) = L_\infty^{can}(\mathfrak{U}) \cap L_\infty(\mathfrak{U}, \Gamma).$$

LEMMA 10. *If* Γ *is the covering group of* $h : \mathfrak{U} \to \Delta$, *then*

$$h^* : L_\infty^{can}(\Delta) \to L_\infty^{can}(\mathfrak{U}, \Gamma)$$

is a bijection.

The proof is a calculation.

LEMMA 11: *If* μ *and* μ_1 *belong to the open unit ball in* $L_\infty^{can}(\Delta)$ *and* $\mu \sim \mu_1$, *then* $\mu = \mu_1$.

This follows from Lemmas 2, 3 and 10.

LEMMA 12. *There exists a holomorphic mapping* Π_Δ *of the open unit ball in* $L_\infty(\Delta)$ *into* $L_\infty^{can}(\Delta)$ *such that* $\Pi_\Delta(0) = 0$, $\Pi_\Delta(\mu) \sim \mu$ *if* $\|\Pi_\Delta(\mu)\| < 1$,

$\|\Pi_\Delta(\mu)\| < k_0$ *if* $\|\mu\| < \theta_0$, *and* $\Pi_\Delta(\mu) = \mu$ *if* $\mu \in L_\infty^{can}(\Delta)$.

Proof: Set $\Pi_\Delta = (h^*)^{-1} \circ \Pi \circ h^*$.

§3. Smoothing Theorem

A *configuration* (D, G) is a pair consisting of an open non-empty set $D \subset \hat{C}$, such that $\hat{C} - D$ contains more than two points, and of a discrete group G of Möbius transformations such that $\gamma(D) = D$ for $\gamma \in G$. We choose such a configuration once and for all.

Note that G acts properly discontinuously on D. To simplify writing, we assume from now on that $\{0, 1, \infty\} \subset \hat{C} - D$. The configuration (D, G) will be called *symmetric* if the conjugation $z \mapsto \bar{z}$ takes D into itself, and if $\gamma(\mathfrak{U}) = \mathfrak{U}$ for all $\gamma \in G$.

If $\mu \in L_\infty(C)$ and $\|\mu\| < 1$. we denote by $z \mapsto w^\mu(z)$ the unique automorphism of $\hat{C}$, with Beltrami coefficient μ, which keeps $0, 1, \infty$ fixed. We denote by $L_\infty(C, G)$ the set of $\mu \in L_\infty(C)$ with

$$\mu[\gamma(z)]\overline{\gamma'(z)}/\gamma'(z) = \mu(z), \qquad \gamma \in G \ .$$

It is a closed linear subspace of $L_\infty(C)$.

LEMMA 13. *The pair* $(w^\mu(D), w^\mu G(w^\mu)^{-1})$ *is a configuration if and only only if* $\mu \in L_\infty(C, G)$.

The proof is standard and easy.

We call $\mu \in L_\infty(C)$ symmetric if

$$\mu(\bar{z}) = \overline{\mu(z)} \ .$$

Such μ form an R-linear closed subspace of $L_\infty(C)$, denoted by $L_\infty^{sym}(C)$.

LEMMA 14. If (D, G) *is symmetric,* $\mu \in L_\infty^{sym}(C, G)$, *and* $\|\mu\| < 1$ *then* $(w^\mu(D), w^\mu G(w^\mu)^{-1}$ *is symmetric.*

The proof of this is obvious.

If $\mu \in L_\infty(C)$, we denote by $\nu = \Pi^D(\mu)$ the element $\nu \in L_\infty(C)$ determined by the conditions

$$\nu \mid C - D = \mu \mid C - D ,$$

$$\nu \mid \Delta = \Pi_{\Delta}(\mu \mid \Delta)$$

for every component Δ of D.

LEMMA 15. Π^{D} *is a holomorphic mapping of the unit ball in* $L_{\infty}(C)$ *into* $L_{\infty}(C)$, *with* $\Pi^{D}(0) = 0$ *and*

$$\Pi^{D} \circ \Pi^{D} = \Pi^{D} .$$

Also, if (D, G) *is a configuration, then*

$$\Pi^{D}[L_{\infty}(C, G)] \subset L_{\infty}(C, G) ,$$

and, if (D, 1) *is symmetric, then*

$$\Pi^{D}[L_{\infty}^{sym}(C)] \subset L_{\infty}^{sym}(C) :$$

Proof: Only the second assertion needs verification. A direct calculation shows that if γ is any Möbius transformation, and if we set, for any $\mu \in L_{\infty}(C)$,

$$(\mu \cdot \gamma)(z) = \mu[\gamma(z)]\overline{\gamma'(z)}/\gamma'(z)$$

then

$$\Pi^{\gamma(D)}(\mu \cdot \gamma) = \Pi^{D}(\mu) \cdot \gamma .$$

If (D, G) is a configuration, $\gamma \in G$ and $\mu \in L_{\infty}(C, C)$, then $\gamma(D) = D$, $\mu \cdot \gamma = \mu$ and therefore $\Pi^{D}(\mu) = \Pi^{D}(\mu) \cdot \gamma$, so that $\Pi^{D}(\mu) \in L_{\infty}(C, G)$. Symmetry is proved similarly.

LEMMA 16. *If* $\mu \in L_{\infty}(C)$ *and* $\|\mu\| < 1$, $\|\mu \mid D\| < \theta_0$, *then* $\|\Pi^{D}(\mu)\| < 1$ *and* $\|\Pi^{D}(\mu) \mid D\| < k_0$.

This fbllows from the definition of θ_0.

We shall call a $\mu \in L_{\infty}(C)$ *canonical with respect to* D *if* $\mu \mid \Delta$ is canonical, in the sense of §2, for every component Δ of D.

We shall call two elements, μ and μ_1, of $L_\infty(C)$, with $\|\mu\| < 1$, $\|\mu_1\| < 1$, *equivalent with respect to* D, and we shall write $\mu \sim \mu_1$, if $w^\mu | C-D = w^{\mu_1} | C-D$, and $\mu|\Delta \sim \mu_1|\Delta$ for every component Δ of D.

THEOREM 1. *Assume that* $\mu \in L_\infty(C)$, $\|\mu\| < 1$, *and* $\|\mu|D\| < \theta_0$. *Set* $\nu = \Pi^D(\mu)$. *Then* $\nu \sim \mu$ *w.r.t.* D.

The theorem asserts that there is a cannonical way of smoothing a given quasiconformal automorphism of C (that is, w^μ of the theorem) into an "equivalent" one in an open set D, provided the dilatation of the given mapping in D is not too large. Note that w^ν is real analytic in D because ν is.

Proof of Theorem 1. We must show only that $w^\mu | C-D = w^\nu | C-D$. Call D tame if each component Δ of D is bounded by finitely many disjoint analytic Jordan curves. Assume that D is tame. For each component Δ of D there exists a quasiconformal homeomorphism $\omega_\Delta: \Delta \to \hat{C}$, with Beltrami coefficient $\nu|\Delta = \Pi_\Delta(\mu|\Delta)$ such that $\omega_\Delta(\Delta) = w^\mu(\Delta)$ and ω_Δ has the boundary values $w^\mu|\partial\Delta$ on the boundary $\partial\Delta$ of Δ. This follows from Lemmas 9 and 12. Hence there exists a topological automorphism W of C such that

$$W|\Delta = \omega_\Delta$$

for every component Δ of D, and

$$W|\hat{C} - D = w^\mu|\hat{C} - D.$$

Since isolated analytic arcs are removable sets for quasiconformal mappings, the mapping W is quasiconformal and its Beltrami coefficient is ν. Since W leaves $0, 1, \infty$ fixed, we conclude that $W = w^\nu$. This proves the theorem for the case of a tame D.

In the general case, construct a sequence of tame open sets D_j such that $D_1 \subset D_2 \subset \cdots \subset D$, $D = D_1 \cup D_2 \cup \cdots$, every component of D_j is of the form $D_j \cap \Delta$ where Δ is a component of D, and $D_j \cap \Delta$ is relatively compact in Δ. Set $\nu^j = \Pi^{D_j}(\mu)$. Then $\|\nu^j\| < \max(\|\mu\|, k_0)$, by Lemma 16. If Δ is a fixed component of D and $\Delta_j = D_j \cap \Delta$, then Δ_j is a domain, and $\phi_j = \lambda^2_{\Delta_j}(\nu_j | \Delta_j)$ is holomorphic. Using Lemma 5 and compactness properties of holomorphic functions, we conclude that a subsequence of $\{\phi_j\}$ converges normally. We conclude next that a subsequence of $\{\nu_j\}$ converges a.e. to an element $\rho \in L_\infty(C)$ such that $\rho | C - D = \mu | C - D$ and $\rho | \Delta \in L^{can}_\infty(\Delta)$ for each component Δ of D. By what was proved above we know that $w^{\nu_j} | \hat{C} - D_j = w^\mu | \hat{C} - D_j$. Therefore, by Lemma 8, $\nu_j | \Delta \sim \mu | \Delta$, so that, by Lemma 1, $\rho | \Delta \sim \mu | \Delta$. We conclude, by Lemmas 11 and 12, that $\rho | \Delta = \Pi_\Delta(\mu)$. Hence $\rho = \nu$. Thus the selection of a subsequence was unnecessary and we have that $\lim \nu_j(z) = \nu(z)$ a.e. Since there is a number $k < 1$ with $\|\nu_j\| \leq k$, we have that $\{w^{\nu_j}(z)\}$ converges to $w^\nu(z)$. Thus $w^\nu | \hat{C} - D = w^\mu | \hat{C} - D$, q.e.d.

§4. Trivial Families

An element $\mu \in L_\infty(C)$ with $\|\mu\| < 1$ will be called *trivial if* $\mu \sim 0$ w.r.t. D. The aim of this section is to construct one-parameter families of trivial Beltrami coefficients.

An element $\mu \in L_\infty(C, G)$ will be called *locally trivial* (with respect to D) if

$$w^{\varepsilon\mu}(z) - z = o(|\varepsilon|), \varepsilon \to 0, \qquad z \in C - D,$$

that is, if

$$\dot{w}^\mu(z) = \left. \frac{\partial w^{\varepsilon\mu}(z)}{\partial \varepsilon} \right|_{\varepsilon = 0} = 0 \text{ for } z \in C - D .$$

(It is known [3] that $w^{\varepsilon\mu}(z)$ is a holomorphic function of ε at $\varepsilon = 0$.)

LEMMA 17. *We have*

$$\dot{w}^\mu(z) = -\frac{z(z-1)}{\pi} \iint_C \frac{\mu(\zeta)\, d\xi\, d\eta}{\zeta(\zeta-1)(\zeta-z)} .$$

This is a known formula [10].

A *quadratic differential* (automorphic form of weight -4) for (D, G) is a holomorphic function $\phi(z)$, $z \in D$, such that

$$\phi(\gamma(z))\gamma'(z)^2 = \phi(z), \quad \gamma \in G .$$

It is called *integrable* if

$$\|\phi\|_A = \iint_{D/G} |\phi(z)|\, dx\, dy < \infty ;$$

integrable quadratic differentials, with the above norm, form a Banach space $A(D, G)$.

For every function $\psi(z)$, $z \in D$, we define the Poincaré theta series

$$(\Theta\psi)(z) = \sum_{\gamma \in G} \psi(\gamma(z))\gamma'(z)^2 .$$

LEMMA 18. *The mapping*

$$\Theta : A(D, 1) \to A(D, G)$$

is a continuous surjection.

This is a known result, see the Appendix to [13] and the literature quoted there.

LEMMA 19. *If* $\psi \in A(D, 1)$ *and* $\mu \in L_\infty(C, G)$, *then*

$$\iint_{D/G} \mu(z)(\Theta\psi)(z)\, dx\, dy = \iint_D \psi(z)\mu(z)\, dx\, dy .$$

The proof of this is a calculation.

LEMMA 20. *For* $z \in C - D$, *set*

$$\psi_z(\zeta) = \frac{1}{\zeta(\zeta - 1)(\zeta - z)} ;$$

then ψ_z D *belongs to* A(D, 1). *Functions of the form*

$$\sum_{j=1}^{N} c_j \psi_{z_j} | D \qquad (z_j \in C - D,\ c_j \in C)$$

are dense in A(D, 1).

Only the second statement needs a proof. This can be obtained by repeating the argument used for the case when D is connected [9].

THEOREM 2. *An element* $\mu \in L_\infty(C, G)$ *is locally trivial if and only if*

$$\mu | C - D = 0$$

and

$$\iint_{D/G} \mu(z)\phi(z)\,dx\,dy = 0 \quad \text{for all } \phi \in A(D, G) .$$

Proof: In view of Lemmas 18 and 19 the orthogonality condition is equivalent to

$$\iint_{D} \mu(z)\psi(z)\,dx\,dy = 0 \qquad \text{for all} \quad \psi \in A(D, 1) .$$

Thus it suffices to prove the theorem for the case G = 1.

Assume that $\mu | C - D = 0$ and the orthogonality condition holds. Then

$$\iint_{C} \mu(\zeta)\psi_z(\zeta)\,d\xi\,d\eta = 0$$

for $z | \in C - D$, so that, by Lemma 17, $\dot{w}^{\mu}(z) - 0$.

Assume next that $\dot{w}^{\mu}(z) = 0$ for $z \in C - D$. Since, by Lemma 17 and by classical potential theory,

$$\frac{\partial \dot{w}^{\mu}(z)}{\partial \bar{z}} = \mu(z) \quad \text{a.e.,}$$

we have that $\mu = 0$ a.e. in C − D. Also, the condition $\dot{w}^{\mu} | C - D = 0$ may be written as

$$\iint_D \mu(\zeta)\psi_z(\zeta)\,d\xi d\eta = 0, \qquad z \in C-D \ .$$

In view of Lemma 20, this implies that

$$\iint_D \mu(\zeta)\,\phi(\zeta)\,d\xi d\eta = 0$$

for all $\phi \in A(D,1)$.

Let $L_\infty^{can}(C,G)$ be the linear subspace of $L_\infty(C,G)$ consisting of canonical elements.

LEMMA 21. *For every $\mu \in L_\infty(C,G)$ there are uniquely determined elements $\nu_1, \nu_2, \nu_3, \ldots,$ of $L_\infty^{can}(C,G)$ such that for $t \in C$, $|t|$ small,*

$$t\mu \sim \nu = t\nu_1 + t^2\nu_2 + t^3\nu_3 + \cdots$$

where the series converges in $L_\infty(C)$, and

$$\nu_1 | C-D = \mu | C-D, \quad \nu_j | C-D = 0, \qquad j > 0.$$

If (D,G) and μ are symmetric, so are all ν_j.

Proof: Set $\nu = \Pi^D(t\mu)$. For small $|t|$ this is a holomorphic function of t with values in $L_\infty^{can}(C,G)$, by Lemma 15; hence there is a convergent expansion $\nu = \nu_0 + t\nu_1 + t^2\nu_2 + \cdots$. Clearly, $\nu_0 = 0$ and the ν_j, $j \geq 1$, have the required properties.

LEMMA 22. *Under the hypothesis of Lemma 21, assume that μ is locally trivial. Then $\nu_1 = 0$.*

Proof: For small $|t|$ we have that $w^\nu(z)$ is a holomorphic function of t; if $z \in C-D$, then $w^\nu(z) = w^{t\mu}(z)$. One computes easily that

$$\left.\frac{\partial w^\nu(z)}{\partial t}\right|_{t=0} = \left.\frac{\partial w^{t\nu_1}(z)}{\partial t}\right|_{t=0} = \dot{w}^{\nu_1}(z)$$

so that $\dot{w}^{\nu_1} | C-D = 0$ and ν_1 is locally trivial. Hence, $\nu_1 | C-D = 0$, by Theorem 2. Also, there exists a holomorphic function $\psi(z)$ in D such that $\nu_1 | \Delta = \lambda_{\Delta}^{-2} \cdot (\bar{\psi} | \Delta)$ for every component Δ of D. Since ν_1 is locally trivial, we conclude, by Theorem 2 and Lemmas 18 and 19 that, for every $\phi \in A(D, 1)$,

$$\iint_D \nu_1(z)\phi(z)\, dx\, dy = \iint_{D/G} \nu_1(z)(\Theta\phi)(z)\, dx\, dy = 0 \; .$$

This implies that for every Δ and every holomorphic function $f(z)$, $z \in \Delta$ with

$$\iint_\Delta |f(z)|\, dx\, dy < \infty$$

we have

$$\iint \lambda_{\Delta}^{-2}(z)\, f(z)\overline{\psi_\Delta(z)}\, dx\, dy \;=\; 0$$

where $\psi_\Delta = \psi | \Delta$. It is well known [13] that this implies that $\psi_\Delta = 0$. Hence $\psi = 0$ and $\nu = 0$.

Theorem 3. *Let* $\mu \in L_\infty(C, G)$ *be locally trivial. For every* $t \in C$, $|t|$ *small, one can find a* $\tau \in L_\infty(C, G)$ *depending on* t, *such that*

$$\|\tau\| = 0(1), \qquad t \to 0$$

and

$$t\mu + t^2\tau \sim 0 \; .$$

If (D, G) *and* μ *are symmetric, and* t *is real,* τ *is symmetric.*

Proof: The ν from Lemma 21 satisfies $\|\nu\| = 0(t^2)$, in view of Lemma 22. Define $\tilde{\nu} \in L_\infty(C)$ by

$$(w^\nu)^{-1} = w^{\tilde{\nu}}$$

(for small $|t|$). Then $\|\tilde{\nu}\| = \|\nu\|$. Define $\sigma \in L_\infty(C)$ by

$$w^\sigma = (w^\nu)^{-1} \circ w^{t\mu} \, .$$

Then σ is trivial. Also, since ν and $t\mu$ belong to $L_\infty(C, G)$ we have, for $\gamma \in G$,

$$w^\sigma \circ \gamma \circ (w^\sigma)^{-1} = (w^\nu)^{-1} \circ w^{t\mu} \circ \gamma \circ (w^{t\mu})^{-1} \circ w^\nu = \gamma$$

so that, by Lemma 13, $\sigma \in L_\infty(C, G)$. Next we compute that

$$\sigma = \frac{t\mu + \hat{\nu}}{1 + \bar{t}\bar{\mu}\hat{\nu}}$$

where

$$\hat{\nu}(z) = \frac{\tilde{\nu}\,[w^{t\mu}(z)]\partial w^{t\mu}(z)/\partial z}{\partial w^{t\mu}(z)/\partial z}$$

Thus, $\|\nu\| = \|\tilde{\nu}\| = 0(|t|^2)$ and $\tau = (\sigma - t\mu)/t^2$ has the required properties. We also note that if (D, G) and μ are symmetric and t real, then σ and τ are symmetric.

§5. Hamilton's Extremality Condition

A Beltrami coefficient $\mu \in L_\infty(C, G)$, $\|\mu\| < 1$ will be called *extremal* with respect to (D, G) if

$$\|\mu \mid D\| \leq \|\tilde{\mu} \mid D\| \quad \text{for} \quad \mu \sim \tilde{\mu}\,, \quad \tilde{\mu} \in L_\infty(C, G)\ .$$

THEOREM 4. *Every* $\mu_0 \in L_\infty(C, G)$, *with* $\|\mu_0\| < 1$, *is equivalent, with respect to* (D, G), *to an extremal one.*

Proof: Set $k = \inf \|\tilde{\mu}\|$ where $\tilde{\mu} \in L_\infty(C, G)$, $\|\tilde{\mu}\| \leq \|\mu_0\|$ and $\tilde{\mu} \sim \mu_0$. There is a sequence of Beltrami coefficients $\{\tilde{\mu}_j\}$ with the above properties such that $\lim \|\tilde{\mu}_j\| = k$. Set $w_j = w^{\tilde{\mu}_j}$. By known compactness properties of quasiconformal mappings one may assume, selecting, if need by, a subsequence, that $\lim w_j = w^\mu$, where $\mu \in L_\infty(C, G)$, $\|\mu\| < 1$, and the convergence is normal. For every $\gamma \in G$ there is a Möbius transformation $\tilde{\gamma}$ such that $w_j \circ \gamma \circ w_j^{-1} = \tilde{\gamma} = w^{\mu_0} \circ \gamma \circ (w^{\mu_0})^{-1}$. It follows that $w^\mu \circ \gamma \circ (w^\mu)^{-1} = \tilde{\gamma}$, so that, by Lemma 13, we have $\mu \in L_\infty(C, G)$. By the lower semicontinuity of the dilatation of a quasiconformal mapping with respect to normal convergence [6] we have $\|\mu\| \leq k$, so that $\|\mu\| = k$.

Clearly, $w^{\mu}(z) = \lim w_j(z) = w^{\mu_0}(z)$ for $z \in C-D$. Now let Δ be a component of D, $h: \mathfrak{U} \to \Delta$ a holomorphic universal covering, Γ the covering group of h, and h^* the operator defined in §2. Set $h^*(\mu_0|\Delta) = \nu_0$, $h^*(\tilde{\mu}_j|\Delta) = \nu_j$ and $h^*(\mu|\Delta) = \nu$. Then there are holomorphic universal coverings

$$\tilde{h}_0: \mathfrak{U} \to w^{\mu_0}(\Delta),\quad \tilde{h}_j: \mathfrak{U} \to w^{\tilde{\mu}_j}(\Delta),\quad \tilde{h}: \mathfrak{U} \to w^{\mu}(\Delta)$$

with

$$\tilde{h}_0 \circ w_{\nu_0} = (w^{\mu_0}|\Delta) \circ h,\quad \tilde{h}_j \circ w_{\nu_j} = (w^{\tilde{\mu}_j}|\Delta) \circ h,\quad \tilde{h} \circ w_{\nu} = (w^{\mu}|\Delta) \circ h\ .$$

We may assume, selecting if need by, a subsequence, that $\lim w_{\nu_j} = w_\sigma$ uniformly on every bounded subset of $\mathfrak{U}$, where $\sigma \in L_\infty(\mathfrak{U})$, $\|\sigma\| < 1$. Since (the continuous extensions of) w_{ν_j} coincides with (the continuous extension of) w_{ν_0} on R, the same is true of w_σ. Thus $\sigma \sim \nu_0$. On the other hand, $\hat{h} = \lim \tilde{h}_j$ is holomorphic, and $\hat{h} \circ w_\sigma = (w^{\mu}|\Delta) \circ h$, so that $\sigma = \nu$. Thus $\nu \sim \nu_0$, so that $\mu|\Delta \sim \mu_0|\Delta$. Since Δ was any component of D, we established that $\mu \sim \mu_0$.

An element $\mu \in L_\infty(C, C)$ will be said to satisfy the *Hamilton condition*, with respect to (D, G), if

$$\|\mu|D\| \leq \|\mu+\nu|D\|$$

for every locally trivial $\nu \in L_\infty(D, G)$. We can now establish

THEOREM 5. *Every extremal Beltrami coefficient satisfies the Hamilton condition.*

Proof: Assume that there is a locally trivial ν such that

$$\|\mu-\nu|D\| = k_1 < k = \|\mu|D\|\ .$$

We shall show that μ is not extremal.

By Theorem 3, one can find, for every small $t > 0$, a $\tau \in L_\infty(C, G)$ with $\|\tau\| = O(1)$, such that $t\nu + t^2\tau \sim 0$. Now define $\mu_0 \in L_\infty(C, G)$ by

$$w^{\mu_0} = w^{\mu} \circ (w^{t\nu + t^2\tau})^{-1} \quad .$$

For small t, we have μ_0 μ. Also,

$$w^{\mu} = w^{\mu_0} \circ w^{t\nu + t^2\tau}$$

and we compute (cf. the proof of Theorem 3) that

$$\hat{\mu}_0 = \frac{\mu - t\nu - t^2\tau}{1-(t\bar{\nu} + t^2\bar{\tau})\mu}$$

where

$$\hat{\mu}_0(z) = \mu_0[w^{t\nu + t^2\tau}(z)] \frac{\overline{\partial w^{t\nu+t^2\tau}(z)/\partial z}}{\partial w^{t\nu+t^2\tau}(z)/\partial z}$$

Thus, for almost all $z \epsilon D$,

$$\begin{aligned}
|\hat{\mu}_0(z)| &= \frac{|\mu(z) - t\nu(z) - t^2\tau(z)|}{|1-[t\nu(z) + t^2\tau(z)]\bar{\mu}(z)|} \\
&= |\mu(z) - t\nu(z)(1-|\mu(z)|^2)| + 0(t^2) \\
&= |[1-t(1-|\mu(z)|^2)]\mu(z) + t(1-|\mu(z)|^2)[\mu(z)-\nu(z)]| + 0(t^2) \\
&\leq [1-t(1-|\mu(z)|^2)]k + t(1-|\mu(z)|^2)k_1 + 0(t^2) \\
&= k - t(1-k^2)(k-k_1) + 0(t^2)
\end{aligned}$$

uniformly in z. Thus $\|\mu_0| D\| < k$ for small $t > 0$, and μ is not extremal.

Assume now that (D, G) is symmetric. A symmetric Beltrami coefficient $\mu \epsilon L_\infty(C, G)$ with $\|\mu\| < 1$ will be called *symmetrically extremal* if $\|\mu | D\| \leq \|\mu_0| D\|$ for all symmetric $\mu_0 \sim \mu$, $\mu_0 \epsilon L_\infty(C, G)$. We say that μ satisfies the *symmetric Hamilton condition* if $\|\mu | D\| \leq \|\mu + \nu | D\|$ for every symmetric locally trivial $\nu \epsilon L_\infty(D, G)$. Repeating the previous arguments, we obtain the following results.

THEOREM 6. *Every symmetric Beltrami coefficient is equivalent to a symmetrically extremal one.*

THEOREM 7. *Every symmetrically extremal Beltrami coefficient satisfies the symmetric Hamilton condition.*

We remark that the Hamilton condition can be restated in a different form (which is, in fact, Hamilton's original formulation).

LEMMA 23. *An element* $\mu \in L_\infty(C, G)$ *satisfies the Hamilton condition if and only if*

$$\sup \left| \iint_{D/G} \mu(z)\phi(z)\,dx\,dy \right| = \|\mu \mid D\|$$

where the supremum is taken over all $\phi \in A(D, G)$ *with* $\|\phi\|_A = 1$.

Proof: Denote the left side in the above formula by k. Clearly, $k \leq \|\mu \mid D\|$. In all cases k is the norm of the linear functional

$$\ell(\phi) = \iint_{D/G} \phi(z)\mu(z)\,dx\,dy$$

on $A(D, G)$. By the theorems of Hahn-Banach and F. Riesz, there exists a $\hat{\mu} \in L_\infty(C, G)$ with $\hat{\mu} \mid C - D = \mu \mid C - D$, $\|\hat{\mu} \mid D\| = k$ and

$$\ell(\phi) = \iint_{D/G} \phi(z)\hat{\mu}(z)\,dx\,dy \ :$$

Clearly, $\hat{\mu} - \mu$ is orthogonal over D/G to all $\phi \in A(D, G)$ and thus locally trivial by Theorem 2. If $\|\mu \mid D\| > k$, then

$$\|\mu + (\hat{\mu} - \mu) \mid D\| < \|\mu \mid D\|$$

and μ does not satisfy the Hamilton condition. The reverse implication is proved similarly.

If (D, G) is symmetric, we call a quadratic differential $\phi(z)$ *symmetric* if $\phi(\bar{z}) = \overline{\phi(z)}$, antisymmetric if $\phi(\bar{z}) = -\overline{\phi(z)}$. The symmetric integrable quadratic differentials form the real Banach space $A_+(D, G) \subset A(D, G)$. We also set $D_+ = D \cap \mathfrak{U}$.

LEMMA 24. *If* (D, G) *and* $\mu \in L_\infty(C, G)$ *are symmetric, then* μ *is locally trivial if and only if*

$$\iint_{D_+/G} \mu(z)\phi(z)\,dx\,dy = 0$$

for all $\phi \in A_+(D, G)$, *and* $\mu \mid C - D = 0$.

Proof: If $\phi \in A(D, G)$, set

$$\phi_+(z) = \frac{1}{2}\phi(z) + \frac{1}{2}\overline{\phi(\bar{z})}, \qquad \phi_-(z) = \phi(z) - \phi_+(z) .$$

Then $\phi_+ \in A_+(D, G)$, ϕ_- is antisymmetric, and $\phi = \phi_+$ if $\phi \in A_+(D, G)$. If μ is symmetric, then

$$\iint_{D/G} \phi(z)\mu(z)\,dx\,dy = 2\iint_{D_+/G} \phi_+(z)\mu(z)\,dx\,dy .$$

Hence the lemma follows from Theorem 2.

LEMMA 25. *If* (D, G) *and* $\mu \in L_\infty(C, G)$ *are symmetric, then* μ *satisfies the symmetric Hamilton condition if and only if*

$$\|\mu \mid D_+\| = \sup \operatorname{Re} \iint_{D_+/G} \mu(z)\phi(z)\,dx\,dy$$

where the supremum is taken over all $\phi \in A_+(D, G)$ *with* $\|\phi\|_A = 2$.

The proof is analogous to that of Lemma 23 and need not be spelled out.

§6. Teichmüller's Condition

The connection between Hamilton's condition and the classical extremality condition of Teichmüller is contained in the following observation, due to Hamilton.

LEMMA 26. *If* $\dim A(D, G) < \infty$, *and* $\mu \in L_\infty(C, G)$ *satisfies the Hamilton condition, then either* $\mu = 0$, *or there is a* $\phi \in A(D, G)$ *such that* $\phi \neq 0$

and

$$\|\mu \mid D\| \; |\phi(z)| = \mu(z)\phi(z)$$

a.e. in D.

Proof: The hypothesis implies that the boundary of the unit ball in A(D, G) is compact. Hence, if $\mu \neq 0$ satisfies the Hamilton condition, then there is a $\phi \in A(D,G)$ with $\|\phi\|_A = 1$ such that

$$\|\mu \mid D\| = \iint_{D/G} \mu(z)\phi(z)\,dx\,dy$$

or

$$\iint_{D/G} \{ \|\mu \mid D\| \; |\phi(z)| - \mu(z)\phi(z)\}\,dx\,dy = 0 \;,$$

and, since $\|\mu \mid D\| \; |\phi(z)| \geq |\mu(z)\phi(z)|$ a.e. in D, the conclusion follows.

We say that a $\mu \in L_\infty(C, G)$ satisfies the *Teichmüller condition* with respect to (D, G) of either $\mu = 0$, or there is a $\phi \in A(D,G)$, with $\phi \neq 0$, and a non-negative locally constant function $k(z)$, $z \in D$, such that (i) for every component Δ of D with $\phi \mid \Delta \neq 0$, we have $k \mid \Delta > 0$ and

$$\mu \mid \Delta = (k \mid \Delta)\, |\phi| \Delta| / (\phi \mid \Delta) \;,$$

and (ii) for every component Δ of D with $\phi \mid \Delta = 0$, we have $k \mid \Delta = \mu \mid \Delta = 0$. Note that $\|\mu \mid \Delta\| = k \mid \Delta$ for every component Δ.

THEOREM 8. *If* dim A(D, G) $< \infty$, *and* D *is connected, then every extremal Beltrami coefficient satisfies the Teichmüller condition.*

Proof: The assertion follows from Theorem 5 and Lemma 26, since if $\phi \in A(D,G)$ and $\phi \neq 0$, then $\phi(z) \neq 0$ a.e. in D.

THEOREM 9. *If* dim A(D, G) $< \infty$, *then every Beltrami coefficient is equivalent to an extremal Beltrami coefficient which satisfies the Teichmüller condition.*

Proof: Let us call a union of components of D which is invariant under G a *part* of (D, G). Let $D_1, D_2, \ldots$ be the minimal parts of (D, G). Then $\dim A(D, G) = \Sigma \dim A(D_j, G)$, so that $\dim A(D_j, G) \neq 0$ only for finitely many parts.

Now let $\mu_0 \in L_\infty(C, G)$ with $\|\mu_0\| < 1$ be given. By Theorem 3 there is an extremal μ_1 equivalent to μ, by Theorem 4 this μ_1 satisfies the Hamilton condition. If $\mu_1 = 0$, there is nothing to prove. Otherwise there is, by Lemma 26, a $\phi \in A(D, G)$ with $\phi \neq 0$ and with $k_1|\phi| = (\mu \mid D)\phi$ where $k_1 = \|\mu \mid D\| > 0$. Let D^1 be the largest part of (D, G) such that $\phi \mid \Delta \neq 0$ for every component of D^1, and set $\phi_1 = \phi \mid D^1$. Then

$$\mu_1 \mid D^1 = k_1|\phi_1|/\phi_1 \; .$$

If $D = D^1$, the proof is completed. Otherwise, apply the same procedure to the Beltrami coefficient μ_1 and the configuration $(D-D^1, G)$. This will yield a Beltrami coefficient μ_2 which is equivalent to μ_1 with respect to $(D-D^1, G)$, and therefore to μ_0 with respect to (D, G), which is extremal with respect to $(D-D^1, G)$ and therefore also with respect to (D, G), and such htat either $\mu_2| D-D^1 = 0$, or there is a part D^2 of $(D-D^1, G)$ and an element $\phi_2 \in A(D^2, G)$, $\phi_2 \neq 0$, such that

$$\mu_2 \mid D^2 = k_2|\phi_2|/\phi_2$$

with $k_2 = \|\mu_2| D^2\| > 0$.

If $D = D^1 \cup D^2$, the proof is completed. Otherwise, apply the same reasoning to μ_2 and to the configuration $(D-D^1-D^2, G)$. After a finite number of steps, one obtains the desired result.

LEMMA 27. *If* (D, G) *is symmetric,* $\dim A(D, G) < \infty$, and $\mu \in L_\infty(C, G)$ *is symmetric and satisfies a symmetric Hamilton condition, then either* $\mu = 0$, *or there is a* $\phi \in A_+(D, G)$ *such that* $\phi \neq 0$ *and*

$$\|\mu \mid D\| \; |\phi(z)| = \mu(z)\phi(z)$$

a.e. in D.

The proof is completely analogous to that of Lemma 26; one must only note that if the complex vector space $A(D, G)$ has finite dimension, so has the real vector space $A_+(D, G)$.

It is now easy to establish the analogues of Theorems 8 and 9.

THEOREM 10. *If* (D, G) *is symmetric,* D *is connected, and* $\dim A(D, G) < \infty$, *then every symmetric and symmetrically extremal Beltrami coefficient satisfies the Teichmüller condition.*

THEOREM 11. *If* (D, G) *is symmetric and* $\dim A(D, G) < \infty$, *then every symmetric Beltrami coefficient is equivalent to a symmetric, symmetrically extremal Beltrami coefficient which satisfies the Teichmüller condition.*

We remark that if a symmetric μ satisfies the Teichmüller condition, the ϕ entering in the condition must be symmetric.

§7. Applications to Fuchsian Groups

The preceding theorems contain, in particular, all known and some new existence and characterization theorems for extremal quasiconformal mappings of Riemann surfaces. We make this explicit considering Fuchsian groups, with or without elliptic transformations. The translation of our remarks into the language of Riemann surfaces may be left to the reader.

Let G be a Fuchsian group acting on the upper half plane $\mathfrak{U}$. Every normalized (leaving the points $0, 1, \infty$ fixed) quasiconformal bijection $w: \mathfrak{U} \to \mathfrak{U}$, such that wGw^{-1} is a Fuchsian group, is of the form w^{μ} where μ is a symmetric element of $L_{\infty}(C, G)$ with $\|\mu\| < 1$. Let $\Lambda(G)$ denote the limit set of G, let $\sigma \subset R \cup \{\infty\}$ be a closed set containing $\Lambda(G)$, which is invariant under G, and let us consider the symmetric configuration (D, G) where $D = \hat{C} - \sigma$. For the sake of brevity we assume also that $\{0, 1, \infty\} \subset \sigma$.

We consider the variational problem: given a symmetric $\mu \in L_{\infty}(C, G)$ with $\|\mu\| < 1$, find a symmetrically extremal $\nu \in L_{\infty}(C, G)$ which is equivalent to μ. The latter condition means: $w^{\mu}|\sigma = w^{\nu}|\sigma$. If $\sigma = \Lambda(G)$, equivalence means only that μ and ν induce the same isomorphism of G:

$$w^{\mu}\circ\gamma\circ(w^{\mu})^{-1} = w^{\nu}\circ\gamma\circ(w^{\nu})^{-1} \quad \text{for} \quad \gamma \in G.$$

In all cases, equivalence implies the equality of the induced isomorphisms. If $\sigma = R \cup \{\infty\}$, equivalence means the same as equality of the induced isomorphisms provided that G is of the first kind, that is, if $\Lambda(G) = R \cup \{\infty\}$.

The existence of a symmetrically extremal ν follows from Theorem 7. By Theorem 8 and Lemma 25, an extremal ν has the Hamilton property which in our case reads: either $\nu = 0$ or

$$\|\nu\| = \sup\left\{\operatorname{Re}\iint_{\mathfrak{U}/G} \mu(z)\phi(z)\,dx\,dy \Big/ \iint_{\mathfrak{U}/G} |\phi(z)|\,dx\,dy\right\}$$

where ϕ runs over $A_+(D,G)-\{0\}$. This means that $\phi(z)$, $z \in \mathfrak{U}$, is holomorphic, $\phi \not\equiv 0$, $\phi(\gamma(z))\gamma'(z)^2 = \phi(z)$ for $\gamma \in G$,

$$\iint_{\mathfrak{U}/G} |\phi(z)|\,dx\,dy < \infty$$

and, if I is any open interval on R containing no points of σ, then $\phi(z)$ is continuous and real on I.

Using known properties of Fuchsian groups, one checks that the space $A_+(D,G)$ is finitely dimensional if and only if G is finitely generated and the set $[\sigma-\Lambda(G)]/G$ is finite. In this case, by Theorem 10, ν is of the Teichmüller form: either $\nu = 0$ or

$$\nu = k|\phi|/\phi$$

where k, $0 < k < 1$, is a constant and $\phi \in A_+(D,G)$.

The present approach yields nothing new concerning uniqueness. It is known by Teichmüller's uniqueness theorem (see [1] or [5]) that if $\dim A_+(D,G) < \infty$, then no two distinct symmetrically extremal ν's can be symmetrically equivalent.

Everything said here extends at once to the case when σ contains points off R, provided that σ is symmetric and $\sigma \cap \mathfrak{U}$ is discrete.

§8. Applications to Kleinian Groups

Let G be a Kleinian group, $\Lambda = \Lambda(G)$ its limit set, and $\Omega = \Omega(G) = (\mathbf{C} \cup \{\infty\}) - \Lambda$ its region of discontinuity. We assume that G is *non-elementary,* that is, that Λ is infinite, and that $\{0, 1, \infty\} \subset \Lambda$. A *quasi-conformal deformation* χ of G is an isomorphism of G (onto a Kleinian group G_1) of the form

$$G \ni \gamma \mapsto \chi(\gamma) = w^{\mu} \circ \gamma \circ (w^{\mu})^{-1} \in G_1$$

where $\mu \in L_{\infty}(C, G)$ with $\|\mu\| < 1$ and $\mu | \Lambda = 0$. (The latter condition is meaningful only if Λ has positive two-dimensional measure.) We call μ *minimal* if χ cannot be induced by a Beltrami coefficient ν with $\|\nu\| < \|\mu\|$.

THEOREM 12. *Every quasiconformal deformation* χ *of a Kleinian group* G *is induced by a minimal Beltrami coefficient* μ. *Every minimal* μ *satisfied the Hamilton condition:*

$$\|\mu\| = \sup \left\{ \iint_{\Omega/G} \mu(z)\phi(z)\, dx\, dy \Big/ \iint_{\Omega/G} |\phi(z)|\, dx\, dy \right\}$$

where ϕ *runs over* $A(\Omega, G) - \{0\}$.

Proof: The verification of the first statement is similar to (and simpler than) the proof of Theorem 4. We must only note that two elements, μ and ν, of the open unit ball in $L_{\infty}(C, G)$ induce the same isomorphism χ if and only if $w^{\mu} | \Lambda = w^{\nu} | \Lambda$. The second assertion follows from Theorem 5, since every minimal μ is also extremal with respect to (Ω, G), though not necessarily vice versa. (We write $\|\mu\|$ and not $\|\mu | \Omega\|$ since, by assumption, $\mu | \Lambda = 0$.)

We say that G is of *finite type* if Ω/G has finitely many components, each component is a compact Riemann surface, with perhaps finitely many points removed, and the natural mapping $\Omega \to \Omega/G$ is ramified over at most finitely many points. According to Ahlfors' finiteness theorem (see [13] and the literature quoted there), every finitely generated group is of finite type. If G is of finite type, then $\dim A(\Omega, G) < \infty$, hence Theorems 9 and 12 yield

THEOREM 13. *Let χ be a quasiconformal deformation of a Kleinian group G of finite type. Then χ can be induced by a Beltrami coefficient μ (with $\mu | \Lambda = 0$) which satisfies the Teichmüller condition.*

Under the hypothesis of this theorem, let Δ be a simply connected component of Ω and $h: \mathfrak{U} \to \Delta$ a conformal bijection. Let G_Δ be the subgroup of G which leaves Δ fixed. Then $G_0 = h^{-1} G_\Delta h$ is a Fuchsian group, and, since G is of finite type, G_0 is a finitely generated group of the first kind. Set $\nu = h^*(\mu | \Delta)$, where h^* is the operator defined in §2. If $\mu | \Delta = 0$, then $\nu = 0$. If $\mu | \Delta \neq 0$, then there is a $\phi \in A(\Delta, G_\Delta)$ and a number k such that $\phi \neq 0$, $0 < k < 1$ and $\mu | \Delta = k|\phi|/\phi$. We set, for $z \in \mathfrak{U}$, $\psi(z) = \phi(h(z))h'(z)^2$. Then $\psi \in A(\mathfrak{U}, G_0)$ and $\nu = k|\psi|/\psi$.

We note that $\Delta_1 = w^\mu(\Delta)$ is determined by χ. Also, there exists a conformal bijection $h_1 : \mathfrak{U} \to w^\mu(\Delta)$ such that

$$h_1 \circ w_\nu = (w^\mu | \Delta) \circ h \quad .$$

We define an isomorphism $\hat{\chi}$ of G_0 by setting, for every $\gamma \in G_\Delta$,

$$\hat{\chi}(h^{-1} \circ \gamma \circ h) = h_1^{-1} \circ \chi(\gamma) \circ h_1 \quad ,$$

and we verify that, for $\gamma \in G_0$,

$$\hat{\chi}(\hat{\gamma}) = w_\nu \circ \hat{\gamma} \circ w_\nu^{-1} \quad .$$

By Teichmüller's uniqueness theorem the isomorphism χ determines ν. Hence χ determines $(\mu | \Delta) = (h^*)^{-1}\nu$.

The proceding considerations contain the proof of

THEOREM 14. *Under the hypotheses of Theorem 13, assume that all components of* $\Omega(G)$ *are simply connected. Then* χ *determines* μ *uniquely.*

Applications of Theorems 12 and 14 to the theory of moduli of Kleinian groups will be considered elsewhere [11].

Miller Institute for Basic Research
University of California at Berkeley
and
Department of Mathematics
Columbia University

REFERENCES

[1]. L. V. Ahlfors, On quasiconformal mappings, *J. Analyse Math.* 3 (1953/54), 1-58.

[2] ______, *Lectures on Quasiconformal Mappings*, Van Nostrand, New York, 1966.

[3] ______ and L. Bers, Riemann's mapping theorem for variable metrics, *Ann. of Math.* 72 (1960), 385-404.

[4] ______ and G. Weill, A uniqueness theorem for Beltrami equations, *Proc. Amer. Math. Soc.*, 13 (1962), 975-978.

[5] L. Bers, Quasiconformal mappings and Teichmüller's theorem, *Analytic Function*, by R. Nevanlinna, et al., Princeton University Press, Princeton, N.J., 1960, 89-119.

[6] ______, Equivalence of two definitions of quasiconformal mappings, *Comm. Math. Helvet.* 37 (1962), 148-154.

[7] ______, Automorphic forms and general Teichmüller spaces, *Proc. Conf. on Complex Analysis*, Minnesota, 1964, Springer-Verlag, 105-113.

[8] ______, On moduli of Riemann surfaces, Lectures at Forschunginstitut fur Mathematik, *Eidgenossische Technische Hochschule*, Zurich, Summer 1964 (mimeographed).

[9] ———, An approximation theorem, *Journal d'Analyse Math.* 14 (1965), 1-4.

[10] ———, A non-standard integral equation with applications to quasiconformal mappings, *Acta Math.* 116 (1966), 113-134.

[11] ———, Spaces of Kleinian groups, *Several Complex Variables I, Maryland* 1970 (Lecture Notes in Mathematics 155), Springer Verlag (1970), 9-34.

[12] R. S. Hamilton, Extremal quasiconformal mappings with prescribed boundary values, *Trans. Amer. Math. Soc.* 138 (1969), 399-406.

[13] I. Kra, Eichler cohomology and the structure of finitely generated Kleinian groups, this volume p. 225.

ISOMORPHISMS BETWEEN TEICHMÜLLER SPACES

Lipman Bers and Leon Greenberg*

§1. Introduction

In this paper we prove a theorem about isomorphisms between Teichmüller spaces (Theorem 1 below) which has been announced without proof in [5,6]. The proof, while much shorter than the original version, is still complicated. We were unable to simplify it further, so that there is no justification for delaying publication. We refer to [1,2,5,6,7,8] for information on Teichmüller space theory. The isomorphism theorem is related to a theorem on Fuchsian groups (Theorem 2 below) which can be stated in purely classical terms. Our proofs of the two theorems are closely intertwined. Recently Marden [13] obtained a topological proof of a result equivalent to Theorem 2.

A) Let U denote the upper half-plane, and let G be a discrete (and therefore discontinuous) group of conformal selfmappings of U, that is a *Fuchsian group.* We denote by U_G the upper half-plane from which one has removed all fixed points of elliptic elements of G. Thus G acts freely on U_G and U_G/G is a Riemann surface.

To every Fuchsian group G one attaches a so-called *Teichmüller space* $T(G)$ and a *reduced Teichmüller space* $T^{\#}(G)$. These are metric spaces with a complex and a real analytic structure, respectively; we shall recall the definition in §2.

* This work has been partially supported by the National Science Foundation under Grant NSF – GP – 8735.

THEOREM 1. *Let* G *and* K *be Fuchsian groups. A conformal bijection* $U_G/G \to U_K/K$ *induces, canonically, isomorphisms* $T(G) \to T(K)$ *and* $T^{\#}(G) \to T^{\#}(K)$.

The theorem is known, and easy to prove if G and K have no elliptic elements (cf. §2, E). We are interested in the case where such elements are present in at least one of the two groups.

The effect of a conformal bijection on the so-called modular groups of G and K is described in §7.

B) THEOREM 2. *Let* G *be a Fuchsian group, and let* $z \mapsto w(z)$ *be an orientation preserving homeomorphism of* U_G *onto itself such that* $w \circ \gamma = \gamma \circ w$ *for all* $\gamma \in G$. *Then* w *is homotopic to the identity by a homotopy which commutes with all* $\gamma \in G$ *(that is by a homotopy* $(t, z) \mapsto \Phi_t(z) \in U_G$, $0 \leq t \leq 1$, $z \in U_G$, *with* $\Phi_0(z) = w(z)$, $\Phi_1(z) = z$, *and such that* $\Phi_t \circ \gamma = \gamma \circ \Phi_t$ *for* $0 \leq t \leq 1$ *and all* $\gamma \in G$*)*.

If G contains no elliptic elements, so that $U_G = U$, this result is well known and can be proved by a geometric device due to Ahlfors (see [3], p. 99). In this case Theorem 2 is equivalent to the statement: an automorphism of an orientable surface which preserves orientation and induces the identity automorphism on the fundamental group is homotopic to the identity. Our proof of Theorem 2, in the case $U_G \neq U$, makes essential use of Teichmüller spaces. More precisely, it will turn out that, for every group G, the statements of Theorems 1 and 2 imply each other. On the other hand, one can prove Theorem 1 directly for finitely generated groups, and one can show that Theorem 2 holds for G if it holds for all finitely generated subgroups of G.

§2. Teichmüller spaces and modular groups.

In this section we recall some basic concepts and results from Teichmüller space theory, and fix our notations.

A) We denote by M(1) the open unit ball in the complex Banach space $L_\infty(U)$, and by $\Sigma(1)$ the group of all quasiconformal selfmappings of U. Every $\omega \in \Sigma(1)$ is the restriction of a topological selfmapping of $R \cup \{\infty\} \cup U$. Let $\Sigma^*(1)$ denote the subgroup which leaves the points $0, 1, \infty$ fixed. Every $\omega \in \Sigma(1)$ can be written, uniquely, in the form $\omega = \alpha_1 \circ w_1$ and in the form $w_2 \circ \alpha_2$ where $w_1, w_2 \in \Sigma^*(1)$ and α_1, α_2 are real Möbius transformations (conformal selfmappings of U). Every $w \in \Sigma^*(1)$ is uniquely determined by its *Beltrami* coefficient $\mu(z) = (\partial w/\partial \bar{z})/(\partial w/\partial z)$ which is an element of M(1), and every $\mu \in M(1)$ is the Beltrami coefficient of a unique $w \in \Sigma^*(1)$; we write $w = w_\mu$ and use the bijection $\mu \mapsto w_\mu$ to identity M(1) with $\Sigma^*(1)$. Thus M(1) may be considered as a *group*. It is also a metric space, under the *Teichmüller distance* δ defined as follows: if $w_\mu \circ w_\nu^{-1} = w_\sigma$ and $k = \|\sigma\|_\infty = \text{ess. sup}\, |\sigma(z)|$, then

$$(2.1) \qquad \delta(\mu,\nu) = \delta(w_\mu, w_\nu) = \log(1+k) - \log(1-k) \;.$$

We recall that M(1) is not a topological group.

B) Let G be a Fuchsian group. We denote by $\Sigma(G)$ the set of those $\omega \in \Sigma(1)$ for which $\omega B \omega^{-1}$ of Möbius transformations, and hence a Fuchsian group, and we set $\Sigma^*(G) = \Sigma(G) \cap \Sigma^*(1)$.

Let $L_\infty(U, G)$ denote the closed linear subspace of $L_\infty(U)$ consisting of those $\mu(z)$ for which

$$(2.2) \qquad \mu(\gamma(z))\, \overline{\gamma'(z)}/\gamma'(z) = \mu(z) \qquad \text{for all} \quad \gamma \in G \,,$$

and set $M(G) = M(1) \cap L_\infty(U, G)$. Under the canonical bijection $M(1) \to \Sigma^*(1)$, M(G) is mapped onto $\Sigma^*(G)$.

We call two elements, ω and $\hat{\omega}$, of $\Sigma(G)$ *R-equivalent* if $\omega \mid R = \hat{\omega} \mid R$, *strongly* G*-equivalent if* $\omega^{-1} \circ \hat{\omega}$ commutes with all elements of G, and G*-equivalent* if there is a real Möbius transformation α such that $\hat{\omega}$ and $\alpha \circ \omega$ are strongly G-equivalent.

We denote the R-equivalence class of ω by $[\omega]$, the strong G-equivalence class of ω by $[\omega]_G$ and the G-equivalence class of ω by $(\omega)_G$.

C) The set of all $[\omega]$, $\omega \in \Sigma^*(G)$, is the *Teichmüller space* T(G); the set of all $(\omega)_G$, $\omega \in \Sigma(G)$, is the *reduced Teichmüller space* $T^{\#}(G)$. The mappings $\mu \longmapsto [w_\mu] \mapsto (w_\mu)_G$ define surjections $M(G) \to T(G) \to T^{\#}(G)$. One uses these surjections to define the Teichmüller metrics δ_G in T(G) and $\delta^{\#}_G$ in $T^{\#}(G)$, as well as a *complex analytic* structure in T(G) and a *real analytic* structure in $T^{\#}(G)$. If G is of the *first kind,* that is if the fixed points of elements of G (distinct from the identity) are dense in $\mathbf{R} \cup \{\infty\}$, then $T(G) = T^{\#}(G)$.

It turns out that T(G) and $T^{\#}(G)$ are complete metric spaces and that they can be realized, canonically, as bounded domains in complex or real Banach spaces. If $G_1 \subset G_2$, then there is a canonical embedding $T(G_2) \subset T(G_1)$. The corresponding relation between $T^{\#}(G_1)$ and $T^{\#}(G_2)$ exists only when the index $[G_2 : G_1] < \infty$.

D) An *isomorphism* between two Teichmüller spaces is a bijection which preserves the Teichmüller distance and the analytic structures. Every element $\omega \in \Sigma(G)$ induces *allowable isomorphisms.*

$$\omega^* : T(G) \to T(\omega G \omega^{-1}), \quad \omega^* : T^{\#}(G) \to T^{\#}(\omega G \omega^{-1}) \;,$$

defined as follows:

$$\omega^*([w]) = [a \circ w \circ \omega^{-1}], \; \omega^*((w)_G) = (w \circ \omega^{-1})_{\omega G \omega^{-1}} \; ; \tag{2.3}$$

here a is a real Möbius transformation determined by the condition $a \circ w \circ \omega^{-1} \in \Sigma^*(1)$.

E) If G and K are Fuchsian groups without elliptic elements, so that $U_G = U_K = U$, then every conformal bijection $U/G \to U/K$ is represented by a real Möbius transformation a such that $a G a^{-1} = K$. It induces, therefore, isomorphisms a^* of T(G) and $T^{\#}(G)$, respectively. This remark proves Theorem 1 for groups without elliptic elements.

F) We denote by $\Sigma_0(G)$ the normalizer of G in $\Sigma(1)$, this is the group of all $\omega \in \Sigma(G)$ with $\omega G \omega^{-1} = G$. The group $\Sigma_0(G)$ induces the *modular group* $\Gamma(G)$ of allowable automorphisms of T(G) and the *reduced modular group* $\Gamma^{\#}(G)$ of allowable automorphisms of $T^{\#}(G)$. If G is of the first kind, then $\Gamma^{\#}(G) = \Gamma(G)$.

G) If elements of G (distinct from the identity) have less than three distinct fixed points, G is called elementary. We assume from now on that G is *non elementary.*

We call G *normalized* if $0, 1, \infty$ are among the fixed points. There is usually no loss of generality in assuming that G is normalized, since this can be achieved by conjugating G with a real Möbius transformation.

Remark. *Let* G *be normalized and let* w *and* $\hat{w}$ *be* G*-equivalent elements of* $\Sigma^*(G)$. *Then* w *and* $\hat{w}$ *are strongly* G*-equivalent.*

Indeed, let a be a real Möbius transformation such that $w^{-1} \circ a \circ \hat{w}$ commutes with all $\gamma \in G$. We shall show that $a = \mathrm{id}$. There is a $\gamma_0 \in G$ such that $\gamma_0 \neq \mathrm{id}$ and $\gamma_0(0) = 0$. We may assume, replacing if need by γ_0 by γ_0^{-1}, that $\lim_{n\to\infty} \gamma_0^n(z) = 0$ for every $z \in U$. (Here $\gamma^n = \gamma \circ \gamma \circ \gamma \circ \cdots$, n times.) Now

$$(2.3) \qquad a \circ \hat{w} \circ \gamma_0^n(z) = w \circ \gamma_0^n \circ w^{-1} \circ a \circ \hat{w}(z) \ .$$

Letting $n \to \infty$ and noting that $\hat{w}(0) = w(0) = 0$, we conclude that $a(0) = 0$. Similarly, $a(1) = 1, a(\infty) = \infty$, so that $a = \mathrm{id}$.

The remark implies that if G is a normalized Fuchsian group, then the space $T^{\#}(G)$ may be defined as the set of all $[w]_G$ with $w \in \Sigma^*(G)$, and the canonical mapping $M(G) \to T^{\#}(G)$ is given by $\mu \mapsto [w_\mu]_G$. The elements of $\Gamma^{\#}(G)$ are of the form $[w_\mu]_G \mapsto [a \circ w_\mu \circ \omega]_G$ where $\omega \in \Sigma_0(G)$ and a is a real Möbius transformation such that $a \circ w_\mu \circ \omega \in \Sigma^*(1)$.

§3. Reformulation of the theorems

In this section we introduce some constructions which will be used throughout the rest of the paper, and then reformulate Theorems 1 and 2.

A) Let G be a Fuchsian group and $h: U \to U_G$ a universal covering. (The mapping h is uniquely determined by U_G, except that it may be replaced by $h \circ \alpha$, α a real Möbius transformation.) Let H be the covering group of h, i.e., the set of real Möbius transformations η with $h \circ \eta = h$. We denote by g the natural mapping $U_G \to U_G/G$, and we define f by the commutativity of the diagram

(3.1)
$$\begin{array}{ccc} & U & \\ {}^{h}\swarrow & & \searrow^{f} \\ U_G & \xrightarrow{\quad g \quad} & U_G/G \end{array}$$

Then f is a universal covering. We denote by F the covering group of f. Then F and H are Fuchsian groups and there is an exact sequence

(3.2)
$$1 \to H \hookrightarrow F \xrightarrow{\chi} G \to 1$$

where χ is defined by

(3.3)
$$h \circ \phi = \chi(\phi) \circ h, \quad \phi \in F .$$

The whole construction is determined by the choice of h. We usually assume in what follows that G is normalized and that h has been chosen once and for all, in such a way that F is also normalized.

B) The limit set $\Lambda(G)$ of a Fuchsian group is, by definition, the set of accumulation points of the set of fixed points. For a group G of the first kind, $\Lambda(G) = R \cup \{\infty\}$. Otherwise the group is called of the second kind and $\Lambda(G)$ is a nowhere dense closed subset of $R(G) \cup \{\infty\}$. For a group G of the second kind we set

$$\tilde{U}_G = (U \cup R \cup \{\infty\}) - \Lambda(G) .$$

The group operates discontinuously on $\tilde{U}_G$, so that the mapping g extends to a mapping $\tilde{U}_G \to \tilde{U}_G/G$.

One verifies easily that the three groups G, H, F considered above are either all of the first kind or all of the second kind. In the latter case the mappings h and f can be extended, by continuity to the universal coverings $h : \tilde{U}_H \to \tilde{U}_G$, $f : \tilde{U}_H \to \tilde{U}_G/G$. (By abuse of language we introduce no new names for the extended mappings.)

C) The groups F and H are covering groups of universal coverings, hence they contain no elliptic elements. The following statement is, therefore, a special case of Theorem 1.

THEOREM 1′. T(F) *is canonically isomorphic to* T(G). $T^{\#}(F)$ *is canonically isomorphic to* $T^{\#}(G)$.

On the other hand, Theorem 1′ implies Theorem 1, in view of the remark in §2, E).

D) Next we reformulate Theorem 2.

THEOREM 2′. *Let* w *be an orientation preserving homeomorphism of* U_G *onto itself, with*

$$w \circ \gamma = \gamma \circ w \quad \textit{for all} \quad \gamma \in G . \tag{3.4}$$

Then there is a homeomorphism W *of* U *onto itself such that*

$$h \circ W = w \circ h \tag{3.5}$$

and

$$W \circ \phi = \phi \circ W \quad \textit{for all } \phi \in F . \tag{3.6}$$

We remark that every automorphism w of U_G can be lifted to an automorphism W of U. The relation (3.5) determines W only modulo composition with an element $\eta \in H$. The theorem asserts that if W satisfies (3.4), then there is an $\eta \in H$ such that $\eta \circ W$ commutes with all $\phi \in F$.

E) Theorem 2 implies Theorem 2′.

Indeed, let Φ_t be the homotopy described in Theorem 2 and let W_0 be some homeomorphism of U onto itself satisfying $h \circ W_0 = w \circ h$. For every t, $0 \le t \le 1$, there is a continuous mapping $\Psi_t : U \to U$ such that $h \circ \Psi_t = \Phi_t \circ h$. If we demand that $\Psi_0 = W_0$ and that Ψ_t depend on t continuously, then Ψ_t is uniquely determined and the mapping $(t, z) \mapsto \Psi_t(z)$ is continuous. Since $\Psi_1 = \text{id}$, we have $h \circ \Psi_1 = h$ and there is an $\eta_1 \in H$ such that $\Psi_1 = \eta_1$.

For every t, $0 \le t \le 1$, and every $\phi \in F$ we have, noting (3.3) and the condition that Φ_t commute with all $\gamma \in G$,

$$h \circ \Psi_t \circ \phi = \Phi_t \circ h \circ \phi = \Phi_t \circ \chi(\phi) \circ h = \chi(\phi) \circ \Phi_t \circ h$$

$$= \chi(\phi) \circ h \circ \Psi_t = h \circ \phi \circ \Psi_t$$

so that there is an $\eta \in H$ with

$$\Psi_t \circ \phi = \eta \circ \phi \circ \Psi_t \ .$$

This η seemingly depends on ϕ and on t, but since the dependence on t must be continuous and H is discrete, η is the same for all t. For $t = 1$ we obtain $\eta_1 \circ \phi = \eta \circ \phi \circ \eta_1$ or $\eta = \eta_1 \circ \phi \circ \eta_1^{-1} \circ \phi^{-1}$. For $t = 0$ we conclude that $W_0 \circ \phi = \eta_1 \circ \phi \circ \eta_1^{-1} \circ W_0$. It is clear that $W = \eta_1^{-1} \circ W_0$ satisfies (3.5) and (3.6).

F) Conversely, Theorem 2′ implies Theorem 2.

Indeed, since F has no elliptic elements, one knows that for an automorphism W of U commuting with every $\phi \in F$ there is a homotopy $\Psi_t : U \to U$, $0 \le t \le 1$, with $\Psi_0 = W$, $\Psi_1 = \text{id}$ and $\Psi_t \circ \phi = \phi \circ \Psi_t$ for all t and all $\phi \in F$. For every t, $0 \le t \le 1$, define the mapping $\Phi_t : U_G \to U_G$ by setting $\Phi_t(h(z)) = h(\Psi_t(z))$. One verifies that the definition is unambiguous, that the mapping $(t, z) \longmapsto \Phi_t(z)$, $0 \le t \le 1$, $z \in U_G$, is continuous, and that $\Phi_0 = w$, $\Phi_1 = \text{id}$. Every $\gamma \in G$ may be written as $\gamma = \chi(\phi)$, $\phi \in F$, and we have $\Phi_t \circ \gamma \circ h = \Phi_t \circ h \circ \phi = h \circ \Psi_t \circ \phi = h \circ \phi \circ \Psi_t = \gamma \circ h \circ \Psi_t = \gamma \circ \Phi_t \circ h$. Since h is surjective, $\Phi_t \circ \gamma = \gamma \circ \Phi_t$. This proves the assertion.

G) It is convenient to have still another reformulation of Theorem 2.

THEOREM 2″. *Let* w *be an orientation preserving homeomorphism of* U_G *onto itself which commutes with all elements of* G. *Then the automorphism* ϖ *of* U_G/G *induced by* w *is homotopic to the identity.*

The homeomorphism ϖ is, of course, defined by

$$\varpi \circ g = g \circ w \quad . \tag{3.7}$$

If we assume Theorem 2, then the conclusion of Theorem 2″ follows, since for every t, $0 \leq t \leq 1$, there is a continuous mapping $P_t: U_G/G \to U_G/G$ such that

$$P_t \circ g = g \circ \Phi_t \quad . \tag{3.8}$$

This may be considered as a continuous mapping of $[0,1] \times (U_G/G)$ into U_G/G, and we have $P_0 = \varpi$, $P_1 = \text{id}$.

Concersely, Theorem 2″ implies Theorem 2. Indeed, if P_t, $0 \leq t \leq 1$, is a homotopy of ϖ into the identity, then for every t there is a continuous mapping $\Phi_t: U_G \to U_G$ satisfying (3.8), with $\Phi_0 = w$ and $\Phi_1 \in G$. If we require that the dependence of Φ_t on t be continuous, Φ_t is determined uniquely.

For every $\gamma \in G$ and for every t we have, by (3.8),

$$g \circ \Phi_t \circ \gamma = P_t \circ g \circ \gamma = P_t \circ g = g \circ \Phi_t \quad .$$

Then there is a $\gamma_t \in G$ with $\Phi_t \circ \gamma = \gamma_t \circ \Phi_t$. Since γ_t must depend on t continuously, and G is discrete, $\gamma_t = \gamma_0$, that is $\gamma_t = \gamma$. Hence Φ_t commutes with all elements of G. It follows that $\Phi_1 = \text{id}$. Then Φ_t has the properties required by Theorem 2.

§4. Construction of the canonical isomorphisms

In this section we construct mappings of $T(F)$ and $T^{\#}(F)$ onto $T(G)$ and $T^{\#}(G)$, respectively, which will be proved, in §5 and §7 to be isomorphisms.

A) For $\zeta \in U_G$ there is a $z \in U$ such that $h(z) = \zeta$. If $\mu \in L_\infty(U,F)$, we set

$$(m\mu)(\zeta) = \mu(z)\, h'(z)/\overline{h'(z)}, \qquad \zeta = h(z) \;. \tag{4.1}$$

LEMMA 1. $m : L_\infty(U,F) \to L_\infty(U,G)$ *is a norm preserving linear bijection.*

Proof: Only two things must be checked: the definition (4.1) is meaningful, and $m\mu \in L_\infty(U,G)$. We note that if $h(z) = \zeta$ and $h(z') = \zeta$, then there is an $\eta \in H$ with $z' = \eta(z)$. Since $h \circ \eta = h$, $H \subset F$, and $\mu(\phi(z))\overline{\phi'(z)}/\phi'(z) = \mu(z)$ for $\phi \in F$, we verify easily that (4.1) does not depend on the choice of z.

Since $U - U_G$ is a discrete set, $m\mu$ is a well-defined element of $L_\infty(U)$. A direct calculation shows that $(m\mu)(\gamma(z))\overline{\gamma'(z)}/\gamma'(z) = m\mu(z)$ for $\gamma = \chi(\phi)$, $\phi \in F$.

We shall occasionally write m_h instead of m, to indicate the dependence on h.

B) If $\mu \in M(F)$, then $m\mu \in M(G)$. We shall write $m(w_\mu) = w_{m\mu}$.

LEMMA 2. *Let* $W \in \Sigma^*(F)$, $w \in \Sigma^*(G)$, $G_1 = wGw^{-1}$. *Then* $w = m_h(W)$ *if and only if there is a universal covering* $h_1 : U \to U_{G_1}$ *and a commutative diagram*

$$\begin{array}{ccc} U & \xrightarrow{\;W\;} & U \\ {\scriptstyle h}\downarrow & & \downarrow{\scriptstyle h_1} \\ U_G & \xrightarrow{\;w\;} & U_{G_1} \end{array} \tag{4.2}$$

(By abuse of language we identify w with $w \mid U_G$.)

Proof: Clearly, $w(U_G) = U_{G_1}$. If the diagram is known to be commutative, one computes that the Beltrami coefficient of w is $m\mu$, where μ is the Beltrami coefficient of $h_1 \circ W$, which is also the Beltrami coefficient of W, since h_1 is holomorphic. If it is known that $w = m(W)$, define h_1 by the relation $h_1 \circ W = w \circ h$. One computes that $h_1 = w \circ h \circ W^{-1}$ is holomorphic, hence a universal covering.

Remark. We note that the groups $F_1 = WFW^{-1}$ and $G_1 = wGw^{-1}$ are normalized. Indeed, if $\phi_0 \in F$ leaves 0 fixed, then $W \circ \phi_0 \circ W^{-1} \in F_1$ also has the fixed point 0, and the argument can be repeated for $1, \infty$ and for G_1. Also, the covering group of h_1 is $H_1 = WHW^{-1}$ as one verifies at once.

C) LEMMA 3. *Let* $W, \hat{W} \in \Sigma^*(F)$ *and set* $w = m_h(W)$, $\hat{w} = m_h(\hat{W})$, $F_1 = WFW^{-1}$, $G_1 = wGw^{-1}$. *Then* $\hat{W} \circ W^{-1} \in \Sigma^*(F_1)$ *and there is a universal covering* $h_1 : U \to U_{G_1}$ *such that*

$$m_{h_1}(\hat{W} \circ W^{-1}) = \hat{w} \circ w^{-1} . \tag{4.3}$$

Proof: The first statement is trivial. Set $\hat{G} = \hat{w} G \hat{w}^{-1}$. According to Lemma 2 there is a universal covering $\hat{h} : U \to U_{\hat{G}}$ and a commutative diagram

(4.4)

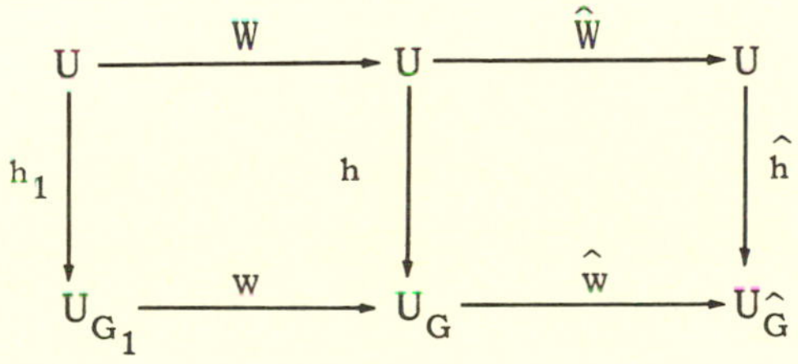

This implies (4.3), again by Lemma 2.

COROLLARY. *The mapping* $m_h : \Sigma^*(F) \to \Sigma^*(G)$ *is an isometry in the Teichmüller metric.*

Proof: We use the notations of Lemma 3. In view of Lemma 1 and of (4.3) the Beltrami coefficients of $\hat{W} \circ W^{-1}$ and of $\hat{w} \circ w^{-1}$ have the same norm.

D) LEMMA 4. *If* $W \in \Sigma^*(F)$ *commutes with all* $\phi \in F$, *then* $w = m(W)$ *commutes with all* $\gamma \in G$.

Proof: By hypothesis h and h_1 in (4.4) are universal coverings with the same covering group $H = WHW^{-1}$. Hence $h_1 = \alpha \circ h$ where $\alpha : U_G \to U_{G_1}$ is a conformal mapping. Since $U - U_G$ and $U - U_{G_1}$ are discrete sets, α is a real Möbius transformation. Now let $\gamma \in G$ be given. Then $\gamma = \chi(\phi)$ for some $\phi \in F$ and using (3.3) and (4.2) we obtain $w \circ \gamma \circ h = w \circ h \circ \phi =$ $h_1 \circ W \circ \phi = \alpha \circ h \circ W \circ \phi = \alpha \circ h \circ \phi \circ W = \alpha \circ \gamma \circ h \circ W =$ $\alpha \circ \gamma \circ \alpha^{-1} \circ \alpha \circ h \circ W = \alpha \circ \gamma \circ \alpha^{-1} \circ h_1 \circ W = \alpha \circ \gamma \circ \alpha^{-1} \circ w \circ h$. Since h is a surjection on U_G we have $w \circ \gamma \circ w^{-1} = \alpha \circ \gamma \circ \alpha^{-1}$. Thus w is G-equivalent to the identity, and (cf. §2, G)) since G is normalized, w is strongly G-equivalent to the identity, so that $\alpha = \mathrm{id}$.

COROLLARY. *If* W *and* $\hat{W}$ *are strongly* F*-equivalent elements of* $\Sigma^*(F)$, *then* $m(W)$ *and* $m(\hat{W})$ *are strongly* G*-equivalent elements of* $\Sigma^*(G)$.

Proof: The hypothesis implies that $\hat{W} \circ W^{-1}$ commutes with all elements of $F_1 = WFW^{-1}$. By Lemmas 3 and 4 it follows that $\hat{w} \circ w^{-1}$ commutes with all elements of $G_1 = wGw^{-1}$. This means that w and $\hat{w}$ are strongly G-equivalent.

E) In view of the Corollary above we can define the mapping $m : T^{\#}(F) \to T^{\#}(G)$ by setting $m[W]_F = [m(W)]_G$ for every $W \in \Sigma^*(F)$.

LEMMA 5. *The mapping* $m : T^{\#}(F) \to T^{\#}(G)$ *is a continuous open surjection.*

Proof: The mapping $m : M(F) \to M(G)$ is an isometric bijection, the natural mappings $M(F) \to T^{\#}(F)$, $M(G) \to T^{\#}(G)$ are continuous and open, and the following diagram is commutative.

$$\begin{array}{ccc} M(F) & \xrightarrow{\quad m \quad} & M(G) \\ \downarrow & & \downarrow \\ T^{\#}(F) & \xrightarrow{\quad m \quad} & T^{\#}(G) \end{array}$$

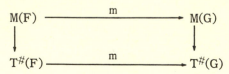

LEMMA 6. *If the mapping* $m : T^{\#}(F) \to T^{\#}(G)$ *is injective, it is an isomorphism.*

This is obvious in view of Lemma 1 and the Corollary to Lemma 3.

F) LEMMA 7. *Let* $W \in \Sigma^*(F)$ *be such that* $W \mid R = \text{id}$, *and set* $w = m(W)$. *Then* $w \mid R = \text{id}$.

(Here and hereafter we identify, by abuse of language, an element of $\Sigma(1)$ with its continuous extension to $U \cup R$.)

Proof: If $W \mid R = \text{id}$, then W is strongly F-equivalent to the identity, and w is strongly G-equivalent to the identity, by Lemma 4. Then $w \mid \Lambda(G) = \text{id}$. If G is of the first kind, the conclusion of the lemma follows. Assume now that G is of the second kind. We must show that, for every component I of $(R \cup \{\infty\}) - \Lambda(G)$, we have $w \mid I = \text{id}$. There is an interval J on $R \cup \{\infty\}$ such that $h(J) = I$; here h stands for the continuous extension of h to $\bar{U}_H$ (cf. §3, B)). We noted in the proof of Lemma 4 that, under our hypothesis, $h \circ W = w \circ h$. For $\zeta \in J$ we have $w \circ h(\zeta) = h \circ W(\zeta) = h(\zeta)$.

COROLLARY. *If* W *and* $\hat{W}$ *are* R*-equivalent elements of* $\Sigma^*(F)$, *then* $w = m(W)$ *and* $\hat{w} = m(\hat{W})$ *are* R*-equivalent.*

Proof: By hypothesis, $\hat{W} \circ W^{-1} \mid R = \text{id}$. By Lemmas 3 and 7, $\hat{w} \circ w^{-1} \mid R = \text{id}$, so that w and $\hat{w}$ are R-equivalent.

G) LEMMA 8. *Let* $W \in \Sigma^*(F)$ *be strongly* F*-equivalent to the identity, and let* $w = m(W)$ *be* R*-equivalent to the identity. Then* $W \mid R = \text{id}$.

Proof: There is nothing to prove if F, H, G are of the first kind. Assume that these groups are of the second kind (cf. §2, B)). The hypothesis implies that $W \mid \Lambda(F) = \text{id}$. Let J be a component of $(R \cup \{\infty\}) - \Lambda(F)$. Then $I = h(J)$ is a component of $R \cup \{\infty\} - \Lambda(G)$, hence an open arc of $R \cup \{\infty\}$. Let H_J be the subgroup of H which maps J onto itself. Then H_J

is the covering group of the unbounded unramified covering $h|J: J \to I$. Hence $H_J = 1$.

The endpoints of J belong to $\Lambda(F)$. Therefore, $W(J) = J$. As in the proof of Lemma 7 we have $h \circ W = w \circ h$. Therefore, for $x \in J$, $h \circ W(x) = w \circ h(x) = h(x)$. Hence $W|J \in H$; therefore $W|J \in H_J$, therefore $W|J = \mathrm{id}$, q.e.d.

COROLLARY TO LEMMA 8. *If W and $\hat{W}$ are strongly F-equivalent elements of $\Sigma^*(F)$, and $w = m(W)$ is R-equivalent to $\hat{w} = m(\hat{W})$, then W and $\hat{W}$ are R-equivalent.*

Proof: By hypothesis, $\hat{W} \circ W^{-1}$ commutes with all elements of $F_1 = WFW^{-1}$ and $\hat{w} \circ w^{-1}|R = \mathrm{id}$. Bt Lemmas 3 and 8 we have $\hat{W} \circ W^{-1}|R = \mathrm{id}$, q.e.d.

H) In view of the corollary to Lemma 7 one can define a mapping $m: T(F) \to T(G)$ by setting $m[W] = [m(W)]$ for every $W \in \Sigma^*(F)$.

LEMMA 9. *If $m: T^{\#}(F) \to T^{\#}(G)$ is injective then $m: T(F) \to T(G)$ is an isomorphism.*

Proof: It follows at once from Lemma 1 and the Corollary to Lemma 3 that if $m: T(F) \to T(G)$ is injective it is an isomorphism.

Assume now that W and $\hat{W}$ are elements of $\Sigma^*(F)$ with $m[W] = m[\hat{W}]$. Then $w = m(W)$ is R-equivalent to $\hat{w} = m(\hat{W})$, hence, a fortiori strongly G-equivalent. If $m: T^{\#}(F) \to T^{\#}(G)$ is injective, then W and $\hat{W}$ are strongly F-equivalent. Now the Corollary to Lemma 8 implies that W and $\hat{W}$ are R-equivalent, so that $[W] = [\hat{W}]$. In this case $m: T(F) \to T(G)$ is injective.

§5. Proof of the isomorphism theorem for finitely generated groups.

In this section we prove Theorem 1′, assuming G to be finitely generated. We begin by proving three lemmas which are valid also for infinitely generated groups.

A) We denote by $\Sigma_0(F,H)$ the subgroup of $\Sigma_0(F)$ consisting of all $\omega \in \Sigma_0(F)$ with

$$\omega \circ \phi \circ \omega^{-1} \circ \phi^{-1} \in H \quad \text{for all } \phi \in F. \tag{5.1}$$

LEMMA 10. *If* $\omega \in \Sigma_0(F,H)$, *then there is a bijection* $\lambda : U_G \to U_G$ *with* $h \circ \omega = \lambda \circ h$. *This bijection commutes with all* $\gamma \in G$.

The proof is obvious.

COROLLARY. *If* $\omega \in \Sigma_0(F,H)$ *is a real Möbius transformation, then* $\omega \in H$.

Proof: If ω is a real Möbius transformation, so is λ, since λ must then be conformal, Hence $\lambda = \text{id}$ and $h \circ \omega = h$.

We denote by $\Gamma^{\#}(F,H)$ the subgroup of $\Gamma^{\#}(F)$ induced by $\Sigma_0(F,H)$, cf. §2, D). It will turn out that this subgroup is trivial.

B) LEMMA 11. *Let* $\omega \in \Sigma_0(F,H)$. *If there is a* $W \in \Sigma^*(F)$ *such that* W *and* $W \circ \omega$ *are* F*-equivalent, then* $\hat{W}$ *and* $\hat{W} \circ \omega$ *are* F*-equivalent for every* $\hat{W} \in \Sigma^*(F)$.

Proof: Assume that W and $W \circ \omega$ are F-equivalent and set $F_1 = WFW^{-1}$, $H_1 = WHW^{-1}$. There is a real Möbius transformation α with

$$\alpha \circ W \circ \phi \circ W^{-1} \circ \alpha^{-1} = W \circ \omega \circ \phi \circ \omega^{-1} \circ W^{-1} \quad \text{for all } \phi \in F \tag{5.2}$$

Since $\omega \circ \phi \circ \omega^{-1} = \eta \circ \phi$ where $\eta \in H$ (with η depending on ϕ), we have $\alpha \circ \phi_1 \circ \alpha^{-1} \circ \phi_1^{-1} \in H_1$ for $\phi \in F_1$. Hence $\alpha \in \Sigma_0(F_1, H_1)$, and by the Corollary to Lemma 10 we have $\alpha \in H_1$. Hence there is a fixed $\eta \in H$ with $\alpha = W \circ \eta \circ W^{-1}$. Now (5.2) shows that ω is strongly F-equivalent to η.

If $\hat{W} \in \Sigma^*(F)$, then there is a real Möbius transformation $\hat{\eta}$ with $\hat{W} \circ \eta = \hat{\eta} \circ \hat{W}$. By what was said above, $\hat{W} \circ \omega$ is strongly F-equivalent to $\hat{W} \circ \eta = \hat{\eta} \circ \hat{W}$ and thus F-equivalent to $\hat{W}$.

COROLLARY. *Elements of* $\Gamma^{\#}(F,H)$ *distinct from the identity have no fixed points in* $T^{\#}(F)$.

C) LEMMA 12. *Let* $W, \hat{W} \in \Sigma^*(F)$. *Then* $w = m(W)$ *is strongly* G-*equivalent to* $\hat{w} = m(\hat{W})$ *if and only if there is a real Möbius transformation* α *and* $a\,\omega \in \Sigma_0(F,H)$ *with* $\hat{W} = \alpha \circ W \circ \omega$.

Proof: Assume $\hat{W} = \alpha \circ W \circ \omega$, where α and ω are as in the statement of the lemma. Note that the mapping λ of Lemma 10 leaves $0, 1, \infty$ fixed, since it commutes with all elements in the normalized group G. Hence $w \circ \lambda \in \Sigma^*(1)$. Now define $\hat{h}$ by the commutativity of the diagram

(5.3)

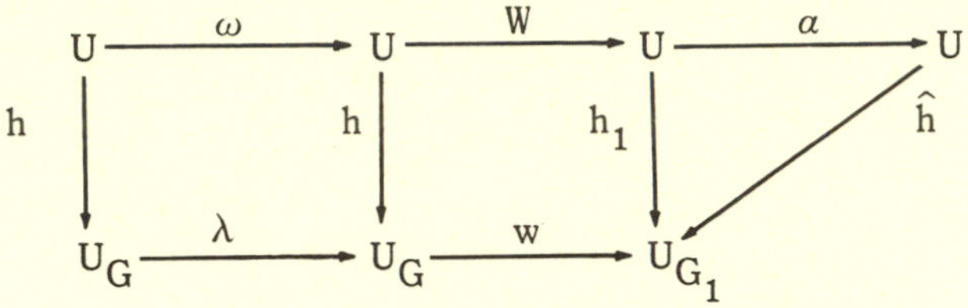

Then $\hat{h}$ is a universal covering and, by Lemma 2, $w \circ \lambda = m(\alpha \circ W \circ \omega) = m(\hat{W}) = \hat{w}$. Hence $w^{-1} \circ \hat{w} = \lambda$, and w is strongly G-equivalent to $\hat{w}$, by Lemma 10.

Assume next that w and $\hat{w}$ are strongly G-equivalent. Then, using the notation of diagram (4.4) we have $G_1 = \hat{G}$. Hence h_1 and $\hat{h}$ are universal coverings, by U, of the same domain $U_{G_1} = U_{\hat{G}}$. Therefore $h_1 = \hat{h} \circ \alpha$ where α is a real Möbius transformation and we have the commutative diagram

(5.4)

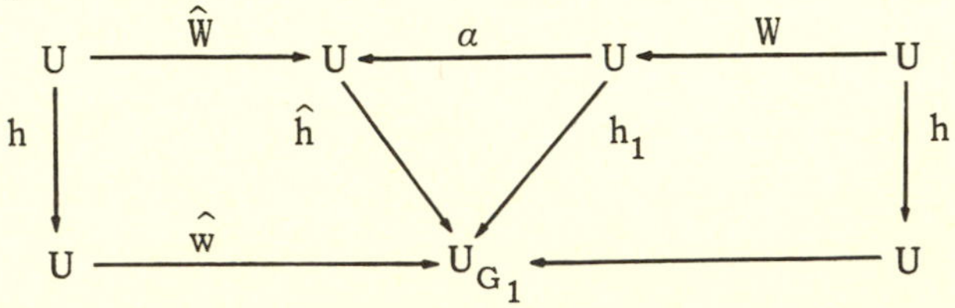

Set $W^{-1} \circ \alpha^{-1} \circ \hat{W} = \omega$; we must show that $\omega \in \Sigma_0(F,H)$. We set $\lambda = w^{-1} \circ \hat{w}$ and note that for every $\phi \in F$ and $\gamma = \chi(\phi)$, we have $\lambda \circ \gamma =$

$\gamma \circ \lambda$. Since $h \circ \omega = \lambda \circ h$ we have $h \circ \omega \circ \phi = \lambda \circ h \circ \phi = \lambda \circ \gamma \circ h = \gamma \circ \lambda \circ h = \gamma \circ h \circ \omega = h \circ \phi \circ \omega$. Hence $\omega \circ \phi \circ \omega^{-1} \circ \phi^{-1} \in H$, q.e.d.

COROLLARY. *Two points of* $T^{\#}(F)$ *have the same image in* $T^{\#}(G)$ *under* m *if and only if there is an element of* $\Gamma^{\#}(F,H)$ *which sends one point into the other.*

D) We assume now that G is finitely generated. It is known that this implies that U_G/G is a Riemann surface obtained from a closed surface of genus p by removing a finite number, say $n_0 + n_1$ disjoing continua of which precisely n_0 are points, with $n_0 \geq 0$, $n_1 \geq 0$ and

$$d = 6p - 6 + 2n_0 + 3n_1 \geq 0 \ .$$

In this case $T^{\#}(G)$ is a real analytic manifold of real dimension d. Since U_G/G and U/F are conformally equivalent, F is also finitely generated and $T^{\#}(F)$ is also a real analytic manifold of dimension d.

It is also known that $T^{\#}(G)$ is simply connected, in fact, homeomorphic to Euclidean space. The result is due to Fricke [9]; it appears also in the Fenchel-Nielsen theory of discontinuous group [10] and in a paper by Keen [12].

Finally, we shall need the fact that the group $\Gamma^{\#}(F)$ acts discontinuously on $T^{\#}(F)$. This is also a classical result by Fricke, reproved in the Fenchel-Nielsen theory and, in the context of Teichmüller space theory, by Kravetz [11].

E) Set $\Gamma^{\#}(F,H) = \Gamma_0$. Since $\Gamma_0 \subset \Gamma^{\#}(F)$, Γ_0 is a discontinuous group of isometries of $T^{\#}(F)$. Its action is free (that is, without fixed points) by the Corollary to Lemma 11. The mapping $m : T^{\#}(F) \to T^{\#}(G)$ is, in view of Lemma 5 (cf. §3, E)) and the Corollary to Lemma 12, an unbounded unramified covering with covering group Γ_0. Since $T^{\#}(G)$ is simply connected, $\Gamma_0 = 1$ and m is a bijection. Theorem 1′ now follows from Lemmas 6 and 9 in Section 4.

§6. Proof of the homotopy theorem

In this section we prove Theorem 2.

A) LEMMA 13. *Let* w_0 *be an orientation preserving automorphism of* U_G *which commutes with all* $\gamma \in G$ *and induces an automorphism* ϖ_0 *of* U_G/G. *Let* ϖ_0 *be homotopic to another automorphism,* ϖ_1, *of* U_G/G. *Then there is an automorphism* w_1 *of* U_G *which commutes with all* $\gamma \in G$ *and which induces* ϖ_1.

Proof: The homotopy which takes ϖ_0 into ϖ_1 can be lifted to U_G. Then we have, for $0 \leq t \leq 1$, commutative diagrams

$$\begin{array}{ccc} U_G & \xrightarrow{\;\Phi_t\;} & U_G \\ {\scriptstyle g}\downarrow & & \downarrow{\scriptstyle g} \\ U_G/G & \xrightarrow[\;P_t\;]{} & U_G/G \end{array}$$

where Φ_t, P_t are continuous mappings, depending continuously on t, and $P_0 = \varpi_0$, $P_1 = \varpi_1$, $\Phi_0 = w_0$. Set $w_1 = \phi_1$. One verifies that w_1 is an automorphism.

For every γ in G and every t, we have that $g \circ \Phi_t \circ \gamma = P_t \circ g \circ \gamma = P_t \circ g = g \circ \Phi_t$. Since γ_t depends continuously on t, it does not depend on t at all. Since $\gamma_0 = \gamma$ by hypothesis we have $w_1 \circ \gamma = \gamma \circ w_1$ for all $\gamma \in G$.

B) LEMMA 14. *Theorem 2″ holds for a group* G *if it holds whenever the mapping* $w : U_G \to U_G$ *is locally quasiconformal. If* G *is finitely generated, then Theorem 2″ holds if it holds for every quasiconformal* w.

Proof: Set $S = U_G/G$, and let w induce the automorphism ϖ of S. It is known (cf. [4]) that ϖ is homotopic to locally quasiconformal automorphism ϖ_1. (Cf. [4]; "locally quasiconformal" means "quasiconformal on every relatively compact subdomain.") In view of Lemma 13 the mapping

ϖ_1 is induced by an automorphism w_1 of U_G such that $\gamma \circ w_1 = w_1 \circ \gamma$ for all $\gamma \in G$. Since g is conformal and $\varpi_1 \circ g = g \circ w_1$, w_1 is locally quasiconformal. If Theorem 2″ holds for w_1, ϖ_1 is homotopic to the identity and so in ϖ.

C) In order to prove the second half of the lemma, we show first that for a finitely generated G the automorphism ϖ is homotopic to a quasiconformal one. Now S can be represented as $S = \tilde{S} - (\delta_1 \cup \cdots \cup \delta_N)$ where $\tilde{S}$ is a compact surface of genus P and $\delta_1, \ldots, \delta_N$ are disjoint continua on $\tilde{S}$; the numbers P and N are subject to the sole restriction that $3P - 3 + N \geq 0$. In order that $\varpi : S \to S$ be homotopic to a quasiconformal automorphism it is sufficient that for every sequence $\{p_\nu\} \subset S$ with $\lim p_\nu \in \delta_i$ and $\lim \varpi(p_\nu) \in \delta_j$, the continua δ_i and δ_j be either both points or both non-degenerate (cf. [4]). We can assert more:

(6.1) $\quad$ if $\quad \lim p_\nu \in \delta_i$, then $\lim \varpi(p_\nu) \in \delta_i$.

Indeed, to every δ_i there belong simple closed curves C on S with the property: the complement of C in $\tilde{S}$ has two components, one of which (called the interior of C) is simply connected and contains δ_i. Consider a sequence $C_1, C_2, C_3, \ldots$ of curves belonging to δ_i, such that C_{n+1} lies in the interior of C_n and no sequence of points $z_n \in C_n$, $n = 1, 2, 3, \ldots$, converges in S. In $\tilde{S}$ every such sequence converges to δ_i.

To every δ_i there belongs a conjugacy class of maximal cyclic subgroups $D_\nu \in G$ such that for every curve C belonging to p_i, each component of $g^{-1}(C)$ is a curve invariant under some D_ν, and, conversely every simple curve C on S with this property belongs to p_i. Now, if a curve $C_0 \subset U_G$ is invariant under a subgroup $D \subset G$, then the curve $w(C_0)$ is invariant under the same subgroup. It follows that if the curve $C \subset S$ belongs to p_i, so does the curve $\varpi(C)$. This remark implies the validity of (6.1).

The proof of Lemma 14 can now be completed by repeating the argument given in B).

D) We proceed to prove Theorem 2′ for a finitely generated group G. Since Theorems 2′ and 2″ are equivalent, we may assume that the mapping w is quasiconformal (see Lemma 14). It is also easy to check that we loose no generality in assuming that the groups G, F, H (see §3, A)) are normalized.

The bijection $w : U_G \to U_G$ can be lifted to U: there is a topoligical selfmapping W_0 of U such that

$$h \circ W_0 = w \circ h \ . \tag{6.2}$$

Since w commutes with every $\gamma \in G$, we have, for $\phi \in F$,

$$h \circ W_0 \circ \phi = w \circ h \circ \phi = x \circ \chi(\phi) \circ h = \chi(\phi) \circ w \circ h$$
$$= \chi(\phi) \circ h \circ W_0 = h \circ \phi \circ W_0$$

so that

$$W_0 \circ \phi \circ W_0^{-1} \circ \phi^{-1} \in H \text{ for all } \phi \in F \ . \tag{6.3}$$

Since h is conformal and w is quasiconformal, W_0 is quasiconformal. Also, if $\phi \in F$ is such that 0 is an attracting fixed point of ϕ, then $w \circ \phi^n(z) = \phi^n \circ w(z)$, so that, letting $n \to \infty$, we have $w(0) = 0$. Similarly, $w(1) = 1$, $w(\infty) = \infty$ and thus $w \in \Sigma^*(G)$. (Here we used the fact that G is normalized.) Now let α be a real Möbius transformation such that $\alpha \circ W_0$ leaves $0, 1, \infty$ fixed. By (6.2) we have $W_0 \in \Sigma(G)$. Hence $W = \alpha \circ W_0 \in \Sigma^*(G)$. Also, $h_1 = h \circ \alpha^{-1} : U \to U_G$ is a universal covering, and $h_1 \circ W = w \circ h$.

By Lemma 2 we have $w = m(W)$. Since we already know that $m : T^{\#}(F) \to T^{\#}(G)$ is injective, and since w is strongly G-equivalent to the identity, W is strongly F-equivalent to the identity (i.e., commutes with all $\phi \in F$). Thus

$$\alpha \circ W_0 \circ \phi \circ W_0^{-1} \circ \alpha^{-1} \circ \phi^{-1} = \text{id}, \text{ for all } \phi \in F \ . \tag{6.4}$$

Together with (6.3) this yields $\alpha \circ \phi \circ \alpha^{-1} \circ \phi^{-1} \in H$ for $\phi \in F$, so that $\alpha \in H$ by the Corollary to Lemma 10. Therefore $h_1 = h \circ \alpha^{-1} = h$, and

$W = \alpha \circ W_0$ has the properties asserted by Theorem 2′.

E) Now let G be infinitely generated. Since G is countable, there exists a sequence $G_1 \subset G_2 \subset G_3 \subset \cdots$ of finitely generated subgroups of G such that $G = G_1 \cup G_2 \cup G_3 \cup \cdots$. For each j let $h_j : U \to U_{G_j}$ be the uniquely determined universal covering which satisfies

$$h_j(i) = h(i), \qquad h_j'(i)\,\overline{h'(i)} > 0 \; . \tag{6.5}$$

We construct the groups H_j and F_j as in Section 3, A), using h_j and G_j instead of h and g. Thus there is an exact sequence

$$1 \to H_j \hookrightarrow F_j \xrightarrow{\chi_j} G_j \to 1 \tag{6.6}$$

with

$$h_j \circ \phi = \chi_j(\phi) \circ h_j \quad \text{for} \quad \phi \in F_j \; . \tag{6.7}$$

LEMMA 15. *We have*

$$\lim_{j \to \infty} h_j(z) = h(z) \quad \text{normally} \tag{6.8}$$

(that is uniformly on compact subsets of U). *Also, for every* $\phi \in F$ *there is an integer* $N > 0$ *and a sequence* ϕ_j, $j = N, N+1, \ldots$ *with* $\phi_j \in F_j$, $\chi_j(\phi_j) = \chi(\phi)$, *and* $\lim_{j \to \infty} \phi_j = \phi$.

The proof is rather routine, but for the sake of completeness we shall carry it out.

F) Proof of Lemma 15. First we assume that $h_j(z)$ converges normally to some function $k(z)$; we shall show that $k(z) = h(z)$.

The function k is holomorphic and $k(i) = h(i)$. Let ζ be a point in U_G. We construct a simply connected domain $D \subset U_G$ containing ζ and $h(i)$. There is a holomorphic mapping $g_j : D \to U$ such that $g_j(h(i)) = i$ and $h_j(g_j(z)) = z$. Since $g_j(D) \subset U$, Schwarz' lemma implies that for every

$D_0 \subset\subset D$ there is a $U_0 \subset\subset D$ (independent of j) with $q_j(D_0) \subset U_0$. (Here $\subset\subset$ denotes relatively compact subsets.) Selecting, if need be, a subsequence, we may assume that $\lim q_j(z) = q(z)$ normally in D, and $k(q(z)) = z$. Hence $k(U) \supset U_G$. Since for every $\zeta_0 \in U - U_G$ we have $h_j(z) \neq \zeta_0$ for $z \in U$ and large j, we must have $k(z) \neq \zeta_0$ (by Hurwitz' theorem). Thus $k(U) = U_G$.

Next let z_0 be any given point in U. Let $D \subset U_G$ be an open disc centered at $h(z_0)$, and let j be so large that $h_j(z_0) \in D$. For such a j there is a holomorphic function $p_j(z)$ in D with $p_j(D) \subset U$, $p_j(h(z_0)) = z_0$ and $h_j(p_j(z)) = z$. As before, a subsequence of $\{p_j(z)\}$ will converge to a holomorphic function $p(z)$ with $k(p(z)) = z$.

We have shown that k is an unbounded and unramified, and thus universal, covering of U_G by U. Therefore $k'(i) \neq 0$, so that, by (6.5), $k'(i)\, h'(i) > 0$. This implies that $k = h$.

We remark next that, by Schwarz' lemma, every $U_0 \subset\subset U$ has the property that $h_j(U_0) \subset U_1 \subset\subset U_0$, where U_1 does not depend on j. Therefore every increasing sequence of integers contains a subsequence $\{j_n\}$ such that $\{h_{j_n}(z)\}$ converges normally. But we have just shown that the limit function of this convergent subsequence must be $h(z)$. Hence the selection of a subsequence was unnecessary and (6.8) holds.

G) It remains to prove the second statement in Lemma 15. We note first that F_j and F act freely on U, so that an element of these groups is determined by its action on a single point.

Let $\phi \neq \mathrm{id}$, an element of F, be given, and set $\gamma = \chi(\phi)$, $Z = \phi(i)$ and $\zeta = \gamma(h(i)) = h(Z)$. There is a sequence $\{Z_j\} \subset U$ with $\lim Z_j = Z$ and $h_j(Z_j) = h(Z)$. There is a $\phi_j \in F_j$ with $\phi_j(i) = Z_j$. We note that this ϕ_j is unique, that $\chi_j(\phi_j) = \gamma$ and hence $h_j \circ \phi_j = \gamma \circ h_j$. We may assume, selecting if need be a subsequence, that there is a real Möbius transformation ψ with $\lim \phi_j = \psi$. Then $h \circ \psi = \gamma \circ h$, so that $\psi \in F$. Since $\psi(i) = Z$, we have $\psi = \phi$, and the selection of the subsequence was unnecessary.

H) Now we prove Theorem 2′ for the group G, assuming that the mapping w is locally quasiconformal in U_G. In view of Lemma 14 this involves no loss of generality.

Since $U - U_G$ is discrete we may extend w, by continuity, to an automorphism of U; we denote the extended mapping by the same letter w. Note that w need not be locally quasiconformal in U.

We use the notations introduced in E). Since Theorem 2′ is already established for finitely generated groups, we know that for every $j = 1, 2, \ldots$, there is an automorphism W_j of U such that

$$(6.9) \qquad h_j \circ W_j = w \circ h_j$$

$$(6.10) \qquad \phi \circ W_j = W_j \circ \phi \quad \text{for all} \quad \phi \in F_j .$$

We now establish

LEMMA 16. *For every domain* $D \subset\subset U$ *there are numbers* N *and* K *such that* $W_j \mid D$ *is* K *quasiconformal for* $j > N$.

(We recall that a K quasiconformal mapping is a homeomorphic solution of a Beltrami equation $\partial W/\partial \bar{z} = \mu(z)(\partial W/\partial z)$ where $|\mu(z)| \leq (K-1)/(K+1)$.)

Proof: The closure of h(D) is compact in $h(U) = U_G$ and there are domains Δ_0 and D_0 such that $h(D) \subset\subset \Delta_0 \subset\subset U_G$, $D \subset\subset D_0$ and $h(D_0) \subset \Delta_0$. Let N be so large that $h_j(D) \subset h(D_0)$ for $j > N$; such an N exists in view of (6.8). Since h_j is conformal everywhere and $w \mid \Delta_0$ is quasiconformal, and therefore K quasiconformal for some K, we conclude from (6.9) that $W_j \mid D$ is K quasiconformal for some K.

Now let $D \subset\subset U$ be a circular disc and let N and K be as in the lemma. Then we can write, for $j > N$,

$$W_j \mid D = \zeta_j \circ \chi_j$$

where χ_j is a K quasiconformal automorphism of D which leaves the center of D fixed, and $\zeta_j : D \to U$ is a conformal mapping. Indeed, we simply choose for χ_j that unique solution of the Beltrami equation satisfied by W_j which maps D and its center onto themselves. It is well known that all χ_j, $j > N$, are equicontinuous. On the other hand the holomorphic functions $\hat{\zeta}_j = (\zeta_j - i)/(\zeta_j + i)$ are uniformly bounded.

Applying the preceding argument to a sequence of discs exhausting U, and selecting if need by a subsequence, we may assume that either (I) the sequence W_j converges normally to a function, or (II) the sequence diverges to ∞, again normally.

Next, case (II) cannot occur. For if it did, choose an element $\phi \in F$ such that $\phi(z) = (as + b)/(cz + d)$, $c \neq 0$, and a sequence $\phi_j \in F_j$ with $\phi_j \to \phi$. Then one may write $\phi_j(z) = (a_j z + b_j)/(c_j z + d_j)$ with $\lim a_j = a$, $\lim b_j = b$, $\lim c_j = c$, $\lim d_j = d$. Now, by (6.10),

$$\frac{a_j W_j(z) + b_j}{c_j W_j(z) + d_j} = W_j\left(\frac{a_j z + b}{c_j z + d}\right) .$$

The left hand side converges to a/c and the right hand side diverges to ∞; this is absurd.

It follows that $\lim\limits_{j \to \infty} W_j(z) = W(z)$ where $W : U \to U$ is a continuous mapping. Since also $h_j \circ W_j^{-1} = w^{-1} \circ h_j$, the same argument shows that W is a topological self-mapping of U. Clearly, $h \circ W = w \circ h$. It also follows from Lemma 13 that W commutes with all $\phi \in F$.

§7. Conclusion

Now we can complete the proof of the isomorphism theorem and discuss the effect of the canonical isomorphisms on the modular groups. We assume again that G and F are normalized.

A) LEMMA 16. *Let* $W \in \Sigma^*(F)$ *and let* $w = m(W)$ *commute with all elements of* G. *Then* W *commutes with all elements of* F.

Proof: By Theorem 2′ there is a $\omega \in \Sigma(F)$ which commutes with all $\phi \in F$ and satisfies $h \circ \omega = w \circ h$. Since F is normalized, ω leaves $0, 1, \infty$ fixed. Thus $\omega \in \Sigma^*(F)$, and $w = m(\omega)$ by Lemma 2. Hence $\omega = W$.

COROLLARY. *If* $W, \hat{W} \in \Sigma^*(F)$ *and* $w = m(W)$ *is strongly* G*-equivalent to* $\hat{w} = m(\hat{W})$, *then* W *is strongly* F*-equivalent to* $\hat{W}$.

Proof: By hypothesis, $\hat{w} \circ w^{-1}$ commutes with all elements of $G_1 = wGw^{-1}$. Hence, by Lemmas 3 and 16, $\hat{W} \circ W^{-1}$ commutes with all elements of $F_1 = WFW^{-1}$. Thus $W^{-1} \circ \hat{W}$ commutes with all elements of F.

B) The corollary asserts that $m : T^{\#}(F) \to T^{\#}(G)$ is an injection. In view of Lemmas 6 and 9 this establishes Theorem 1′.

C) Addition to Theorem 1′. *The canonical isomorphisms* m *induce isomorphisms of* $\Gamma(G)$ *and* $\Gamma^{\#}(G)$ *into subgroups of* $\Gamma(F)$ *and* $\Gamma^{\#}(F)$, *respectively.*

Proof: Let $w \in \Sigma^*(G)$, $\omega \in \Sigma_0(G)$ and let α be a real Möbius transformation such that $\alpha \circ w \circ \omega \in \Sigma^*(G)$. There is a $W \in \Sigma^*(F)$ with $w = m(W)$, so that we have the commutative diagram (4.2). Also, the mapping $\omega : U_G \to U_G$ can be lifted to a quasiconformal selfmapping Ω of U such that $h \circ \Omega = \omega \circ h$. One verifies easily that $\Omega \in \Sigma_0(F)$. Let β be a real Möbius transformation such that $\beta \circ W \circ \Omega \in \Sigma^*(F)$, and set $h_2 = \alpha \circ h_1 \circ \beta^{-1}$, $G_2 = \alpha^{-1} G_1 \alpha$. Then we have the commutative diagram

$$\begin{array}{ccccccc} U & \xrightarrow{\Omega} & U & \xrightarrow{W} & U & \xrightarrow{\beta} & U \\ \downarrow h & & \downarrow h & & \downarrow h_1 & & \downarrow h \\ U_G & \xrightarrow{\omega} & U_G & \xrightarrow{w} & U_{G_1} & \xrightarrow{\alpha} & U_{G_2} \end{array}$$

and h_2 is a universal covering. By Lemma 2 we have that $m(\beta \circ W \circ \Omega) = \alpha \circ w \circ \omega$.

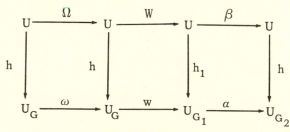

It follows that ω^* and Ω^*, considered either in the modular groups or in the reduced modular groups, are connected by the relations $m \circ \Omega^* = \omega^* \circ m$. In other words, $m^{-1}\Gamma^{\#}(G)m$ and $m^{-1}\Gamma(G)m$ are subgroups of $\Gamma^{\#}(F)$ and $\Gamma(F)$, respectively.

D) The subgroups considered above are, in general, *proper* subgroups.

Example. Let S be a closed Riemann surface of genus 3 and let p_1, p_2, p_3 be 3 distinct points on S. Let G be a Fuchsian group such that U_{G}/G is conformally equivalent to $S - \{p_1, p_2, p_3\}$. Assume that G contains elliptic elements of orders 2, 3 and 5, respectively. Then $\Gamma(F) = \Gamma^{\#}(F)$ is isomorphic to the group of homotopy classes of quasiconformal selfmappings of $S - \{p_1, p_2, p_3\}$, but only selfmappings which leave each point p_ν fixed induce elements of $\Gamma(G) = \Gamma^{\#}(G)$.

E) It can be shown that for a finitely generated group G the index of $m^{-1}\Gamma^{\#}(G)m$ in $\Gamma^{\#}(F)$ is finite.

Department of Mathematics,
Columbia University

Miller Institute for Basic Research,
University of California at Berkeley

Department of Mathematics
University of Maryland

REFERENCES

[1] L. V. Ahlfors, *Lectures on Quasiconformal Mappings*, D. Van Nostrand Co., 1966.

[2] _____ and L. Bers, "Riemann's mapping theorem for variable metrics," *Ann. of Math.* 72, 385-404 (1960).

[3] L. Bers, "Quasiconformal mappings and Teichmüller's theorem," in *Analytic Functions*, 89-119, Princeton University Press, 1960.

[4] _____ "Uniformization by Beltrami equation," *Comm. Pure Appl. Math.* 14, 215-228 (1961).

[5] _____ *On Moduli of Riemann Surfaces* (mimeographed lecture notes), Eidgenössische Technische Hochschule, Zurich, 1964.

[6] _____ "Automorphic forms and general Teichmüller spaces," *Proc. Conf. Compl. Anal.* (Minneapolis 1964), 109-113, Springer-Verlag, 1965.

[7] _____ "Universal Teichmüller space," in *Analytic Methods in Mathematical Physics*, 65-83, Gordon and Breach Science Publishers, 1970.

[8] C. J. Earle, "Reduced Teichmüller spaces," *Trans. Amer. Math. Soc.* 126, 54-63 (1967).

[9] R. Fricke and F. Klein, *Vorlesungen über die theorie der automorpher funktionen*, Vol. 2, B. G. Teubner, 1926.

[10] L. Greenberg, to appear.

[11] S. Kravetz, "On the geometry of Teichmüller spaces and the structure of their modular groups," *Ann. Acad. Sci. Fenn. Ser. AI* No. 278, 1-35 (1959).

[12] L. Keen, "Intrinsic moduli on Riemann surfaces," *Ann. of Math.* 84, 404-420 (1966).

[13] A. Marden, "On homotopic mappings of Riemann surfaces," *Ann. of Math.* 90, 1–8 (1969).

ON THE MAPPING CLASS GROUPS OF CLOSED SURFACES AS COVERING SPACES

by

Joan S. Birman* and Hugh M. Hilden

§1. Introduction

Let T_g be a closed, orientable surface of genus g, and let $\mathcal{H}(T_g)$ be the group of all orientation-preserving homeomorphisms of $T_g \to T_g$. $\mathcal{H}(T_g)$ contains a subgroup $\mathcal{D}(T_g)$ consisting of all homeomorphisms which are isotopic to the identity. The mapping class group $M(T_g)$ of T_g is defined to be the quotient group $\mathcal{H}(T_g)/\mathcal{D}(T_g)$. Alternatively, $M(T_g)$ is known as the Teichmüller modular group, and also as the homeotopy group of T_g (although the latter term usually includes the orientation-reversing homeomorphisms of $T_g \to T_g$). $M(T_g)$ can also be characterized algebraically as the group of all classes of "proper" automorphisms of the fundamental group $\pi_1 T_g$ of the surface, where the inner automorphisms constitute the class of the identity. (A proper automorphism is one which maps the single defining relation in any one-relator presentation of $\pi_1 T_g$ into a conjugate of itself. If one allows orientation-reversing homeomorphisms, the corresponding algebraic characterization would be the totality of automorphism classes of $\pi_1 T_g$, where one admits automorphisms which map the basic relator into either a conjugate or itself or of its inverse.)

It is well-known that every homeomorphism of $T_g \to T_g$ is isotopic to a sequence of special types of homeomorphisms known as twists, and that the $3g-1$ classes which are represented by the twists about the closed paths $u_1, \ldots, u_g, z_1, \ldots, z_{g-1}, y_1, \ldots, y_g$ in Figure 1 generate $M(T_g)$

* The work of the first author was supported in part by a summer research grant from Stevens Institute of Technology.

[7]. A complete set of defining relations for the group $M(T_g)$, however, has not been obtained for $g > 1$. Our main result is to obtain defining relations for $M(T_2)$ and, for $g > 2$, for a sequence of subgroups $M_{s_i}(T_g)$, $i = 0, 1, \ldots, g-1$, which, taken together, generate $M(T_g)$, although each $M_{s_i}(T_g)$ is a proper subgroup of $M(T_g)$ if $g > 2$. The derivation, being based on the fact that T_g can be regarded as a ramified covering of a sphere with an appropriate set of branch points, also casts some light on the relationship between mapping class groups of surfaces and Artin's braid group.

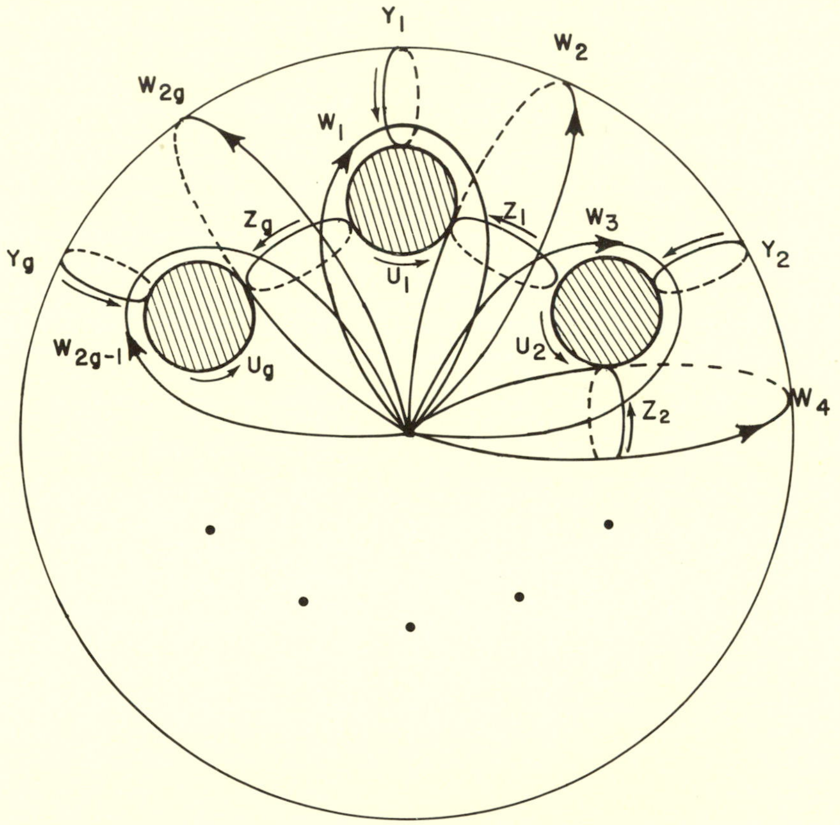

FIG. 1

To relate our work to known results, we remark that Siegel's modular group is known to be a quotient group of $M(T_g)$. Algebraically, Siegel's modular group can be characterized as the group of proper automorphisms of the abelianized fundamental group of T_g, or alternatively as the group $Sp(2g, Z)$ of $2g \times 2g$ proper symplectic matrices with integral entries [15, page 178]. Defining relations for $Sp(2g, Z)$ are known for arbitrary g [13], but the kernel of the homomorphism from $M(T_g)$ to $Sp(2g, Z)$ is not known for $g > 1$. For $g = 1$ the groups $M(T_1)$ and $Sp(2, Z)$ coincide because $\pi_1 T_1$ is abelian, $M(T_1)$ being the group of 2×2 matrices with integral entries and determinant $+1$ [6, page 85]. This group has been studied extensively in the literature. The group $M(T_2)$ was investigated by Bergau and Mennicke [2], who proposed a system of defining relations; however, a gap in their proof (Lemma 5 on p. 430 of [2]) has left open the question of completeness, which is now confirmed by our proof. For $g > 2$ little is known about defining relations in $M(T_g)$, and our results are new.

We begin in Section 2 by investigating the relationship between mapping class groups of T_g and of a sphere with $2g+2$ points removed, denoted $T_{0,2g+2}$, which arises from the fact that T_g can be regarded as a 2-sheeted covering of T_0 with $2g+2$ branch points. We show that an isomorphism exists between $M(T_{0,2g+2})$ and $\tilde{S}(T_g)$, where $\tilde{S}(T_g)$ is a quotient group of a special mapping class group $S(T_g)$, in which admissible maps are restricted to those homeomorphisms of $T_g \to T_g$ which preserve fibres in the covering space projection from $T_g \to T_{0,2g+2}$. Using known generators and defining relations for $M(T_{0,2g+2})$, as determined in [14], we thus obtain a presentation for $\tilde{S}(T_g)$ which arises in a natural geometric manner.

In Section 3 the relationship between $S(T_g)$ and $M(T_g)$ is investigated. It is shown that $S(T_g)$ is isomorphic to each of the subgroups $M_{s_i}(T_g)$, $i = 0, 1, \ldots, g-1$, where $M_{s_i}(T_g)$ is defined to be the subgroup of $M(T_g)$ generated by the $2g+1$ mapping classes represented by the twists about the closed paths $u_1, \ldots, u_g, z_{1+i}, \ldots, z_{g-1+i}, y_{1+i}, y_{g+i}$ in Figure 1, where all subscripts are modulo g. We will use the symbol $M_s(T_g)$ to denote any one of these subgroups, e.g., $M_{s_0}(T_g)$. The establishment of this isomorphism is the central idea in our development. It rests on a proof of the fact that if a homeomorphism of $T_g \to T_g$ preserves fibres in a particular 2-sheeted covering of T_0, and is deformable to the identity map, then in fact it is deformable to the identity map via an isotopy which is fibre-preserving. It then follows (Section 4) that defining relations for $\tilde{S}(T_g)$ can be used to obtain defining relations for $S(T_g)$ and $M_s(T_g)$.

Section 5 contains some partial results on the relationship between $M_s(T_g)$ and the full mapping class group $M(T_g)$ for $g \geq 3$. In Section 6 some connections with other work are discussed briefly.

Since the group $M(T_1)$ is well-known, and also because some of our results (noteably Lemma 4.1 and Th. 6) require special treatment for the case $g = 1$, we restrict our attention for the most part to $g \geq 2$.

§2. The special mapping class group

Let T_g be a surface of genus g, embedded in E^3 in the manner illustrated in Figure 2. The embedding is chosen in such a way (see Figure 2) that:

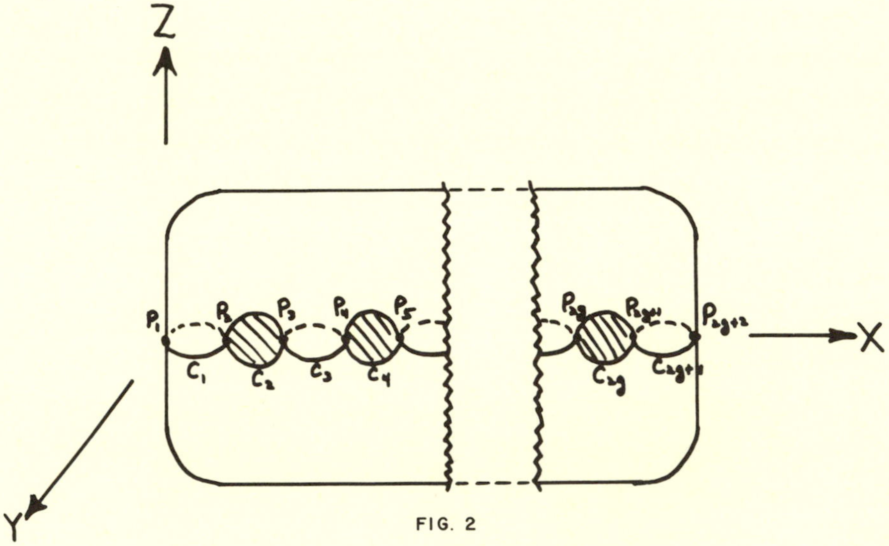

FIG. 2

(a) T_g is invariant under reflections in the xz and xy planes.

(b) $T_g \cap \{(x, y, z)/-1 \leq y \leq 1\}$ is a set of g right circular cylinders of radius 1 with axes parallel to the y axis. Let $c_1, c_3, \ldots, c_{2g+1}$ denote the intersections of T_g with xy plane, and $c_2, c_4, \ldots, c_{2g}$ the intersections of T_g with the xz plane. Each c_i is a circle.

Define an equivalence relation on T_g by $(x, y, z) \sim (+x, -y, -z)$. The natural projection $\pi: T_g \to T_g/\sim$ is easily seen to be 2-to-1 except for $2g+2$ points on the x axis which we shall call $P_1, \ldots, P_{2g+2}$ in order of increasing x coordinates. The surface $T_g/\sim$ is homeomorphic to a sphere T_0, the images of the exceptional points $P_1, \ldots, P_{2g+2}$ under π being a set of exceptional points $Q_1, \ldots, Q_{2g+2}$ on T_0.

Let $\iota: T_g \to T_g$ be defined by $\iota(x, y, z) = (x, -y, -z)$. (Thus ι interchanges the points in each fibre.) We say that a homeomorphism $H: T_g \to T_g$ is "symmetric" if it has the property:

$$H(\iota(z)) = \iota(H(z)) \tag{1}$$

where z now denotes an arbitrary point on the surface T_g. We remark that the involution ι is a symmetric homeomorphism. Also, that every symmetric homeomorphism keeps the special points $P_1, \ldots, P_{2g+2}$ fixed as a set, although not necessarily individually fixed. Of particular interest will be the fact that certain of the twists which generate $M(T_g)$ can be represented by symmetric homeomorphisms (for genus 1 and 2 every homeomorphism is isotopic to a symmetric homeomorphism), as will be demonstrated in Section 2.

Let $S(T_g)$ be the "symmetric" mapping class group of T_g, that is the group $\mathcal{S}(T_g)$ of all orientation-preserving symmetric homeomorphisms modulo the subgroup $\mathcal{SD}(T_g) \subseteq \mathcal{S}(T_g)$ of those symmetric homeomorphisms which are isotopic to the identity map within the group $\mathcal{S}(T_g)$. (We will describe this situation by saying that a map is "symmetrically isotopic to 1"). Let $\mathcal{I} \in S(T_g)$ denote the element represented by the involution ι. Then $\mathcal{I}$ has order 2 and lies in the center of $S(T_g)$. Define $\tilde{S}(T_g)$ to be the factor group of $S(T_g)$ modulo the cyclic subgroup of order 2 generated by $\mathcal{J}$. We claim:

THEOREM 1: $\tilde{\tilde{S}}(T_g)$ *is isomorphic to the mapping class group* $M(T_{0,2g+2})$ *of the sphere with* $2g+2$ *points removed.*

Proof: First we show that $M(T_{0,2g+2})$ is a homomorphic image of $\tilde{\tilde{S}}(T_g)$; second that the homomorphism has an inverse.

Let $H: T_g \to T_g$ be symmetric. Consider any point $\xi \in T_0$. Then $\pi^{-1}(\xi)$ will be a pair of points $z, \iota(z) \in T_g$. Since H is symmetric, the points $H(z)$, $H(\iota(z))$ are mapped by π into a single point on T_0 which we denote $H^*(\xi)$. Thus every symmetric homeomorphism H induces a homeomorphism $H^*: T_0 \to T_0$ defined by $H^* = \pi H \pi^{-1}$. Since H keeps the special points $P_1, \ldots, P_{2g+2}$ fixed as a set, and $\pi: P_i \to Q_i$ for each $i = 1, \ldots, 2g+2$, it follows that H^* will keep $Q_1, \ldots, Q_{2g+2}$ fixed as a set. Moreover, if H is symmetrically isotopic to 1, then the isotopy H_t will induce an isotopy H_t^* defined by $H_t^* = \pi H_t \pi^{-1}$. Thus the projection π induces a homomorphism from $S(T_g)$ to $M(T_{0,2g+2})$, and also from $\tilde{\tilde{S}}(T_g)$ to $M(T_{0,2g+2})$, since $\pi H \pi^{-1} = \pi i H \pi^{-1}$.

Now, if Γ is any simple closed curve on $T_{0,2g+2}$, then Γ lifts to a closed curve if and only if Γ encloses an even number of special points. The property of enclosing an even number of special points is preserved under homeomorphisms, the property of a closed curve lifting to a closed curve is preserved under homotopy, and $\pi_1(T_{0,2g+2})$ is generated by equivalence classes of simple closed curves. This implies that a closed curve Γ lifts to a closed curve if and only if its image under an arbitrary homeomorphism of $T_{0,2g+2}$ lifts to a closed curve.

Now we claim every homeomorphism $H^*: T_{0,2g+2} \to T_{0,2g+2}$ can be lifted to a homeomorphism $H: T_g \to T_g$ which represents a unique element of $\tilde{\tilde{S}}(T_g)$. Let H^* be such a map. Choose any particular point $z_0 \in T_{g,2g+2}$, where $T_{g,2g+2}$ is the surface T_g with the points $P_1, \ldots, P_{2g+2}$ removed. We note that the pair $(\pi, T_{g,2g+2})$ are a covering space over $T_{0,2g+2}$. If $\xi_0 = \pi(z_0)$, then $\pi^{-1}(H^*(\xi_0))$ will consist of two points, and we pick one arbitrarily, denoting it $H(z_0)$. Now choose any $y \in T_{g,2g+2}$, and let γ be a curve from z_o to y. Then $\pi(\gamma)$ is a curve from ξ_o to $\pi(y)$, and $H^*(\pi(\gamma))$ is a curve from $H^*(\xi_o)$ to $H^*(\pi(y))$. We define $H(y)$ to be the endpoint of the unique lift of $H^*(\pi(\gamma))$ whose initial point is $H(z_o)$. It is easily checked that H is well-defined and a homeomorphism.

Now, suppose H^* is isotopic to the identity, via an isotopy which we denote by H_t^*. For any particular point $\xi \in T_{0,2g+2}$ the isotopy H_t will define a path on $T_{0,2g+2}$ joining $H^*(\xi)$ to ξ. By the path lifting property for covering spaces, this path can be lifted to a path joining one of the points in the set $\pi^{-1}(H^*(\xi))$ to one of the points in the set $\pi^{-1}(\xi)$ which is unique modulo the choice of the initial lift $\pi^{-1}(H^*(\xi_o))$. However, whereas H_t^* pulls each point ξ back to the identity, the induced isotopy H_t might pull each point $H(z)$ back to $\iota(z)$ instead of z. Thus a homomorphism exists from $M(T_{0,2g+2})$ onto $\tilde{S}(T_g)$, but not necessarily onto $S(T_g)$.

Since the homomorphism from $M(T_{0,2g+2})$ onto $\tilde{S}(T_g)$ is invertible, it follows that $\tilde{S}(T_g)$ and $M(T_{0,2g+2})$ are isomorphic groups, which completes the proof of Theorem 1.

Now consider the mapping class group $M(T_{0,2g+2})$. This group was studied by W. Magnus [14], who showed that $M(T_{0,2g+2})$ admits the presentation*:

generators: $\sigma_1, \ldots, \sigma_{2g+1}$

defining relations:

(2.1) $\sigma_i \leftrightarrows \sigma_j \qquad 1 \leq i \leq 2g-1 \qquad |i-j| \geq 2$;

(2.2) $\sigma_i \sigma_{i+1} \sigma_i = \sigma_{i+1} \sigma_i \sigma_{i+1} \qquad 1 \leq i \leq 2g$;

(2.3) $(\sigma_1 \sigma_2 \cdots \sigma_{2g+1})^{2g+2} = 1$

(2.4) $\sigma_1 \sigma_2 \cdots \sigma_{2g} \sigma_{2g+1} \sigma_{2g+1} \sigma_{2g} \cdots \sigma_2 \sigma_1 = 1$

The generators σ_i, $1 \leq i \leq 2g+1$ in the above presentation can be represented by the following self-homeomorphisms of $T_{0,2g+2}$. Let D be the disc $r \leq 2$ in the Euclidean plane, parametrized by (r, θ). Define a homeomorphism $h: D \to D$ by $h(r, \theta) = (r, \theta - \pi r)$. Let h_i be an embedding of D into $T_{0,2g+2}$, where $h_i(D)$ includes Q_i and Q_{i+1} but is disjoint from all Q_j, $j \neq i, i+1$. Suppose also that $h_i(1,0) = Q_i$ and $h_i(1,\pi) = Q_{i+1}$. Then $h_i h h_i^{-1}$ extended to the identity map outside $h_i(D)$ defines a self-homeomorphism of $T_{0,2g+2}$ which represents σ_i. The effect of this map is to interchange of the points Q_i and Q_{i+1}, while leaving the other points Q_j fixed.

Our next task is to determine what types of self-homeomorphisms of T_g are defined by the proceedure we described in the proof of Theorem 1 for lifting self-homeomorphisms of $T_{0,2g+2}$ to self-homeomorphisms of T_g? In fact, it will turn out that the representatives of the σ_i lift to precisely the types of twist maps which are known to generate $M_s(T_g)$.

A twist τ_c about a simple closed curve c on the surface T_g is defined by considering a neighborhood of c on T_g which is homeomorphic to a cylinder. Cut the surface T_g along one base of this cylinder, twist the free end of the cylinder through 2π, and then glue it back together again. The resulting map of $T_g \to T_g$ is the twist τ_c.

In order to study how the maps which represent the σ_i lift to T_g, we go back to the representation of T_g illustrated in Figure 2, and show that a twist about the curve c_i on T_g projects to a representative of σ_i on $T_{0,2g+2}$.

We note that in Figure 2, if i is even the curve c_i is the unit circle about $(2i-1,0,0)$ in the xz plane. We recall that c_i lies on a right circular cylinder of radius 1 whose axis coincides with the y axis. For convenience, introduce coordinates (y,θ) on this cylinder. A twist τ_{c_i} about c_i is given by the identity map outside the cylinder, with $\tau_{c_i}(y,\theta) = (y,\theta+\pi(y+1))$. It is easy to check that τ_{c_i} preserves fibres with respect to the projection π. Also that τ_{c_i} maps $P_k \to P_k$ if $k \neq i, i+1$, while $P_i \to P_{i+1}$ and $P_{i+1} \to P_i$, and that τ_{c_i} projects to the elementary twist on $T_{0,2g+2}$ interchanging Q_i and Q_{i+1}. A similar argument holds for each curve $c_1, \ldots, c_{2g+1}$. Thus we have:

THEOREM 2. *The maps which represent the generators* $\sigma_1, \ldots, \sigma_{2g+1}$ *of* $M(T_{0,2g+2})$ *lift to twists* $\tau_{c_1}, \ldots, \tau_{c_{2g+1}}$ *about the curves* $c_1, \ldots, c_{2g+1}$ *on* T_g, *and these twists represent elements* $\tilde{\sigma}_1, \ldots, \tilde{\sigma}_{2g+1}$ *of* $\tilde{S}(T_g)$.

Theorems 1 and 2 then combine to give:

THEOREM 3. $\tilde{S}(T_g)$ *admits a presentation:*

generators: $\tilde{\sigma}_1, \ldots, \tilde{\sigma}_{2g+1}$

(3.1) $\tilde{\sigma}_1 \leftrightarrows \tilde{\sigma}_j$ $\qquad 1 \leq i \leq 2g-1, \quad |i-j| \geq 2$

(3.2) $\tilde{\sigma}_i \tilde{\sigma}_{i+1} = \tilde{\sigma}_i \tilde{\sigma}_i \tilde{\sigma}_{i+1}$ $\qquad 1 \leq i \leq 2g$

(3.3) $(\tilde{\sigma}_1 \tilde{\sigma}_2 \cdots \tilde{\sigma}_{2g} \tilde{\sigma}_{2g+1})^{2g+2} = 1$

(3.4) $\tilde{\sigma}_1 \tilde{\sigma}_2 \cdots \tilde{\sigma}_{2g} \tilde{\sigma}_{2g+1}^2 \tilde{\sigma}_{2g} \cdots \tilde{\sigma}_2 \tilde{\sigma}_1 = 1$

§3. The relationship between $M(T_g)$ and $S(T_g)$

Putting aside the group $\tilde{S}(T_g)$ for the moment, we fix our attention on the relationship between $S(T_g)$ and $M(T_g)$. Since any self-homeomorphism of T_g which represents an element of $S(T_g)$ clearly also represents an element of $M(T_g)$, and since maps which are symmetrically isotopic to 1 and thus represent the identity in $S(T_g)$ also represent the identity in $M(T_g)$, therefore a natural homomorphism exists from $S(T_g)$ to $M(T_g)$. The question we now consider is the extent to which the converse of this statement might be true.

If we restrict our attention to the subgroup $M_s(T_g)$ generated by the mapping classes represented by $\tau_{c_1}, \ldots, \tau_{c_{2g+1}}$ we note at the outset that every element in this subgroup can be represented by a symmetric homeomorphism which is an appropriate product of $\tau_{c_1}, \ldots, \tau_{c_{2g+1}}$. Much more subtle is the following theorem, which we will develop via a sequence of pre-theorems in the following pages:

THEOREM 7. *If* $G: T_g \to T_g$ *is a symmetric homeomorphism which is isotopic to* 1, *then* G *is symmetrically isotopic to* 1.

As a Corollary to Theorem 7 we will establish:

COROLLARY 7.1: $M_s(T_g)$ *is isomorphic to* $S(T_g)$.

The proof of Theorem 7 begins with:

THEOREM 4: *Let* G *be a symmetric homeomorphism of* T_g, *isotopic to* Id, *such that* G *leaves* $P_1, \ldots, P_{2g+2}$ *fixed. There is a second isotopy between* G *and* Id *such that the entire isotopy leaves* $P_1, \ldots, P_{2g+2}$ *fixed.*

Proof: Let G_t be the original isotopy. Then $p_j(t) = G_t(P_j)$ is in general not constant, although it is a closed loop since $G(P_j) = P_j$. The next two lemmas allow us to construct a sequence of isotopics $K_t^1, \ldots, K_t^{2g+2}$ of G with Id, each one leaving one more special point fixed than its predecessor.

LEMMA 4.1. *Let* G_t *be an isotopy such that* $G_0 = \text{Id}$, $G_1 = G$, *where* $G \in \mathcal{S}(T_g)$ *and* $G(P_j) = P_j$, $j = 1, \ldots, 2g+2$. *Then the orbit of each special point* $p_j(t) = G_t(P_j)$ *is homotopic to the constant curve* P_j.

Furthermore if $G_t(P_k) = P_k$ *for* $0 \leq t \leq 1$, $k < j$, *a homotopy* $h_s(t)$ *may be chosen so that* $h_1(t) = p_j(t)$, $h_0(t) = P_j$ *and* $h_s(t) \neq P_k$, $0 \leq s \leq 1$, $k < j$.

Proof: The idea of the proof is to show that $p_j(t)$ is homotopic to the product of two curves r_1 and r_2 such that r_1 lies on c_{j-1} and r_2 lies on c_j. It will then be shown that as a consequence of the symmetry of G, the curves $p_j(t)$ and $p_j(i(t))$ are isotopic, which is impossible for curves of type $r_1 r_2$ unless $p_j(t) \cong 0$.

In the following construction refer to Figure 3.

1. To make the technical argument simpler we shall assume that T_g is a polyhedron in R^3 such that c_{2j-1} lies in the xy plane and c_{2j} in the xz plane for each j. Let j be any integer, $1 \leq j \leq 2g+1$. Let $\gamma_j(\theta)$ be a parametrization of c_j, where $\gamma_j(0) = \gamma_j(1) = P_j$, $\gamma_j(1/2) = P_{j+1}$.

2. Let C_1 be a right circular cylinder in R^3 parametrized by (t, θ),

$0 \leq t \leq 1$ and let $f: C_1 \to T_g$ be defined by $f(t, \theta) = G_t(\gamma_{j-1}(\theta))$. We need to approximate f by another function $\tilde{f}$ so that a certain set is a 1-manifold. This will become clear in 4.

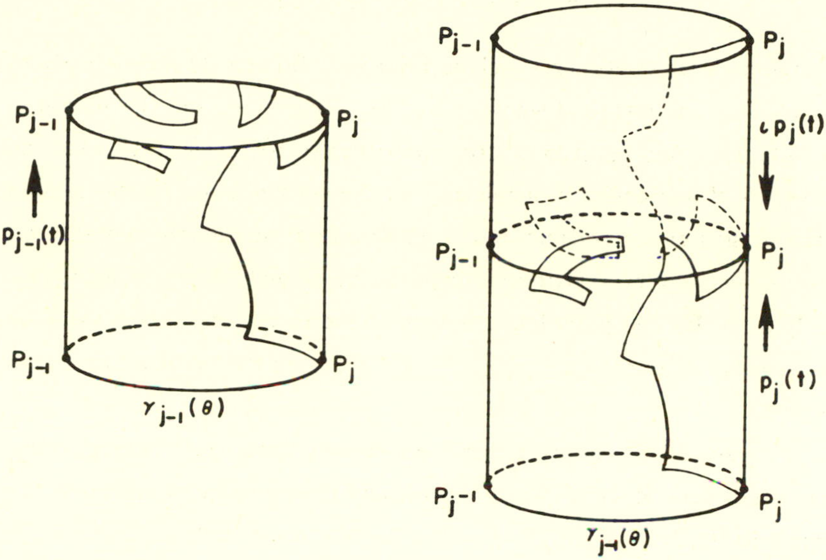

FIG. 3

3. Let $\varepsilon > 0$ be such that $|q(s) - r(s)| < \varepsilon$ implies $q \sim r$ for any curves q and r going from one point on T_g to another. Let $\tilde{f}$ be an approximation to f such that:

a) $|f(t,\theta) - \tilde{f}(t,\theta)| < \varepsilon$ for $0 \leq t \leq 1$, $0 \leq \theta \leq 2\pi$.

b) $\tilde{f}(1,\theta) = \iota f(1,-\theta)$; $f(0,\theta) = \iota\tilde{f}(0,-\theta)$

(This is reasonable, since f has this property by the symmetry of G.)

c) $\tilde{f}$ is piecewise linear. In the triangulation of C_1 assume that $(1,\pi)$ and $(0,\pi)$ are each the vertices of only two triangles.

d) except for $(1,\pi)$ and $(0,\pi)$, $\tilde{f}$ maps no vertex on c_j, and $\tilde{f}$ maps no 2-simplex into a degenerate 2-simplex.

4. Let C_2 be a right circular cylinder parametrized by (t,θ), $0 \leq t \leq 2$.

We define a map g of $C_2 \to T_g$ as follows:

$$g(t,\theta) = \tilde{f}(t,\theta) \quad \text{if} \quad 0 \le t \le 1$$

$$g(t,\theta) = \iota f(2-t,-\theta) \quad \text{if} \quad 1 \le t \le 2\,.$$

3b implies g is continuous and well defined. 3c and 3d above imply that $g^{-1}(c_j)$ is a one manifold whose boundary is precisely the two points $\{(0,\pi),(2,\pi)\}$. (Any point in $g^{-1}(c_j)$ is either $(0,\pi),(1,\pi)$ or $(2,\pi)$ or not the vertex of a triangle. Considering all the possibilities for the location of the images of the vertices in the quadrilateral or triangle containing the point, subject to the restrictions imposed by 3c and 3d and using 1, one easily sees that the point has a neighborhood in $g^{-1}(c_j)$ which is an open line segment or in the case of $(0,\pi)$ and $(2,\pi)$ is the endpoint of a line segment).

$(0,\pi)$ and $(2,\pi)$, being the only boundary points, must belong to the same component M of $g^{-1}(c_j)$. By an easy connectedness argument M is symmetric ($(t,\theta) \in M$ iff $(2-t,-\theta) \in M$) and must contain $(1,\pi)$ or $(1,0)$. M contains $(1,\pi)$.

5. All that we need from 4, is the fact that there is a path $r_1(s)$ on C_2 from $(0,\pi)$ to $(1,\pi)$ such that $g(r_1(s)) \in c_j$. Since $\pi_1(C_2) = Z_\infty$, $r_1(s)$ is homotopic to a curve $r_2(s) = (0, 4ks+\pi)$, $0 \le s \le 1/2$, k some integer, followed by $r_3(s) = (2s-1,\pi)$, $1/2 \le s \le 1$. This implies $g(r_1) \simeq g(r_3)\cdot g(r_2)$. (Here these are closed curves with base point $= g(0,\pi) = g(1,\pi) = P_j$). Since $|p_j(t) - g(r_3, t/2 + 1/2)| < \varepsilon$, and $g(r_2)$ lies within ε of c_{j-1}, it follows that $p_j(t)$ is homotopic to a curve lying on c_{j-1} followed by a curve lying on c_j. Now, for any closed curve η with base point p_j, we find that $p_j \cdot \eta \cdot p_j^{-1} \simeq G(\eta)$, the homotopy being defined by $G_t(s)$. Replacing η by $\iota\eta$ we see that $p_j \cdot \iota\eta \cdot p_j^{-1} \simeq G(\iota\eta)$. Applying ι to both sides and using the face that G is symmetric we obtain $\iota p_j \cdot \eta \cdot (\iota p_j)^{-1} \simeq p_j \cdot \eta \cdot p_j^{-1}$. Thus $(\iota p_j)^{-1} \cdot p_j \cdot \eta \simeq \eta \cdot (\iota p_j)^{-1} \cdot p_j$. But for $g \ge 2$ the center of $\pi_1 T_g$ is trivial, so $p_j \simeq \iota p_j$. This is possible for a curve of type p_j, which first lies entirely on c_{j-1} and then on c_j, only if $p_j \simeq 0$.

6. To prove the second part of the theorem suppose $G_t(P_k) = P_k$ for $0 \leq t \leq 1$ and $k < j$. We repeat the previous construction with certain changes which are necessary to meet the additional requirement. Pick $\varepsilon_1 > 0$ such that:

a) For any closed curve $g(t)$ with base point P_j, $|g(t) - p_j(t)| < \varepsilon_1$ implies $g \sim p_j$ through a homotopy never taking on the values $P_1, \ldots, P_{k-1}$. (Note that $p_j(t) = G_t(P_j) \neq P_k$ because G_t is 1-1 and $G_t(P_k) = P_k$).

b) $|f(t,\theta) - P_k| > \varepsilon_1$ for $(t,\theta) \in C_1$, $k \leq j-2$.

c) Pick δ such that $|f(t,\theta) - c_j| > \varepsilon_1$ if $-\delta \leq \theta \leq +\delta$ (recall $f(t,0) = P_{j-1}$).

d) Pick $\varepsilon > 0$ not bigger than ε_1 and such that $|f(t,\theta) - P_{j-1}| > \varepsilon$ if $-\pi \leq \theta \leq -\delta$ or $\delta \leq \theta \leq \pi$.

e) Let $\tilde{f}$ as defined in 3 be an ε-approximation to f.

7. In 5, since $g(r_1(s)) \in c_j$ by 6c, $r_1(s) \in C_1 - \{(t,\theta)/-\delta < \theta < \delta\}$. $\pi_1[C_1 - \{(t,\theta)/-\delta < \theta < \delta\}]$ is trivial so $r_1(s) \simeq r_2(s) = (s,\pi)$ where the whole homotopy takes place in $C_1 - \{(t,\theta)/-\delta < \theta < \delta\}$. Hence $g(r_1(s)) \simeq g(r_2(s))$ where by 6b and 6d the whole homotopy never takes the values P_k for $k < j$. Also $g(r_2(s)) \simeq p_j(s)$ by 6a, where again the homotopy never takes the values P_k for $k < j$. Finally $g(r_1(s))$ lies on c_j and, being homotopic to P_j, is homotopic to zero. Again since $g(r_1(s))$ lies on c_j we can find a homotopy to zero taking all its values on c_j and therefore not taking the values P_k, $k < j$.

LEMMA 4.2. *Let $q(t)$ be a curve homotopic to the constant curve P_j through the homotopy $h(s,t)$. Suppose $h(s,t) \neq P_k$ for $k < j$, $0 \leq s, t \leq 1$. Then there is an isotopy F_r such that $F_0 = F_1 = \mathrm{Id}$, $F_t(q(t)) = P_j$ and $F_t(P_k) = P_k$, for $k < j$ and $0 \leq t \leq 1$.*

The isotopy F_t is merely a push along the curve $h(s, t)$ (s varies, t is fixed—see Figure 4). The only problem (and that is technical) is in constructing an isotopy F_t which is continuous in t. For technical convenience we assume T_g is a differentiable manifold embedded in R^3.

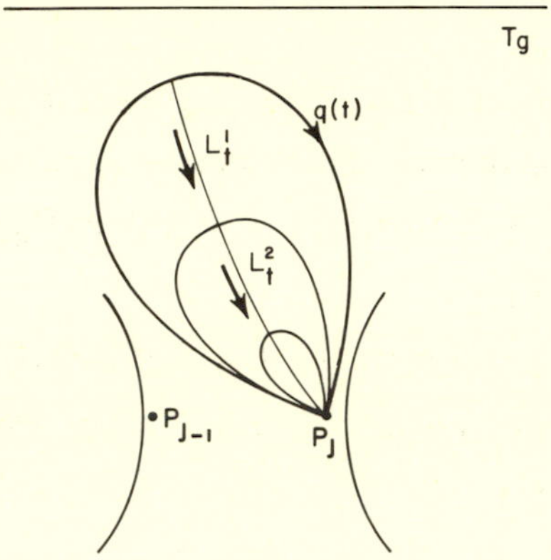

FIG. 4

Proof: We construct the isotopy in several steps..

1. Pick $\varepsilon > 0$ so that:

a) At any point Q on T_g we may define a homeomorphism, H_Q, of the disc of radius 3ε in the tangent plane, centered at Q, (called $D^{3\varepsilon}(Q)$) into T_g by projecting each point in $D^{3\varepsilon}(Q)$) parallel to the normal at Q, into the nearest point on T_g.

b) $|H_Q(R) - H_Q(S)| < 2|R-S|$ for any pair of points R, S in $D^{3\varepsilon}(Q)$.

c) $|h(s,t) - P_k| > 6\varepsilon$ for $0 \le s,\ \ t \le 1,\ k < j$.

2. We define a set of isotopies of $D^{3\varepsilon}(Q)$. Let f(r) be a C^1 function of r such that $0 \le f(r) \le 1$ with $f(r) = 1$ for $r \le \varepsilon$, and $f(r) = 0$ for $r \ge 2\varepsilon$. For any $P \in D^{3\varepsilon}(Q)$, $P \neq Q$, define the vector field $V_{P,Q}(R) = f(|R-Q|)\dfrac{Q-P}{|Q-P|}$. Let M_t be the one parameter family of diffeomorphisms of $D^{3\varepsilon}(Q)$ that takes a point R into c(t; R) where c is an integral curve for the vector field $V_{P,Q}$ taking the value R at $t = 0$. Finally let $G_{P,Q} = M_{|P-Q|}$. If $P = Q$ we define $G_{Q,Q} = \mathrm{Id}$. We note that $G_{P,Q}$ depends continuously on P and that $G_{P,Q}(P) = Q$ if $|P-Q| < \varepsilon$.

3. Pick N so large that $|s-s_0| < 1/N$ implies $|h(s,t)-h(s_0,t)| < \varepsilon$ for $0 \le t \le 1$. For each $i \le N$ we define an isotopy L_t^i as follows:

a) $L_t^i = \text{Id}$ outside of $H_Q D^{3\varepsilon}(Q)$ where Q is taken as $h(1-\frac{i}{N}, t)$.

b) For the point R on T_g in $H_Q D^{3\varepsilon}(Q)$ define $L_t^i(R) = H_Q G_{P,Q} H_Q^{-1}(R)$ where P is taken as $h(1-\frac{i-1}{N}, t)$.

L_t^i is continuous in it's argument because $G_{P,Q} = \text{Id}$ outside of $D^{2\varepsilon}(Q)$. L_t^i is continuous in t because $G_{P,Q}$ is continuous in P and the tangent plane at Q moves continuously with Q. L_t^i pushes the curve $h(1-\frac{i-1}{N}, t)$ into the curve $h(1-\frac{i}{N}, t)$. Also, by 1b and 1c, L_t^i doesn't move any P_k such that $k < j$.

4. Let $F_t(R) = L_t^N(L_t^{N-1}(\ldots(L_t^1(R))\ldots))$. Then F_t is the required isotopy.

Proof of Theorem 4: Lemmas 4.1 and 4.2 together imply that we may define a sequence of $2g+2$ isotopics $K_t^o, \ldots, K_t^{2g+2}$, each one leaving one more special point fixed than its predecessor. Suppose K_t^n leaves $P_1, \ldots, P_n$ fixed. By Lemma 4.1 $P_{n+1}(t) \simeq P_{n+1}$ through a homotopy $h_s(t) \neq P_j$ for $j = 1, \ldots, n$. By Lemma 4.2 there is an isotopy F_t such that $F_t(P_j) = P_j$ for $j = 1, \ldots, n$, $F_t(p_{n+1}(t)) = P_{n+1}$, and $F_o = F_1 = \text{Id}$. Define $K_t^{n+1} = F_t K_t^n$. Then $K_t^{n+1}(P_j) = P_j$, $j = 1, \ldots, n+1$. Thus K^{2g+2} leaves all the special points fixed.

Lemma 4.1 can also be proved by a second method, which is mainly algebraic. However this latter proof does not give the information that the homotopy $h(t,s)$ may be chosen to avoid all P_j such that $j < k$. The proof is: Consider the orbit $G_s(P_j)$ of the point P_j under the isotopy G_s, where $G_o = G$ and $G_1 = \text{Id}$. The loop $G_s(P_j)$ represents some element in the fundamental group $\pi_1 T_g$ with base point at P_j, which we will show is necessarily the identity. Let $\eta(t)$ be any other P_j-based loop. Define the function $g(s,t)\colon I^2 \to T_g$ by $g(s,t) = G_s(\eta(t))$. Looking at the restriction of $g(s,t)$ to ∂I^2 we obtain that $\eta(t) \simeq G_t(P_j)G(\eta(t))G_t^{-1}(P_j)$.

Replacing $\eta(t)$ by $\iota(\eta(t))$ the same argument gives that $\iota(\eta(t)) \simeq G_t(P)\,G(\iota(\eta(t)))G_t^{-1}(P_j)$. But since G is symmetric, $G(\iota(\eta(t))) = \iota(G(\eta(t)))$. Thus $\iota(\eta(t)) \simeq G_t(P_j)\iota(G(\eta(t)))G_t^{-1}(P_j)$, or equivalently, $\eta(t) \simeq \iota(G_t(P_j))G(\eta(t))\iota(G_t^{-1}(P_j))$. Combining these two, it follows that $\iota(G_t^{-1}(P_j))G_t(P_j)G(\eta(t)) \simeq G(\eta(t))\iota(G_t^{-1}(P_j))G_t(P_j)$. But since $\eta(t)$ was taken to be any P_j-based loop, it follows that $\iota(G_t^{-1}(P_j))G_t(P_j)$ must belong to the center of $\pi_1(T_g)$. Since $\pi_1(T_g)$ is centerless for $g \geq 2$, this implies that $G_t(P_j)$ and $\iota G_t(P_j)$ represent the same element of $\pi_1 T_g$.

We want to show that $G_t(P_j)$ can only represent the identity element of $\pi_1 T_g$. This will follow if we can show there are no non-trivial elements in $\pi_1 T_g$ which are fixed under the automorphism induced by ι. To establish this, we note that $\pi_1 T_g$ is generated by the lopps $v_1, \ldots, v_{2g}$ illustrated in Figure 5. In terms of these generators the single defining relation in $\pi_1 T_g$ is:

(4) $$(v_1 v_2^{-1} v_3 v_4^{-1} \ldots v_{2g-1} v_{2g}^{-1})(v_1^{-1} v_2 v_3^{-1} v_4 \ldots v_{2g-1}^{-1} v_{2g}) = 1$$

The effect of the involution ι is to induce an automorphism ι^* in $\pi_1 T_g$ which maps each $v_i \to v_i^{-1}$. Thus if

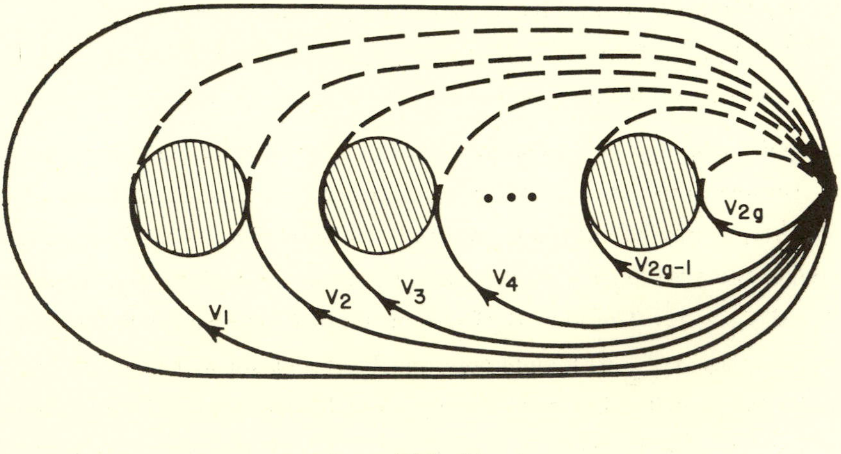

FIG. 5

$$\text{(5)} \qquad W = v_{\mu_1}^{\varepsilon_1} v_{\mu_2}^{\varepsilon_2} \cdots v_{\mu_r}^{\varepsilon_r}$$

is a word in $\pi_1 T_g$, its image under ι^* will be

$$\text{(y)} \qquad \iota^* W = v_{\mu_1}^{-\varepsilon_1} v_{\mu_2}^{-\varepsilon_2} \cdots v_{\mu_r}^{-\varepsilon_r}$$

The condition $W = \iota^* W$ thus implies:

$$\text{(7)} \qquad (W)(\iota^* W)^{-1} = (v_{\mu_1}^{\varepsilon_1} v_{\mu_2}^{\varepsilon_2} \cdots v_{\mu_r}^{\varepsilon_r})(v_{\mu_r}^{\varepsilon_r} \cdots v_{\mu_2}^{\varepsilon_2} v_{\mu_1}^{\varepsilon_1}) = 1$$

Now Dehn's algorithm for the solution of the word problem in the group $\pi_1 T_g$ [11] implies that a freely reduced word cannot be equal to 1 in $\pi_1 T_g$ unless it contains a segment which is more than ½ of the basic relator or a cyclic permutation of the basic relator. The three possibilities are that such a segment is in the word W, or in $\iota(W)^{-1}$, or in the interface between the two. The last is impossible, because the interface contains a term $v_{\mu_r}^{2\varepsilon_r}$ which cannot appear in any cyclic permutation of the basic relator or its inverse. The first two possibilities imply that W is itself 1. Therefore $G_t(P)$ can only represent the identity element of $\pi_1(T_g)$, and the (second) proof of Lemma 4.1 is complete.

The next step in the proof is to establish:

THEOREM 5. *If* $G \in \mathcal{S}(T_g)$ *is such that there exists an isotopy* G_s *of* G *with* Id *such that* $G_s(P_j) = P_j$ *for* $0 \le t \le 1$ *and* $j = 1, \ldots, 2g+2$, *then* $G \in \mathcal{S}\mathfrak{D}(T_g)$.

Proof: We begin by introducing some notation. It will be convenient to divide the curves c_i in Figure 2 into segments, each with boundary points P_i and P_{r+1}. We will label these segments $a_1, \ldots, a_{2g+1}$ and $b_1, \ldots, b_{2g+1}$ as illustrated in Figure 6. In addition, we will need the curves c_{2g+2}, a_{2g+2} and b_{2g+2}. Also, we will need curves d_i, e_i which differ by a small amount from a_i, b_i for each i. We will use Greek letters $\alpha_i(t)$, $\beta_i(t)$, $\gamma_i(\theta)$, $\delta_i(t)$, $\varepsilon_i(t)$ to indicate a parametrization of a_i, b_i, c_i, d_i or e_i respectively, where the orientation of each curve is indicated. The argument θ is in general used for the special case where the path is closed (θ a real number mod 2π), otherwise t. The curves are all defined in a symmetric

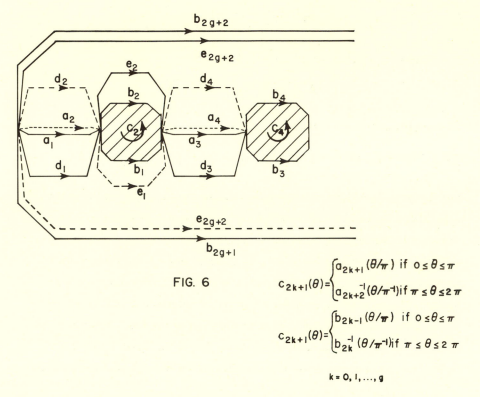

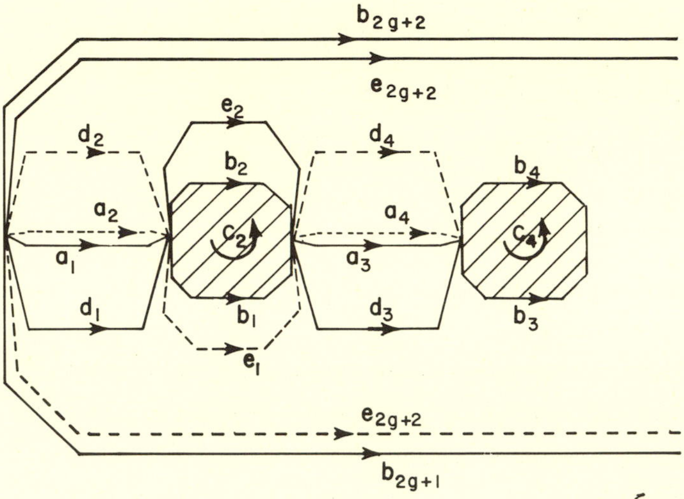

FIG. 6

$$c_{2k+1}(\theta) = \begin{cases} a_{2k+1}(\theta/\pi) & \text{if } 0 \le \theta \le \pi \\ a_{2k+2}^{-1}(\theta/\pi - 1) & \text{if } \pi \le \theta \le 2\pi \end{cases}$$

$$c_{2k+1}(\theta) = \begin{cases} b_{2k-1}(\theta/\pi) & \text{if } 0 \le \theta \le \pi \\ b_{2k}^{-1}(\theta/\pi - 1) & \text{if } \pi \le \theta \le 2\pi \end{cases}$$

$k = 0, 1, \ldots, g$

manner, as illustrated. Thus, the involution ι as defined in Section 1 operates as follows:

$$\iota(c_i(\theta)) = c_i(2\pi - \theta), \text{ and } \iota(a_{2i-1}(t)) = a_{2i}(t) .$$

1. The proof of the theorem revolves around the behavior of the curves $G(\delta_i(t))$ and $G(e_i(t))$. Since G is isotopic to the identity there is a sequence of isotopies which push these curves back to their original positions. It turns out that since G is symmetric we can choose these isotopies to be symmetric isotopies. We begin with several technical simplifications. Once these are out of the way, we can get on to the main part of the proof which comprises subsection 4 and includes Lemmas 5.2, 5.3 and 5.4.

2. We shall assume that T_g and $T_{0,2g+2}$ are triangulated polyhedra in R^3 such that:

a) "s" is a simplex in T_g iff "ιs" is a simplex in T_g.

b) The curves $\delta_j(t)$ and $\varepsilon_j(t)$ are piecewise linear.

c) The curves a_j lie in the xy plane, the curves b_j in the xz plane.

d) The map $\pi: T_g \to T_{0,2g+2}$ is linear with respect to the given triangulations

3. In this subsection we show that G is symmetrically isotopic to a homeomorphism H such that:

a) H is piecewise linear and symmetric.

b) H is piecewise linearly isotopic to the identity.

c) H and the entire isotopy of H with Id leave a neighborhood of each P_j pointwise fixed.

d) The curves $H(\delta_k(t))$ and $H(\varepsilon_k(t))$ cross each a_j and b_j finitely many times. If $H(\delta_k(t))$ or $H(\varepsilon_k(t))$ intersects an a_j or b_j at $t = t_0$, then it crosses that a_j or b_j at $t = t_0$.

(We eliminate the possibility that the curves touch and come back on the same side.)

It follows that we may assume G had these properties in the first place. We need the following Lemma from [8].

LEMMA 5.1. *Let* M *be a closed, triangulated, polyhedral* 2-*manifold embedded in* R^3. *There is an* $\varepsilon > 0$ *such that any homeomorphism* K *satisfying* $|P - K(P)| < \varepsilon$ *for every* $P \in M$ *is isotopic to* Id. *Furthermore, if* K *leaves neighborhoods* $N_1, \ldots, N_n$ *of points* $P_1, \ldots, P_n$ *pointwise fixed, the entire isotopy may be chosen to leave (smaller) neighborhoods of* $P_1, \ldots, P_n$ *pointwise fixed. Finally, if* K *is piecewise linear, the entire isotopy may be chosen to be piecewise linear, and to leave neighborhoods of the* $P_1, \ldots, P_n$ *pointwise fixed.*

This is a special case of a theorem of Fischer (Th. 6, [8]). The second and third parts of the lemma aren't stated explicitly by Fischer, but the isotopy which is constructed in his proof satisfies the condition on the neighborhoods if the mesh of the triangulation used in the proof is small enough. Also it is clear that if K is piecewise linear the isotopy may be chosen to be piecewise linear.

To find a homeomorphism H, symmetrically isotopic to G and with properties 3a – 3d we project G down to a homeomorphism $\tilde{G} \in \mathcal{H}(T_{0,2g+2})$, where $\tilde{G} = \pi G \pi^{-1}$ and $\tilde{G}(Q_j) = Q_j$. We find a homeomorphism $\tilde{K}$ that agrees with $\tilde{G}$ except on a disc neighborhood of each Q_j and leaves other (smaller) neighborhoods of each Q_j pointwise fixed. (That such a $\tilde{K}$ exists follows from Theorem 5 of [8].) $\tilde{K}$ is isotopic to $\tilde{G}$ because any two maps of a two-cell minus a point into a two-cell minus a point which agree on the boundary are isotopic [1]. Now approximate $\tilde{K}$ to within the $\varepsilon > 0$ of Lemma 5.1 for $M = T_o$ by a piecewise linear homeomorphism $\tilde{H}$ leaving neighborhoods of the Q_j pointwise fixed and such that the curves $\tilde{H}(\pi(\delta_k(t)))$ and $\tilde{H}(\pi(\varepsilon_k(t)))$ (which are piecewise linear) cross each $\pi(a_j)$ or $\pi(b_j)$ only finitely many times and such that if $\tilde{H}(\pi(\delta_k(t)))$ or $\tilde{H}(\pi(\varepsilon_k(t)))$ touch $\pi(a_j)$ or $\pi(b_j)$ at $t = t_o$ they cross at $t = t_o$. By Lemma 5.1, $\tilde{H}$ is isotopic to $\tilde{K}$ which is isotopic to $\tilde{G}$. This isotopy lifts to an isotopy between G and H, where H satisfies properties 3a, 3c, and 3d.

To construct a piecewise linear isotopy between H and Id satisfying 3b and 3c, find N such that $|s - s_o| < 1/N$ implies $|H_s(P) - H_{s_o}(P)| < \varepsilon/3$ for any $P \in T_g$, where H_s is the isotopy between H and Id and ε is the ε of Lemma 5.1 for T_g. Let $K_{k/N}$ be a piecewise linear $\varepsilon/3$ approximation to $H_{k/N}$ leaving a neighborhood of each P_j pointwise fixed. Let $K_1 = H$; $K_o = Id$. Then $|K_{k/N}(P) - K_{k+1/N}(P)| < \varepsilon$ for any $P \in T_g$, so by Lemma 5.1 $K_{k/N} K_{k+1/N}$ is piecewise linearly isotopic to Id through an isotopy satisfying 3d. Thus we find a piecewise linear isotopy between $K_{k/N}$ and $K_{k+1/N}$ for $k = 0, \ldots, N-1$. When we fit these isotopies together we have constructed an isotopy of H with Id satisfying 3b and 3c.

4. Let $p(t)$ be a curve on T_g from P_k to P_{k+1} such that:

a) $p(t) \neq P_\ell$ for $0 < t < 1$, $\ell = 1, \ldots, 2g+2$;

b) $p(t)$ crosses each a_j or b_j finitely many times.

We now define the k-symbol associated with p, denoted $\sigma^k(p)$. Let F be a homeomorphism of T_g onto a differentiable manifold $\tilde{T}_g$ such that if $g(t)$

is a piecewise linear curve on T_g then $F(g(t))$ is piecewise differentiable on $\tilde{T}_g$ with one-sided derivatives at every t. Fix an orientation on $\tilde{T}_g$ and call it positive. If $p(t)$ crosses a_j at $t = t_o$ and the ordered basis for the tangent space $\{F(p)'(t_o), F(\alpha_j)'(p(t_o))\}$ is positively oriented we assign the mark A_j^+ to the crossing. Note that it doesn't matter which of the two one-sided derivatives we use for $F(p)'(t_o)$ since the curve $p(t)$ crosses a_j. We use the symbols B_j^+ or B_j^- to indicate crossings of b_j, the negative exponent being used if the basis is negatively oriented. The symbol $\sigma^k(p)$ is simply a list of marks assigned to the crossings of a_j or b_j in the order in which the crossings occur. The points $t = 0$ and $t = 1$ are not counted as crossings, and if $p(t)$ touches but does not cross a curve, no mark is assigned to this event. See Figure 7 for an illustration. We say a symbol is reducible to 0 if it can be reduced to no symbol by cancelling adjacent marks of the form $A_j^+ A_j^-$, $B_j^+ B_j^-$, $A_j^- A_j^+$ or $B_j^- B_j^+$.

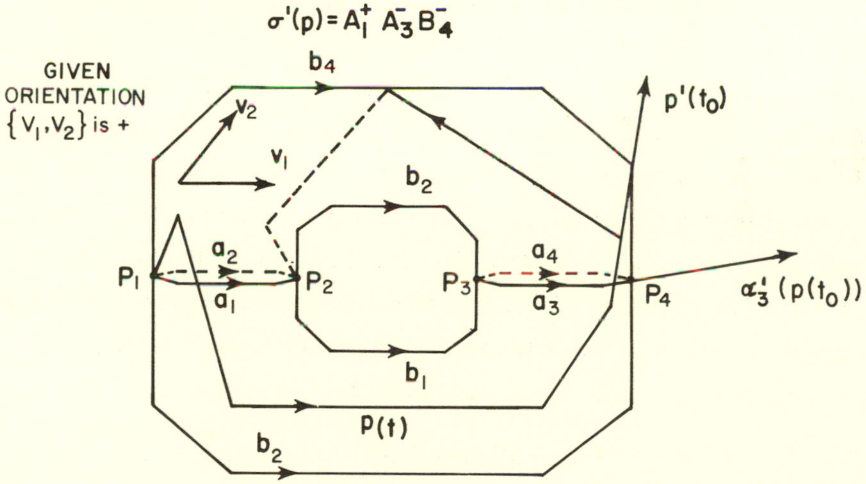

Figure 7

LEMMA 5.2. *If* G *is a homeomorphism of* T_g *satisfying* 3a–3d *then* $\sigma^{2k+1}(G(\delta_{2k+1}(t)))$, $\sigma^{2k+1}(G(\delta_{2k+2}(t)))$, $\sigma^{2k+2}(G(\varepsilon_{2k+1}(t)))$ *and* $\sigma^{2k+2}(G(\varepsilon_{2k+2}(t)))$ *are all reducible to zero,* $k = 0, \ldots, g$.

Proof: We shall prove only that $\sigma^{2k+1}(G(\delta_{2k+1}(t)))$ reduces to 0. The proof for the others is similar. The curve $G(\delta_{2k+1}(t))$ is homotopic to δ_{2k+1} through a piecewise linear homotopy defined by the equation

$$h_s(t) = G_s(\delta_{2k+1}(t))$$

where G_s leaves a neighborhood of P_{2k+1} and P_{2k+2} pointwise fixed. Since G_s leaves neighborhoods of P_{2k+1} and P_{2k+2} pointwise fixed and $\delta_{2k+1}(t)$ doesn't lie on a_j or b_j for $t = 0, 1$, it follows that $h_s(t)$ has the property:

a) There is an $\varepsilon > 0$ such that $h_s(t)$ doesn't lie on a_j or b_j for $0 < \varepsilon < t$ or $1 - \varepsilon < t < 1$.

Now, $h_s(t)$ is piecewise linear. By changing the value of $h_s(t)$ slightly on the vertices of a triangulation of $[0, 1] \times [0, 1]$ we can insure that h maps no line segment of form $[(s, t_0), (s, t_1)]$ into an a_j or b_j and that therefore $h_s(t)$ intersects each a_j or b_j only finitely many times. We can do this without losing property "a" above.

Since $h_s(t)$ is continuous and piecewise linear in s and t, there is a neighborhood of s in which the symbol $\sigma(h_s)$ for h_s can only change if an intersection which is not a crossing becomes a double crossing. See Figure 8. This can change $\sigma(h_s)$ only by adding adjacent pairs of marks of form $A_j^- A_j^+$, $A_j^+ A_j^-$, $B_j^- B_j^+$, or $B_j^+ B_j^-$ which in no way affects reducibility to zero. The sets $\{s/\sigma^{2k+1}(h_s)$ is reducible to zero$\}$ and $\{s/\sigma^{2k+1}(h_s)$ is not reducible to zero$\}$ are both open and form a disconnection of $[0, 1]$. Since the first set contains $s = 0$ it must contain $s = 1$.

LEMMA 5.3. *If* $G \in \mathcal{S}(T_g)$ *satisfies* 3a-3d *then* G *is symmetrically isotopic to an* $H \in \mathcal{S}(T_g)$ *such that*

$$\sigma^{2k+1}(H(\delta_{2k+1}(t))), \ldots, \sigma^{2k+2}(H(\varepsilon_{2k+2}(t)))$$

are all equal to zero, $k = 0, \ldots, g$.

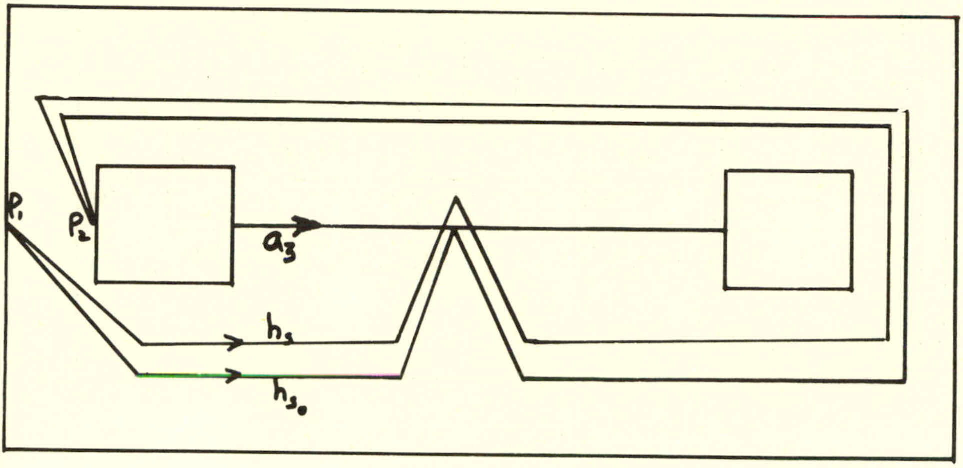

Figure 8

Proof: We shall construct a symmetric isotopy between G and a homeomorphism $K \in \mathcal{S}(T_g)$ such that K satisfies 3a-3d and the total number of marks in the symbols $\sigma^{2k+1}(K(\delta_{2k+1}(t))), \ldots, \sigma^{2k+2}(K(e_{2k+2}(t)))$ is less than the total number of marks in the corresponding symbols for G. This process can then be iterated to give the desired $H \in \mathcal{S}(T_g)$.

Pick a curve, say $G(\delta_{2k+1}(t))$ such that $\sigma^{2k+1}(G(\delta_{2k+1}(t))) \neq 0$. Then since, by the previous lemma, $\sigma^{2k+1}(G(\delta_{2k+1}(t)))$ is reducible to zero, adjacent marks of form, say $A_j^- A_j^+$, must appear in this symbol. Hence there are points t_1, t_2 with $t_1 < t_2$ such that $G(\delta_{2k+1}(t))$ crosses a_j, at $t = t_1$, and then crosses a_j again, at $t = t_2$, in the opposite direction, and such that $G(\delta_{2k+1}(t))$ doesn't cross any a_ℓ or b_ℓ for $t_1 < t < t_2$. (Refer to Figure 9.) Also, by 3d, $G(\delta_{2k+1}(t))$ doesn't touch any a_ℓ or b_ℓ for $t_1 < t < t_2$. It follows, since G is a homeomorphism and $G(\delta_{2k+1}(t))$ doesn't intersect itself, that $\{G(\delta_{2k+1}(t)) | t_1 \leq t \leq t_2\}$ together with the subarc of a_j whose endpoints are $G(\delta_{2k+1}(t_1))$ and $G(\delta_{2k+1}(t_2))$ form the boundary of a two cell which we shall call X_1. Since X_1 is contained in

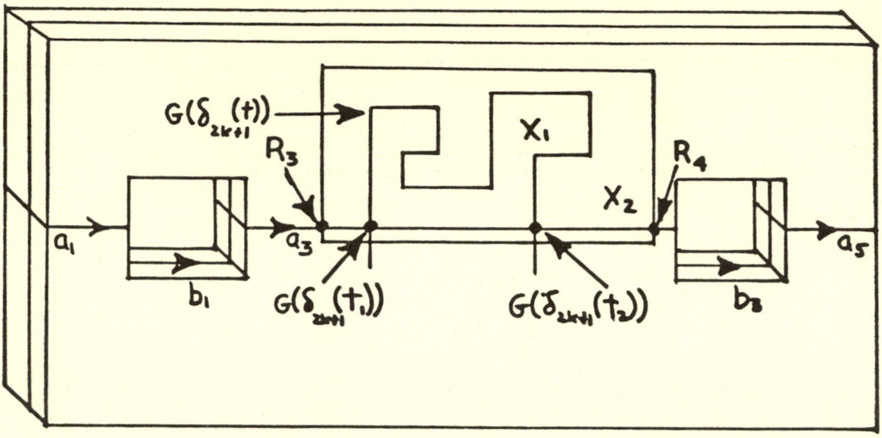

Figure 9

one "quadrant" of T_g it is easy to see that $X_1 \cap \iota X_1 = \emptyset$. By the disjointness (except for the P_k) of the curves $G(\delta_j(t))$ and $G(\gamma_j(t))$ we find another polygonal two-cell X_2 with the properties:

a) $X_2 \cap \iota X_2 = \emptyset$;

b) $X_1 \subset$ interior X_2;

c) $X_2 - X_1 \cap$ the quadrant of T_g containing X_1 contains no points of any $G(\delta_j(t))$ or $G(\gamma_j(t))$ $j = 1, \ldots, 2g+2$;

d) $X_2 \cap b_\ell = \emptyset$, all ℓ; $X_2 \cap a_\ell = \emptyset$, $\ell \neq j$;
$X_2 \cap a_j =$ two points, R_3 and R_4.

Let M be a piecewise linear homeomorphism of X_2 with the unit cube taking the segment $[R_3, R_4]$ into the line segment $[(-1,0),(0,1)]$ and X_1 into $\{(x,y) | y \geq 0\}$. Let I_s be a piecewise linear isotopy of the unit square such that

$$I_o = \text{Id}, \ I_1(M(X_1) \subset \{(x,y)| y < 0\}, \ I_1\{(x,y)| y < 0\} \subset \{(x,y)| y < 0\},$$

and I_s leaves the boundary of the unit square fixed.

Now using I_s we define an isotopy K_s of T_g such that

e) $K_s = \text{Id}$ on $T_g - (X_2 \cap \iota X_2)$;

f) $K_s = M^{-1} I_s M$ on X_2;

g) $K_s(P) = \iota K_s(\iota(P))$ on $i X_2$.

Let $K = K_1$. We see that every crossing of a $K(\delta_\ell(t))$ or a $K(\varepsilon_\ell(t))$ is also a crossing of a $G(\delta_\ell(t))$ or $G(\varepsilon_\ell(t))$ but that $K(\delta_{2k+1}(t))$ no longer crosses at $t = t_1$ or t_2. Then K satisfies 3a-3d and has fewer total marks than G.

LEMMA 5.4. *If* $H \in \mathcal{S}(T_g)$ *is such that* H *satisfies* 3a-3d *and* $\sigma^{2k+1}(H(\delta_{2k+1}(t))), \ldots, \sigma^{2k+2}(H(\varepsilon_{2k+2}(t)))$ *are all equal to zero,* $k = 0, \ldots, g$, *then* $H \in \mathcal{SD}(T_g)$.

Proof: The curves $a_1, b_1, a_3, b_3, \ldots, a_{2g+1}, b_{2g+1}$ bound a two cell containing the curves $d_1, d_3, \ldots, d_{2g+1}$ and $H(d_1), H(d_3), \ldots, H(d_{2g+1})$. Lemmas 1 and 2 of Smith [18] imply that if X is a two cell and $p_1(t)$ and $p_2(t)$ are two simple curves in X such that $p_1(0) = p_2(0)$, $p_1(1) = p_2(1)$, and $p_1(0)$ and $p_1(1)$ belong to the boundary X and $p_1(t)$ and $p_2(t)$ belong to the interior of X, $0 < t < 1$, then there is an isotopy K_s of X such that K_s leaves boundary X fixed and $K_1(p_1(t)) = p_2(t)$. Applying this several times we find an isotopy K_s of the two cell bounded by $a_1, \ldots, b_{2g+1}$ which is the identity on the boundary and such that $K_1(H(\delta_{2k+1}(t))) = \delta_{2k+1}(t)$, $k = 0, \ldots, g$. If we apply $\iota K_s \iota$ to the two cell bounded by $a_2, b_2, \ldots, b_{2g+2}$ simultaneously we obtain a symmetric isotopy of T_g taking $H(\delta_k(t))$ into $\delta_k(t)$ for $k = 1, \ldots, 2g+2$. In a similar way we obtain a symmetric isotopy taking $H(\varepsilon_k(t))$ into $\varepsilon_k(t)$ for $k = 1, \ldots, 2g+2$. In summary H is symmetrically isotopic to a homeomorphism which we will call G, such that $G = \mathrm{Id}$ on the curves d_j and ε_j, $j = 1, \ldots, 2g+2$. But the curves $d_1, \varepsilon_1, \ldots, d_{2g+1}$ bound a two cell and G restricted to this two cell is isotopic to the identity through an isotopy F_s leaving the boundary fixed. If we perform $\iota F_s \iota$ simultaneously we obtain a symmetric isotopy of G with a homeomorphism equal to Id on two of the four quadrants of T_g. Finally, one more symmetric isotopy applied to the other two quadrants brings us to the identity.

The proof of Theorem 5 now follows immediately from Lemmas 5.2, 5.3, and 5.4.

To prove Theorem 7, we need one more theorem:

THEOREM 6. *If* $G \in \mathcal{S}(T_g) \cap \mathcal{D}(T_g)$ *and genus* $T_g > 1$ *then* $G(P_j) = P_j$ *for* $j = 1, \ldots, 2g+2$.

Proof: Refer to Figure 10 in this proof.

1. We shall take T_g to be a triangulated polyhedron. By an argument analogous to that used in Theorem 3, following the statement of Theorem 5, we may assume that G is piecewise linear, that the curves $G(\gamma_j(\theta))$ cross each c_k, $k \neq j$, a finite number of times, and that G_s, the isotopy of G with Id, is piecewise linear.

The formula $h_s^j(\theta) = G_s(\gamma_j(\theta))$ defines a piecewise linear homotopy of $G(\gamma_j(\theta))$, with $\gamma_j(\theta)$. (This is a homotopy of simple closed curves where no base point is held fixed.) By an argument analogous to that used in the proof of Lemma 5.2, we may assume that $h_s^j(\theta)$ crosses each c_k, $k \neq j$, a finite number of times for each fixed s. Again as in the argument used in the proof of Lemma 5.2, for small changes in s, the only change in the number of crossings can be the addition of a double crossing. Hence the number of crossings modulo 2 is a constant function of s. This shows that:

a) $G(\gamma_j(\theta))$ crosses c_k an odd number of times if $k = j-1$ or $j+1$.

b) $G(\gamma_j(\theta))$ crosses c_k an even number of times, if $k \neq j-1$, j, and $j+1$.

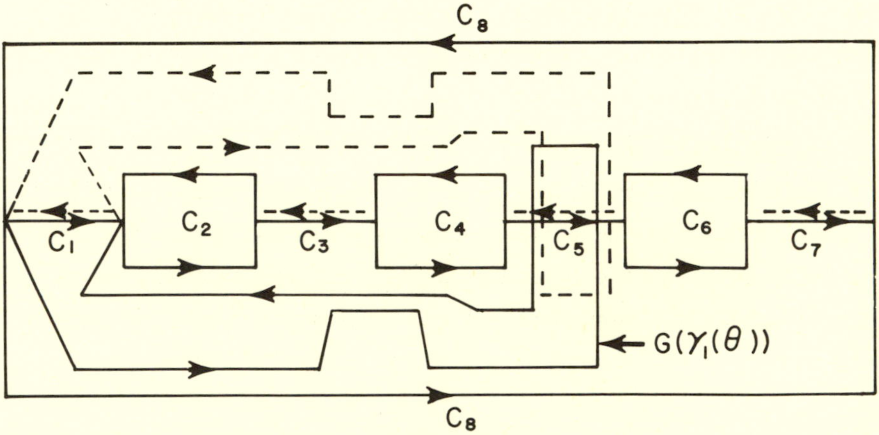

Figure 10

2. The curve $G(\gamma_j(\theta))$ is symmetric, as is $\gamma_j(\theta)$. Hence $G(\gamma_j(\theta_o))$ is a crossing of c_k iff $G(\gamma_j(\pi-\theta_o))$ is a crossing of c_k. Hence except for $\theta = 0$ and $\theta = \pi$ crossings occur in pairs, and not at special points. Since G is symmetric, G must map $\gamma_j(0) = P_j$ and $\gamma_j(1) = P_{j+1}$ into special points. From 1a, it follows that:

a) $G(P_j) = P_{j-1}$ or P_j or P_{j+1} or P_{j+2}

b) $G(P_{j+1}) = P_{j-1}$ or P_j or P_{j+1} or P_{j+2}

Since this argument in no way depends on which curve we choose, we can cyclically permute indices in b to obtain:

c) $G(P_j) = P_{j-2}$ or P_{j-1} or P_j or P_{j+1}

a., and c., imply

d) $G(P_j) = P_{j-1}$ or P_j or P_{j+1}.

Suppose $G(P_j) = P_{j+1}$. Since $G(\gamma_{j-1}(\theta))$ crosses c_{j+1} an even number of times, we must have $G(P_{j-1}) = P_{j+2}$. But this violates d., hence $G(P_j) \neq P_{j+1}$. By a similar argument we can show $G(P_j) \neq P_{j-1}$. This proves Theorem 6.

A second, purely algebraic, proof of Theorem 6 can be sketched as follows. We claim that a homomorphism exists from the group $M_s(T_g)$ onto the symmetric group Σ_{2g+2} on $2g+2$ letters. This homomorphism is defined by identifying the twist τ_{c_i} with the permutation of the points $P_1, \ldots, P_{2g+2}$ which interchanges P_i and P_{i+1}. The fact that this identification is a homomorphism follows from results in a recent paper by W. Magnus and A. Pelluso [17]. It is shown there that a certain subset of the generators of the symplectic group $Sp(2g, Z)$ can be mapped homomorphically onto Σ_{2g+2}. Since $Sp(2g, Z)$ is a homomorphic image of $M(T_g)$ [15], it follows that an appropriately chosen subgroup of $M(T_g)$ can be mapped homomorphically onto Σ_{2g+2}. If one identifies explicitly the subgroup of $Sp(2g, Z)$ in the Magnus-Pelluso proof, it can be shown to be exactly the image of $M_s(T_g)$ under the natural homomorphism from $M(T_g) \to Sp(2g, Z)$. It then follows that a relator in $M(T_g)$ must map into a relator in Σ_{2g+2}, and therefore every symmetric homeomorphism which is isotopic to 1 must keep the points $P_1, \ldots, P_{2g+2}$ individually fixed.

Theorem 7 now follows immediately by combining Theorems 4, 5, and 6.

It remains to prove Corollary 7.1, that is to show that $M_s(T_g)$ is isomorphic to $S(T_g)$. Since the generators of $M_s(T_g)$ can be represented by symmetric maps, and since, by Theorem 7, each representative of the class of the identity in $M_s(T_g)$ is also a representative of the class of the identity in $S(T_g)$, a homomorphism exists from $M_s(T_g)$ to $S(T_g)$. Moreover, by the discussion at the beginning of Section 3, a homomorphism exists from $S(T_g)$ to $M_s(T_g)$. If we can establish that both homomorphisms are onto and mutually inverse then the proof will be complete. To establish this, it is sufficient to show that the set of homeomorphisms $\tau_{c_1}, \ldots, \tau_{c_{2g+2}}$ which represent the generators of $M_s(T_g)$ also represent the generators of $S(T_g)$.

Now, by Theorem 3 the homeomorphisms $\tau_{c_1}, \ldots, \tau_{c_{2g+1}}$ are known to represent the generators of $\tilde{S}(T_g)$. Since the kernel of the homomorphism from $S(T_g) \to \tilde{S}(T_g)$ is the cyclic group generated by the mapping class of the involution ι, it follows that $\iota, \tau_{c_1}, \ldots, \tau_{c_{2g+1}}$ may be taken as representatives of the full set of generators of $S(T_g)$. However, we claim that:

(8) $\qquad \iota$ is isotopic to $\tau_{c_1} \cdots \tau_{c_{2g}} \tau_{c_{2g+1}} \tau_{c_{2g+1}} \tau_{c_{2g}} \cdots \tau_{c_1}$.

To establish (8), we make use of the fact that the group $M(T_g)$ has a faithful representation in terms of classes of automorphisms of $\pi_1 T_g$, where the inner automorphisms constitute the class of the identity. The automorphisms of $\pi_1 T_g$ induced by the generators of $M(T_g)$ have been worked out explicitly by several authors, e.g., L. Goeritz [10], P. Bergau and J. Mennicke [2] for genus 2, and J. Birman [5] for genus g. The method used to obtain these automorphisms is described in any of the above references. However, since none of these references give the information in the most convenient form for present purposes, we repeat the results, which have been reworked in terms of the system of canonical curves $v_1, \ldots, v_{2g}$ illustrated in Fig. 5 and $w_1, \ldots, w_{2g}$ in Fig. 1. The single defining relation for $\pi_1 T_g$ is given in terms of the v_i's by Equation (4), or equivalently in terms of the w_i's by:

(9) $$(w_1 w_2 w_1^{-1} w_2^{-1})(w_3 w_4 w_3^{-1} w_4^{-1}), \ldots, (w_{2g-1} w_{2-} w_{2g-1}^{-1} w_{2g}^{-1}) = 1 .$$

The generators $v_1, \ldots, v_{2g}$ are related to $w_1, \ldots, w_{2g}$ by:

(10) $$v_{2i-1} = (w_{2i-1} w_{2i}^{-1})(w_{2i+1} w_{2i+3} \cdots w_{2g-1})$$

(11) $$v_{2i} = (w_{2i}^{-1})(w_{2i+1} w_{2i+3} \cdots w_{2g-1})$$

where $1 \leq i \leq 2g$. The twists $\tau_{u_i}, \tau_{z_i}, \tau_{y_i}$ about the curves u_i, z_i, y_i in Figure 1 induce the following automorphisms in $\pi_1 T_g$:

(12) $$\tau_{u_i}: w_{2i} \to w_{2i} w_{2i-1}$$

(13) $$\tau_{z_i}: w_{2i-1} \to w_{2i-1} w_{2i}^{-1} w_{2i+1} w_{2i+2} w_{2i+1}^{-1}$$

$$w_{2i} \to w_{2i+1} w_{2i+2}^{-1} w_{2i+1}^{-1} w_{2i} w_{2i+1} w_{2i+2} w_{2i+1}^{-1}$$

$$w_{2i+1} \to w_{2i+1} w_{2i+2}^{-1} w_{2i+1}^{-1} w_{2i} w_{2i+1}$$

(14) $$\tau_{y_i}: w_{2i-1} \to w_{2i-1} w_{2i}^{-1}$$

where $1 \leq i \leq g$ and all subscripts for the w_i's are modulo 2g. Any generator of $\pi_1 T_g$ not listed explicitly in Equations (12)-(14) is assumed to be mapped into itself by that particular automorphism.

We may of course identify $\tau_{c_1}, \ldots, \tau_{c_{2g+1}}$ with $\tau_{y_1}, \tau_{u_1}, \tau_{z_1}, \tau_{u_2}, \tau_{z_2}, \ldots \tau_{u_{g-1}}, \tau_{z_{g-1}}, \tau_{u_g}, \tau_{y_g}$ respectively. It is then a straightforward (albeit tedious) calculation to verify that $\tau_{c_1} \cdots \tau_{c_{2g}} \tau_{c_{2g+1}} \tau_{c_{2g+1}} \tau_{c_{2g}} \cdots \tau_{c_1}$ induces an automorphism which takes $v_i \to v_i^{-1}$ for each ι (modulo an inner automorphism of $\pi_1 T_g$) Since the involution ι has precisely this effect on the generators $v_1, \ldots, v_{2g}$ of $\pi_1 T_g$, it follows that ι and

$\tau_{c_1} \cdots \tau_{c_{2g}} \tau_{c_{2g+1}} \tau_{c_{2g+1}} \tau_{c_{2g}} \cdots \tau_{c_1}$ represent the same element of $S(T_g)$, as claimed, and the proof of Corollary 7.1 is complete.

§4. Defining relations for $M_s(T_g)$

Using the results of Sections 2 and 3, we are now able to establish our main Theorem:

THEOREM 8. *Let* $\tau_1, \ldots, \tau_{2g+1}$ *be mapping classes in* $M_s(T_g)$ *which are represented by the twists* $\tau_{c_1}, \ldots, \tau_{c_{2g+1}}$ *respectively. Then* $M_s(T_g)$ *admits the presentation:*

generators: $\tau_1, \ldots, \tau_{2g+1}$

defining relations:

(15.1) $\tau_i \leftrightarrows \tau_j \qquad 1 \le i \le 2g-1, \quad |i-j| \ge 2$;

(15.2) $\tau_i \tau_{i+1} \tau_i = \tau_{i+1} \tau_i \tau_{i+1} \qquad 1 \le i \le 2g$;

(15.3) $(\tau_1 \tau_2 \cdots \tau_{2g+1})^{2g+2} = 1$;

(15.4) $(\tau_1 \cdots \tau_{2g} \tau_{2g+1} \tau_{2g+1} \tau_{2g} \cdots \tau_1)^2 = 1$;

(15.5) $(\tau_1 \cdots \tau_{2g} \tau_{2g+1} \tau_{2g+1} \tau_{2g} \cdots \tau_1) \leftrightarrows \tau_1$.

For $g = 2$, $M_s(T_g)$ *coincides with the full mapping class group* $M(T_2)$*.

Proof: The validity of relations 15 is easily verified by making use of the representation of $M(T_g)$ as $\mathrm{Aut}\, \pi_1 T_g / \mathrm{Inn}\, \pi_1 T_g$, as given in Equations (12), (13), (14). To establish completeness, we note that $S(T_g)$ is isomorphic to $M_s(T_g)$ by Corollary 7.1. Therefore $\tilde{S}(T_g)$ is a quotient group of

* The presentation for $M(T_2)$ obtained in Theorem 8 coincides with the presentation given by Bergau and Mennicke in [2]. The methods used are, however, distinct. A gap in their proof had left open the question of completeness, which is now established by our method.

$M_s(T_g)$. The kernel of the homomorphism from $M(T_g) \to \tilde{S}(T_g)$ is the normal subgroup generated by $\mathcal{J}$. Since the twists τ_{c_i}, $1 \le i \le 2g+1$ represent both the generators τ_i of $M_s(T_g)$ and the generators $\tilde{\sigma}_i$ of $\tilde{S}(T_g)$, the homomorphism from $M_s(T_g) \to \tilde{S}(T_g)$ is given by $\tau_i \to \tilde{\sigma}_i$, $1 \le i \le 2g+1$, and $\mathcal{J} \to \text{id}$. Now, in view of Equation (8), we know that in $M(T_g)$:

$$(16) \qquad \mathcal{J} = \tau_1 \cdots \tau_{2g}\tau_{2g+1}\tau_{2g+1}\tau_{2g} \cdots \tau_1$$

A straightforward calculation shows that the commutativity of $\mathcal{J}$ with the elements τ_j, where $j = 2, 3, \ldots, 2g+1$ is a consequence of relations 15.1 and 15.2, so that only relation 15.5 is needed to express the action of $\tau_1, \ldots, \tau_{2g+1}$ on the generator of the kernel. Relations 15.1, 15.2, and 15.3 are the lifts of relations 3.1, 3.2, and 3.3 to $M_s(T_g)$, while relation 15.4 expresses the fact that the right hand side of relation 3.4 lifts to $\mathcal{J}$, which is an element of order 2 in $M_s(T_g)$. Therefore Equations 15 contain all the information needed to define the extension from $\tilde{S}(T_g)$ to $M_s(T_g)$, and hence are a complete set of defining relations.

§5. The full mapping class group

If $g = 2$ the group $M_s(T_g)$ is the full mapping class group, so that relations 15 give a full set of defining relations for $M(T_2)$. For $g > 2$, however, $M_s(T_g)$ is a proper, non-normal subgroup of $M(T_g)$, and our results are more limited.

Using the representation of T_g given in Figure 1, we note [12] that $M(T_g)$ is generated by elements $U_1, \ldots, U_g, Z_1, \ldots, Z_g, Y_1, \ldots, Y_g$ which are represented by the twists $\tau_{u_1}, \ldots, \tau_{u_g}, \tau_{z_1}, \ldots, \tau_{z_g}, \tau_{y_1}, \ldots, \tau_{y_g}$, where any one of the Z-type generators can be omitted. In this representation the subgroup $M_s(T_g)$ is generated by $U_1, \ldots, U_g, Z_1, \ldots, Z_{g-1}, Y_1, Y_g$ or $g-1$ other sets of generators obtained from these by a cyclic permutation of the handles. All such relations are easily obtained from one such set by introducing a cyclic permutation of the handles. The expression for the element

θ which accomplishes this, as a function of the generators of $M(T_g)$, is:

(17)
$$\theta = (Y_1U_1Y_1)^{-4} X_1X_2 \cdots X_{g-1}$$
$$X_i = (Y_iU_iY_i)(Y_iU_iZ_iU_{i+1}Y_{i+1})^3(Y_iU_iY_i)^{-1}$$
$$1 \leq i \leq g-1$$

The fact that the right-hand side of Equation (17) is indeed a cyclic permutation of the handles can be verified by calculation, again using the representation of $M(T_g)$ as a group of automorphism classes, as given by Equations (12)-(14). Of course:

(18)
$$\theta^g = 1 \ .$$

The generators of $M(T_g)$ will be related by the equations:

(19)
$$U_{i+1} = \theta^{-1}U_i\theta$$
$$Y_{i+1} = \theta^{-1}Y_i\theta$$
$$Z_{i+1} = \theta^{-1}Z_i\theta$$

where all subscripts are modulo g. Equations (19), in combination with the relations obtained by replacing $\tau_1, \ldots, \tau_{2g+1}$ by $Y_1, U_1, Z_1, U_2, Z_2, \ldots U_{g-1}, Z_{g-1}, U_g, Y_g$ respectively in Equation (15.1)-(15.4) then give all relations in $M(T_g)$ which arise by regarding T_g as a 2-sheeted covering of T_0. We conjecture that these relations, plus relations (17) and (18), are a complete set of defining relations for $M(T_g)$.

§6. Additional results

An interesting tie-in with known work concerns a matrix representation for the braid group B_n originally discovered by Burau, and recently investigated anew by W. Magnus and A. Pelluso [17]. It is shown by Magnus and Pelluso that the Burau representation for B_{2g+2} induces a representation for B_{2g+2} in terms of matrices which generate Siegel's modular group $Sp(2g, Z)$ for $g = 1$ and 2, but a proper subgroup of $Sp(2g, Z)$ for $g \geq 3$. Now, it follows from the present analysis that B_{2g+2} can be mapped homomorphically to $M(T_g)$, the mapping being into for $g \geq 3$. Since $Sp(2g, Z)$ is known to be a homomorphic image of $M(T_g)$ it follows that a homomorphism exists from B_{2g+2} to $Sp(2g, Z)$ which is into if $g \geq 3$. Thus one might expect exactly the type of representation which is established in [17], and indeed the proof given there is motivated by the same underlying geometric situation in terms of covering spaces which was considered here.

We point out also that the analysis presented here offers both a generalization and a geometric interpretation of the well-known but previously unexplained fact that the modular group, i.e., $M(T_1)$, is a homomorphic image of Artin's braid group B_4. Since $M(T_{0,2g+2})$ is a homomorphic image of B_{2g+2} (the geometric basis for this homomorphism is established in [14, also 4], and since $M_s(T_g)$ is now known to be a homomorphic image of $M(T_{0,2g+2})$, it follows that $M_s(T_g)$ is a homomorphic image of B_{2g+2} for every $g \geq 1$.

As a final remark, we note that the isomorphism established in Theorem 7 between $M_s(T_g)$ and $S(T_g)$ has a rather unexpected implication, namely that the mapping class group of the $(2g+2)$-punctured surface of genus g has a subgroup which is isomorphic to the subgroup $M_s(T_g)$ of the mapping class group of the closed surface. Since $M(T_g)$ is a homomorphic image of $M(T_g$ - n points) for every g and n, one would really not expect this to be the case. For example, for $g = 2$ we obtain that $M(T_2)$ is both a homomorphic image of $M(T_2$ -6 points) and a subgroup of $M(T_2$ -6 points), where the former property is easily understood but the latter much more subtle.

Stevens Institute of Technology

Acknowledgment

The authors wish to express their thanks to Professor Wilhelm Magnus, who stimulated their original interest in this problem, and is never too busy to listen to or discuss Mathematics. Also, to Professor Seymore Lipschutz, who pointed out the relevance of reference 11 to the proof of Lemma 4.1. Finally, to Professor J. Mennicke, for helpful comments.

REFERENCES

[1] J. W. Alexander, "On the deformation of an n-cell," *Proc. N.A.S.* 9 (1923), 406-7.

[2] P. Bergau and J. Mennicke, "Uber topologische Abbildungen der Brezelflache vom Geschlecht 2," *Math. Zeitschr.* 74 (1960), 414-435.

[3] J. Birman, "On Braid Groups," *Comm. on Pure and App. Math.*, XXII (1969), 41-72.

[4] ______, "Mapping Class Groups and Their Relationship to Braid Groups," *Com. on Pure and App. Math.* 22 (March, 1969), 213-238.

[5] ______, "Automorphisms of the fundamental group of a closed, orientable 2-manifold," *Proc. of Amer. Math Soc.*, 21, No. 2 (1969), 351-4.

[6] H. Coxeter and W. Moser, *Generators and Relations for Discrete Groups*, Springer-Verlag (1965).

[7] M. Dehn, "Die Gruppe der Abbildungsklassen," *Acta. Math.*, 69 (1938), 135-206.

[8] G. M. Fisher, "On the group of all homeomorphisms of a manifold," *Transactions of the A.M.S.*, 97 (1960), 193-212.

[9] R. Gillette and J. Van Buskirk, "The word problem and its consequences for the braid groups and mapping class groups of the 2-sphere," *Trans. Amer. Math. Soc.*, 131, No. 2 (May, 1968), 277-296.

[10] L. Goeritz, "Die Abbildungen der Brezelflachen und der Vollbrezel vom Geschlecht 2," *Hamb. Abh.*, 9 (1933), 244-259.

[11] M. Greendlinger, "Dehn's algorithm for the word problem," *Comm. Pure and Appl. Math.*, 13 (1960), 67-83.

[12] M. Hall, Jr., *The Theory of Groups*, MacMillan Co., (1964), Section 15.4.

[13] H. Klingen, "Charakterisierung der Siegelschen Modelgruppe durch ein endliches System definierender Relationen," *Math. Ann.*, 144 (1961), 64-82.

[14] W. Magnus, "Uber Automorphismen von Fundamentalgruppen berandeter Flachen," *Math. Ann.*, 109 (1934), 617-646.

[15] W. Magnus, A. Karass and D. Solitar, *Combinatorial Group Theory*, Interscience, New York, 1966.

[16] W. Magnus and A. Pelluso, "On knot groups," *Comm. on Pure and Appl. Math.*, XX (1967), 749-770.

[17] ———, "On a theorem of V. I. Arnold," *Comm. Pure and Appl. Math.* XXII, (1969), 683-692.

[18] H. L. Smith, "On the continuous representations of a square upon itself," *Ann. Math.*, 19 (1918-19), 137-41.

FOOTNOTE

* The presentation for $M(T_{0,2g+2})$ in equation (2) is not quite the same as that given in the original reference [14], however it follows from the latter by a sequence of calculations which can be described briefly as follows. Denoting equation numbers from reference [14] by a bar, we note that (2.1) and (2.2) are the same as $(\overline{13a})$, and (2.3) is the same as the last equation in the set $(\overline{19})$. Now, equation $(\overline{14})$ in conjunction with (2.1), (2.2) and (2.3) imply equation (2.4). The remaining equations in the set $(\overline{19})$ are redundant, since they are a consequence of (2.1), (2.2), (2.3) and (2.4). We omit details of these calculations because we believe they are well known [e.g., see 9] and present no real difficulty.

SCHWARZIAN DERIVATIVES AND MAPPINGS ONTO JORDAN DOMAINS

Peter L. Duren

Abstract

Some years ago, Z. Nehari proved that if f(z) is analytic in $|z| < 1$ and its Schwarzian derivative satisfies

$$|\{f(z), z\}| \leq 2(1-r^2)^{-2}, \qquad r = |z| < 1 \ ,$$

then f(z) is univalent in $|z| < 1$. More recently, L. Ahlfors and G. Weill (*Proc. Amer. Math. Soc.*, 13 (1962), 975-978) showed that if

$$|\{f(z), z\}| \leq k(1-r^2)^{-2}$$

for some $k < 2$, then f(z) maps the unit disk onto a Jordan domain. The following theorem sharpens this result.

THEOREM. *Let* $\lambda(r)$ *be continuously differentiable and non-decreasing on the interval* $0 \leq r < 1$, *with* $0 < \lambda(r) \leq 1$ *and*

$$\int_0^1 (1-r)^{-\lambda(r)}\,dr < \infty \ .$$

Suppose f(z) *is analytic in* $|z| < 1$ *and*

$$|\{f(z), z\}| \leq 2\lambda(r)(1-r^2)^{-2}, \qquad r = |z| < 1 \ .$$

Then f(z) *maps* $|z| < 1$ *conformally onto a Jordan domain (on the Riemann sphere).*

The proof is based on an idea of O. Lehto ("Homeomorphisms with a given dilatation," to appear) for constructing a homeomorphic solution to a Beltrami equation $w_{\bar{z}} = \mu w_z$ in which $|\mu(z)|$ is not bounded away from 1, but is allowed to approach 1 "slowly" near a "small" exceptional set. Details will be published elsewhere (*Ann. Acad. Sci. Fenn.*).

Note added in proof: In place of this paper, a simpler proof of a stronger theorem will be published jointly with O. Lehto.

ON THE MODULI OF CLOSED RIEMANN SURFACES WITH SYMMETRIES*

Clifford J. Earle

§1. Introduction

This article has two main sections. The first, §2, concerns the Teichmüller theory of closed Riemann surfaces with automorphisms. It is closely related to the author's joint research with J. Eells (see [3] and [4]) and A. Schatz [5]. In fact, §5 of [5] treats the same situation for compact bordered surfaces. Our present treatment is a bit more thorough since we introduce the relative Teichmüller spaces T(X, H). Such spaces were first considered by A. Kuribayashi [9].

While §2 was in progress, the paper of Alling and Greenleaf on Klein surfaces and real algebraic function fields appeared [1]; it became clear that one should study the moduli of Klein surfaces. For that purpose one needs to understand the quotient of Teichmüller space by the appropriate modular group. We attempt that study in §3. The situation is more complicated than one might hope. For instance, we call attention to Theorem 2: There are closed surfaces which are equivalent to their conjugates but cannot be defined by real polynomials. For the most part we have offered only indications of proofs. Our results are too incomplete to require a full presentation.

Since we allow our surfaces to have conjugate holomorphic automorphisms, they are represented by Fuchsian groups which are subgroups of reflection groups (discrete groups of non-Euclidean motions which include direction-reversing maps). Such groups have been studied by R. Sibner [12]

---

* This research was partly supported by the National Science Foundation through Grant GP -11767.

and the author [2], but a complete account of their Teichmüller theory—at least for non-finitely generated groups—has yet to be given.

The author is grateful to R. Accola, D. Mumford, and B. O'Byrne for helpful suggestions.

§2. Teichmüller theory with symmetries

2.1. We start with a closed oriented smooth surface X of genus $g \geq 2$. M(X) is the space of smooth conformal structures on X. Diff(X) is the group of all (C^∞) diffeomorphisms of X. It acts on M(X) in a natural way: $\mu \cdot f$ is the pullback of the conformal structure μ by the diffeomorphism f.

Two important normal subgroups of Diff(X) are $\mathrm{Diff}^+(X) = \{f \in \mathrm{Diff}(X);$ f preserves orientation$\}$, $\mathrm{Diff}_0(X) = \{f \in \mathrm{Diff}(X);$ f is homotopic to the identity$\}$. The corresponding quotient spaces $R(X) = M(X)/\mathrm{Diff}^+(X)$ and $T(X) = M(X)/\mathrm{Diff}_0(X)$ are the *Riemann space* and *Teichmüller space of* X. The points of R(X) are the Riemann surfaces which are homeomorphic to X. The points of T(X) are called marked Riemann surfaces. Of course R(X) is the quotient of T(X) by the group

$$\Gamma^+(X) = \mathrm{Diff}^+(X)/\mathrm{Diff}_0(X) \ .$$

$\Gamma^+(X)$ is called Teichmüller's modular group.

The C^∞ topology makes M(X) and Diff(X) Fréchet manifolds and Diff(X) a topological group. Further, there is a natural complex structure on M(X) such that $\mathrm{Diff}^+(X)$ is a group of biholomorphic maps. There is a unique complex structure on the manifold T(X) such that the quotient map from M(X) to T(X) is holomorphic.

2.2. Now we consider a finite subgroup $H \subset \mathrm{Diff}(X)$. Since X carries an H-invariant Riemannian metric, the fixed point set

$$M(X)^H = \{\mu \in M(X);\ \mu \cdot h = \mu \quad \text{for all} \quad h \in H\}$$

is not empty. In other words, X can be made a Riemann surface so that H is a group of conformal and anti-conformal maps. We want to study all such

Riemann surfaces. To that end we form the normalizer $N^+(H)$ of H in $\mathrm{Diff}^+(X)$, which consists of all f in $\mathrm{Diff}^+(X)$ such that there is a commuting diagram

$$\begin{array}{ccc} X & \xrightarrow{\;\;f\;\;} & X \\ \downarrow & & \downarrow \\ X/H & \dashrightarrow & X/H \end{array} .$$

Since μ belongs to $M(X)^H$ if and only if it induces a conformal structure on X/H, $M(X)^H$ is invariant under the action of $N^+(H)$. We therefore define the relative Riemann and Teichmüller spaces

$$R(X,H) = M(X)^H/N^+(H) \ ,$$
$$T(X,H) = M(X)^H/N_o(H) \ ,$$

where $N_o(H) = N^+(H) \cap \mathrm{Diff}_o(x)$. These spaces parametrize the Riemann surfaces (or marked Riemann surfaces) on which H is a group of conformal and anti-conformal maps. They acquire topological and real-analytic (complex analytic if $H \subset \mathrm{Diff}^+(X)$) structures from $M(X)^H$, which is a real (or complex) analytic submanifold of $M(X)$. The relative modular group

$$\Gamma^+(X,H) = N^+(H)/N_o(H)$$

acts on $T(X,H)$, producing $R(X,H)$ as quotient.

2.3 A. Kuribayashi [9] studied $T(X,H)$ when H is a cyclic subgroup of $\mathrm{Diff}^+(X)$. He showed that $T(X,H)$ is an analytic manifold homeomorphic to Euclidean space. We shall see that the same is true for every finite group H. The crucial step is to identify $T(X,H)$ with a subset of $T(X)$. For that purpose we need to study the obvious mapping of $T(X,H)$ into $T(X)$ produced by the inclusion of $M(X)^H$ in $M(X)$.

LEMMA. *Suppose* $f \in \mathrm{Diff}_o(X)$ *and* $\mu \in M(X)^H$. *Then* $\mu \cdot f \in M(X)^H$ *if and only if* f *commutes with every element of* H.

Proof: Suppose μ and $\mu \cdot h \in M(X)^H$. Then

$$\mu \cdot hf = \mu \cdot f = \mu \cdot fh \quad \text{for each } h \in H,$$

so $\mu \cdot hfh^{-1}f^{-1} = \mu$ for each $h \in H$. If $f \in \mathrm{Diff}_0(X)$, then $(hfh^{-1})f^{-1} \in \mathrm{Diff}_0(X)$ also. But $\mathrm{Diff}_0(X)$ acts freely on $M(X)$, so $hf = fh$. The other half of the lemma is obvious.

COROLLARY. *The obvious map of* $T(X,H)$ *into* $T(X)$ *is injective.*

To complete the story we need a better description of the image of $T(X,H)$ in $T(X)$, and we need to show that $T(X,H)$ is analytically equivalent with that image. As a first step we note that the image of $T(X,H)$ is $\Phi(M(X)^H)$, where $\Phi : M(X) \to T(X)$ is the quotient map. We remark also that if $\mu \in M(X)$ is H-invariant, then $\Phi(\mu)$ is $\theta(H)$-invariant, where

$$\theta : \mathrm{Diff}(X) \to \mathrm{Diff}(X)/\mathrm{Diff}_0(X) = \Gamma(X)$$

is the quotient map. Thus Φ maps $M(X)^H$ into the fixed point set $T(X)^{\theta(H)}$. We now appeal to a classical theorem.

PROPOSITION.

(a) $T(X)^{\theta(H)}$ *is a closed connected subset of* $T(X)$, *homeomorphic to a Euclidean space.*

(b) *Further,* $T(X)^{\theta(H)}$ *is an analytic submanifold of* $T(X)$, *and*

$$\Phi : M(X)^H \to T(X)^{\theta(H)}$$

is an analytic foliation map (with local analytic right inverses).

Part (a) is due to Saul Kravetz [8] and part (b) to H. E. Rauch [11]. The proposition has several immediate consequences:

COROLLARY 1. Φ *maps* $M(X)^H$ *onto* $T(X)^{\theta(H)}$.

COROLLARY 2. *The obvious map of* $T(X,H)$ *onto* $T(X)^{\theta(H)}$ *is an analytic equivalence.*

To prove Corollary 1 we note that $T(X)^{\theta(H)}$ is the union of the sets $\Phi(M(X)^{H'})$ over all finite groups H' with $\theta(H') = \theta(H)$. Further, $\Phi(M(X)^H)$ and $\Phi(M(X)^{H'})$ are disjoint unless they coincide, which they do if and only if $H' = fHf^{-1}$ for some $f \in \mathrm{Diff}_0(X)$. From part (b) of the proposition we conclude $\Phi(M(X)^H)$ is both open and closed in $T(X)^{\theta(H)}$, and from part (a) that it equals $T(X)^{\theta(H)}$. To prove Corollary 2 we remark that part (b) of the proposition makes the natural map an analytic equivalence of $T(X, H)$ with its image.

2.4. As an application of the above facts we can prove

THEOREM 1. *The quotient map*

$$\Phi : M(X)^H \to T(X, H) \tag{1}$$

defines a trivial principal fibre bundle with structure group $N_0(H) = N(H) \cap \mathrm{Diff}_0(X)$. *Moreover, the group* $N_0(H)$ *is contractible.*

Proof: The proposition implies that Φ is open and continues with local cross-sections. It is already known that (1) defines a principal fibre bundle when H is the trivial group [4]. In view of the local cross-sections, the general case follows. The bundle (1) is trivial because the base space $T(X, H)$ is contractible. The total space $M(X)^H$ is also contractible (see the corresponding theorem for compact X with boundary in §5 of [5]); hence so is the group $N_0(H)$.

COROLLARY. *Let* Y *be a closed non-orientable surface whose oriented* 2*-sheeted closed covering surface* X *has genus* ≥ 2. *Let*

$$\mathrm{Diff}_0(Y) = \{f \in \mathrm{Diff}(Y);\ f \textit{ is homotopic to the identity}\} .$$

Then $\mathrm{Diff}_0(Y)$ *is contractible.*

In fact, $\mathrm{Diff}_0(Y) = N_0(H)$, where $H \subset \mathrm{Diff}(X)$ is the cover group of the covering $X \to Y$. A different proof of this corollary was sketched in [3].

2.5. There is also an application to the theory of moduli of real algebraic function fields in one variable. Alling and Greenleaf [1] have pointed out the important correspondence between those function fields and the compact

"Klein surfaces." A compact Klein surface is either a closed Riemann surface or (essentially uniquely) the quotient of a closed Riemann surface by a group H of order two whose generator reverses orientation. The Teichmüller spaces T(X, H) for such groups H parametrize the real algebraic function fields in one variable. Of course T(X, H) is transcendental in character. One ought to put some sort of algebraic structure on the Riemann space R(X, H). We shall take some steps in that direction in §3.

2.6. When the group H preserves orientation, both T(X) and T(X, H) have natural interpretations as Teichmüller spaces of Fuchsian groups. We shall take a brief look at them here because the relative modular group $\Gamma^+(X, H)$ takes an instructive form. We use the same notation as Bers and Greenberg in their paper in these Proceedings.

Choose an H-invariant complex structure μ on X and a μ-holomorphic universal covering $\pi: U \to X$. (As, usual, U is the upper half-plane with its standard complex structure.) The covering group G is Fuchsian, and π induces a bijection between T(X) and T(G). Under that bijection, T(X, H) corresponds to the Teichmüller space T(G′), where G′ is a Fuchsian group, G is a normal subgroup of G′, and G′/G = H. The Riemann space T(X) corresponds to the quotient of T(G) by the modular group $\Gamma(G)$. But the relative Riemann space R(X, H) is not always the quotient of T(G′) by the modular group $\Gamma(G')$. That is so because $\Gamma(G')$ need not map T(G) into itself. The correct statement is that R(X, H) is the quotient of T(G′) by the group $\Gamma(G') \cap \Gamma(G)$.

There are simple examples of Fuchsian groups G and G′ such that G is normal in G′but $\Gamma(G')$ is not contained in $\Gamma(G)$. In fact one can choose for G′ the universal covering group of a closed surface of genus 2 and for G a subgroup of index 2 which covers a closed surface of genus 3.

§3. The structure of the Riemann space R(X, H)

3.1. When we considered the relative Teichmüller space T(X, H), we discovered two simple facts. First, T(X, H) was naturally injected in T(X).

Second, the image of T(X, H) in T(X) was the fixed point set of a subgroup $\theta(H)$ of $\Gamma(X)$. The analogous statements about R(X, H) are both false.

For example, the Riemann space R(X) has a conjugate-holomorphic involution J which carries each Riemann surface to its conjugate surface. J is the generator of the quotient group $\Gamma(X)/\Gamma^+(X)$, which of course acts on R(X). If H is a subgroup of Diff X such that H has order two and its generator reverses orientation, then every H-invariant complex structure μ on X corresponds to a fixed point of J in $R(X) = M(X)/\Gamma^+(X)$. One might suppose that all fixed points of J have that form. However, that conjecture is false. J can have fixed points which correspond to a Riemann surface which has no conjugate-holomorphic involution. We reformulate that statement as

THEOREM 2. *There are closed Riemann surfaces of genus* 5 *which are equivalent to their conjugates but cannot be defined by real polynomials.*

Proof: Let X be a smooth closed oriented surface of genus 5. First we show that X has an orientation reversing diffeomorphism f of order 4 such that f^2 has no fixed points. The following simple argument is due to Brian O'Byrne, Let X be the surface shown in Figure 1. The map $f: X \to X$ is the composite of a 90° rotation and reversal of front and back. Thus f^2 is a 180° rotation, and f^4 is the identity.

Figure 1.

Now let H be the cyclic subgroup of Diff(X) generated by f. We claim that there is some μ in M(X) whose isotropy subgroup in Diff(X) is precisely H. Since the action of the modular group $\Gamma(X)$ on T(X) is properly discontinuous we need to prove only that the dimension of T(X, H) exceeds that of T(X, G) whenever H is a subgroup of G. First we remark that for all such groups G

$$\dim_R T(X, G) = \dim_C T(X, G')$$

where $G' = G \cap \mathrm{Diff}^+(X)$ (and G' has index two in G). So we need to show that $T(X, H')$ has greater dimension than $T(X, G')$ when H is a subgroup of G. That is easy to verify, since we know that X/H' is a closed surface of genus 3 and that $T(X, H')$ has dimension $3 \cdot 3 - 3 = 6$.

Finally, any Riemann surface X_μ such that the isotropy group of μ is H satisfies the conditions of Theorem 2. In fact, f is a conjugate automorphism of X_μ, so X_μ is equivalent to its conjugate surface. But X_μ has no conjugate automorphism of order two, so it cannot be defined by a real polynomial.

REMARKS.

(a) It is easy to show that

$$\dim_R T(X, G) = \dim_C T(X, G')$$

whenever $G' = G \cap \mathrm{Diff}^+(X) \neq G$.

(b) It can happen that $T(X, G_1) = T(X, G_2)$ even though G_1 is a proper subgroup of G_2, but that is a rare occurrence. It happens only if the genus of X/G_1' and the ramification of the map $X \to X/G_1'$ satisfy certain conditions. A list of the possible genera and ramifications has been made by Leon Greenberg ([6], Theorems 3A and 3B).

(c) There are obvious modifications of our example to fit any closed surface whose genus is congruent to one modulo 4. There are even simpler examples of genus two. For instance, the Riemann surface

$$y^2 = x(x^2 - \alpha^2)(x^2 + t\alpha^2 x - \alpha) ,$$

with $t > 0$ and $\alpha = \exp(2\pi i/3)$, has an anti-conformal automorphism $(x, y) \to (-1/\bar{x}, i\bar{y}/\bar{x}^3)$ of order 4 but none of order 2 unless $t = 1$.

3.2. The reason why $R(X,H)$ is harder to study than $T(X,H)$ is that $\mathrm{Diff}^+(X)$ does not act freely on $M(X)$. To avoid that difficulty we define, for any integer $n > 2$, the groups

$$\mathrm{Diff}_n(X) = \{f \in \mathrm{Diff}\, X;\ f \text{ induces identity map on } H_1(X, Z/nZ)\},$$

$$\mathrm{Diff}_n^+(X) = \mathrm{Diff}_n(X) \cap \mathrm{Diff}^+(X)\ ,$$

$$N_n(H) = \text{normalizer of } H \text{ in } \mathrm{Diff}_n^+(X)\ ,$$

and the Teichmüller spaces

$$T_n(X) = M(X)/\mathrm{Diff}_n^+(X)$$

and
$$T_n(X,H) = M(X)^H/N_n(H)\ .$$

It is clear that $T_n(X)$ and $T_n(X,H)$ are finite branched coverings of $R(X)$ and $R(X,H)$. Further, Serre (see the Appendix of [7]) has shown that $\mathrm{Diff}_n^+(X)$ acts *freely* on $M(X)$. We therefore have this

LEMMA. *The obvious map of* $T_n(X,H)$ *into* $T_n(X)$ *is injective.*

That is proved by the same method as Lemma 2.3. Proceeding as in §2.3, we let

$$\theta_n : \mathrm{Diff}(X) \to \mathrm{Diff}(X)/\mathrm{Diff}_n^+(X) \doteq \Gamma_n(X)$$

be the quotient map. We wish to study the fixed point set of $\theta_n(H)$ in $T_n(X)$ and to compare it with $T_n(X,H)$. Corresponding to Proposition 2.3 and its corollaries we have

THEOREM 3.

(a) $T_n(X)^{\theta_n(H)}$ *is a closed analytic submanifold of* $T_n(X)$, *not necessarily connected.*

(b) $\Phi_n : M(X) \to T_n(X)$ *maps* $M(X)^H$ *onto a connected component of* $T_n(X)^{\theta_n(H)}$. *That component is analytically equivalent to* $T_n(X,H)$.

(c) *Each component of* $T_n(X)^{\theta_n(H)}$ *is of the form* $T_n(X,H')$ *where* H' *is a finite group and* $\theta_n(H') = \theta_n(H)$.

Theorem 3 is a fairly direct consequence of the results in §2.3 because the projection from $T(X)$ to $T_n(X)$ is a covering map. The fixed point set $T_n(X)^{\theta_n(H)}$ can be disconnected when $g = 1$ and $n = 2$; for the cases $g > 1$ and $n > 2$ nothing is proved.

3.3. We have shown that $R(X, H)$ is the quotient of a manifold $(T_n(X, H))$ by a finite group. We can show even more. In fact, as Mumford has shown [10], $T_n(X)$ is not only a complex manifold but a (quasi-projective) variety, on which $\Gamma_n(X)$ acts as a group of (real) automorphisms. Therefore the fixed point set $T_n(X)^{\theta_n(H)}$ is a (real) subvariety of $T_n(X)$. (It is a complex subvariety if and only if $H \subset \mathrm{Diff}^+(X)$; otherwise $\theta_n(H)$ contains conjugate automorphisms.) $T_n(X, H)$ is itself a piece of the variety $T_n(X)^{\theta_n(H)}$. A similar statement holds for $R(X, H)$.

THEOREM 4. *Let* $\Gamma_n(H)$ *be the normalizer of* $\theta_n(H)$ *in the group* $\theta_n(\mathrm{Diff}^+(X)) = \Gamma_n^+(X)$. *Then*

(a) $\Gamma_n(H)$ *is a group of automorphisms of* $T_n(X)^{\theta_n(H)}$,

(b) *The quotient space* $T_n(X)^{\theta_n(H)}/\Gamma_n(H)$ *is the disjoint union of Riemann spaces* $R(X, H')$. *The union is over the* $\mathrm{Diff}^+(X)$ *conjugacy classes of finite groups* H' such that $\theta_n(H') = \theta_n(H)$.

Proof: (a) is obvious. To prove (b) we first determine the $\Gamma_n(H)$-orbit of $T_n(X, H)$. Let $x \in T_n(X, H)$ and $\gamma \in \Gamma_n(H)$. Then we can write $x = \Phi_n(\mu)$, $\gamma = \theta_n(f)$, and $x \cdot \gamma = \Phi_n(\mu \cdot f)$, where μ is fixed by H and $\mu \cdot f$ is fixed by H'.

Since γ normalizes $\theta_n(H) = \theta_n(H')$, we can find a map $\alpha : H \to H'$ so that $f\alpha(h)f^{-1}h^{-1} \in \text{kernel } \theta_n = \mathrm{Diff}_n^+(X)$ for all $h \in H$. Now, if $h \in H$,

$$\mu \cdot f\alpha(h)f^{-1} = (\mu \cdot f) \cdot \alpha(h)f^{-1} = (\mu \cdot f) \cdot f^{-1} = \mu = \mu \cdot h,$$

so $f\alpha(h)f^{-1}h^{-1}$ fixes μ. Since $\mathrm{Diff}_n^+(X)$ acts freely on $M(X)$,

$$\alpha(h) = f^{-1}hf \quad , \tag{2}$$

and the groups H and H′ are conjugate. Conversely, it is easy to see that if H and H′ are conjugate and $\theta_n(H) = \theta_n(H')$, then $T_n(X, H)$ and $T_n(X, H')$ lie in the same $\Gamma_n(H)$-orbit.

We next show that the image of $T_n(X, H)$ in the quotient space is R(X, H). Let x, γ, μ, and f be as above. We need to prove that if $\mu \cdot f$ is fixed by H then f normalizes H. But that is an immediate consequence of (2) because α maps H onto H′ = H.

To prove (b) we apply the above reasoning to each connected component of $T_n(X)^{\theta_n(H)}$. The image of each component is a space R(X, H′), and two components $T_n(X, H')$ and $T_n(X, H'')$ have the same image if and only if H′ and H″ are conjugate subgroups.

REMARKS.

(a) I have no idea whether the quotient space in Theorem 4 has more than one component.

(b) When H has order two and its generator reverses orientation, the points of R(X, H) are in bijective correspondence with certain real algebraic function fields, as we noted in §2.5. The above quotient space is a space of moduli for these function fields. Each component corresponds to the function fields whose Klein surface X/H has given topological type. Even in this case the number of components is unknown. One ought to show that for suitable choice of n there is just one component.

REFERENCES

[1] N. Alling and N. Greenleaf, Klein surfaces and real algebraic function fields, *Bull. Amer. Math. Soc.*, 75 (1969), 869-872.

[2] C. Earle, Teichmüller spaces of groups of the second kind, *Acta Math.*, 112 (1964), 91-97.

[3] C. Earle and J. Eells, A fibre bundle description of Teichmüller theory, *J. Differential Geometry*, 3 (1969), 19-43.

[4] ———, Deformations of Riemann surfaces, *Lectures in Modern Analysis and Applications I*, Springer-Verlag, 1969, pp. 122-149.

[5] C. Earle and A. Schatz, Teichmüller theory for surfaces with boundary, *J. Differential Geometry,* 4(1970), 169-186.

[6] L. Greenberg, Maximal Fuchsian groups, *Bull. Amer. Math. Soc.,* 69 (1963), 569-572.

[7] A. Grothendieck, Techniques de construction en géométrie analytique X, *Séminaire H. Cartan,* Paris, 1960-61, Exp. 17.

[8] S. Kravetz, On the geometry of Teichmüller spaces and the structure of their modular groups, *Ann. Acad. Sci. Fenn.,* 278 (1959), 1-35.

[9] A. Kuribayashi, On analytic families of compact Riemann surfaces with non-trivial automorphisms, *Nagoya Math. J.,* 28 (1966), 119-165.

[10] D. Mumford, Abelian quotients of the Teichmüller modular group, *J. Analyse Math.,* 18 (1967), 227-244.

[11] H. E. Rauch, A transcendental view of the space of algebraic Riemann surfaces, *Bull. Amer. Math. Soc.,* 71 (1965), 1-39.

[12] R. J. Sibner, Symmetric Fuchsian groups, *Amer. J. Math.,* 90 (1968), 1237-1259.

Cornell University.

AN EIGENVALUE PROBLEM FOR RIEMANN SURFACES

Leon Ehrenpreis

The first part of this lecture is expository; in the second part we shall present some new ideas.

Part I. *The two dimensional picture.* We denote by H the upper half plane: $\tau = x + iy$, $y > 0$, and we denote by G the group SL(2, R) of 2 by 2 real matrices of determinant 1. Actually, we shall be a little sloppy in in what follows in that we shall not distinguish between SL(2, R) and the factor $SL(2, R)/\pm I$, denoting both of them by G. We are sure that this will not cause any difficulty to the reader.

Since G acts transitively on H, we can identify H with the quotient G/K where K is the subgroup of G leaving a convenient point of H fixed. We choose for this point the point $\tau = i$. It is easily seen that

$$K = \left\{\begin{pmatrix} \cos\theta & \sin\theta \\ -\sin\theta & \cos\theta \end{pmatrix}\right\} \quad . \tag{1}$$

Thus K is just the circle group.

There exists on H a differential operator

$$D = y^2\left(\frac{\partial^2}{\partial x^2} + \frac{\partial^2}{\partial y^2}\right) \tag{2}$$

which commutes with the action of G. D is essentially unique; in fact, any linear differential operator on H which commutes with G is a polynomial in D. Looking at things in another way, D is the Laplace-Beltrami operator for the Poincaré metric on H.

Let S be a Riemann surface whose universal covering is H. Then we can identify

$$S = \Gamma/H = \Gamma\backslash G/K$$

where Γ is the fundamental group of S, considered as a subgroup of G. Thus, from a group-theoretic point of view, S appears as a space of double cosets of G. Γ is a discrete subgroup of G.

Since D commutes with the action of G, D commutes with Γ. Thus D defines a differential operator on S which we shall continue to denote by D. Again, D can be described intrinsically on S: We put on S a metric of constant negative curvature (−4) and D is the Laplace-Beltrami operator for this metric.

Main Problem. Study the eigenfunctions and eigenvalues of D.

The problem can be put in a somewhat more precise form: There is a natural Hilbert space setting for our problem, namely the space $L^2(S)$ of square integrable functions on S with respect to the area element associated with the metric of constant negative curvature. From the point of view of the upper half-plane H, this is the space of functions f on $\Gamma\backslash H$ with the norm

$$\|f\|^2 = \int_\Omega |f(\tau)|^2 \frac{dx\,dy}{y^2} \tag{3}$$

where Ω is a fundamental domain for H. D defines a self-adjoint operator on $L^2(S)$ and we are interested in its spectral properties. We shall restrict our considerations to the case when the area of S is finite.

It should be pointed out that there is a linear second order differential operator $\tilde{D}$ on G known as the Casimir operator, which commutes with left and right multiplication of G such that $\tilde{D}$ agrees with D on H, that is,

$$\tilde{D}f = Df \tag{4}$$

if $f(gk) = f(g)$ for all $k \in K$ (so f is really a function on $G/K = H$). In general, the eigenvalue problems for $\tilde{D}$ on $\Gamma\backslash G/K$ are rather similar. The reason for this is that if f is any function on $\Gamma\backslash G$ we can write

(5) $$f(g) = \Sigma\, f_n(g)$$

where

(6) $$f_n(\gamma gk) = \chi_n(k)\, f_n(\gamma g) \ .$$

Here χ_n are the characters of K which, by (1), is just the circle group. The expansion (5) is just the Fourier series expansion of f under the right action of K. In general, the nature of the eigenvalue problems for the classes of functions defined by (6) for each fixed n is essentially independent of n and is thus roughly the same as the Riemann surface S (which corresponds to (6) for the trivial character which we denote by χ_0). There is, however, one important difference between χ_0 and χ_n for $n \neq 0$. which we shall discuss below.

From a geometric point of view, $\Gamma\backslash G$ is the tangent sphere bundle over $\Gamma\backslash G/K$, that is, the bundle whose fibres are the unit tangent vectors to S.

We shall make the following normalization for the eigenvalues: We write

(7) $$D\,u(\tau, s) = -s(1-s)\,u(\tau, s)$$

or, more generally

(8) $$\tilde{D}\,u(g,s) = -s(1-s)\,u(g,s) \ .$$

Since D and $\tilde{D}$ are self-adjoint, $s(1-s)$ must be real. Thus, s is real or Re $s = \frac{1}{2}$. By abuse of language, we shall refer to the set with multiplicity of s which corresponds to the eigenvalues of D or $\tilde{D}$ as the "spectrum."

The operator $-D$ is non-negative. This means that the spectrum of D is contained in the union of Re $s = \frac{1}{2}$ and $0 \leq s \leq 1$. Since we are assuming that S is of finite area, the constants are always eigenfunctions of D so 0 and 1 are always in the spectrum.

If we fix n and consider the restriction of $\tilde{D}$ to the class of functions defined by (6) then $-\tilde{D}$ is bounded from below. It can be seen that, in addition to Re $s = \frac{1}{2}$ and the interval $0 \leq s \leq 1$, only *integral values* of s can appear in the spectrum. The dimension of the eigenspace corresponding to an integral point $s = \ell$ is the same as the dimension of a suitable space of holomorphic and anti-holomorphic differentials on S. (By "differentials" we mean "differentials of higher order" that is, expressions which are locally of the form $h(z)dz^g$.) Since the dimensions of the spaces of differentials of all orders determine the topological type (that is, the genus and the number of punctures) of S we conclude:

PROPOSITION 1. *The spectrum of* $\tilde{D}$ *determines the topological type of* S.

Problem 1. Does the spectrum of D or $\tilde{D}$ determine the conformal type of S?

A partial solution to Problem 1 is given by Gelfand in [2]. This asserts that if S is compact and we consider a one-parameter family S_t of Riemann surfaces with $S_0 = S$ and the spectrum of $\tilde{D}$ on S_t being independent of t then the S_t are all conformally the same.

Suppose that S is compact. Thus Γ consists of hyperbolic elements. By the *norm* of an hyperbolic element we mean the larger of its eigenvalues. The Selberg trace formula (see [4]) shows that the determination of the spectrum of $\tilde{D}$ on Γ G is equivalent to the determination of the norms of the elements of Γ. Thus Problem 1 becomes

Problem 1′. Do the norms of the elements of Γ determine Γ as a subgroup of G up to conjugation by an element of G?

Problem 1′ can be stated directly in terms of S, since the norm of an element $\gamma \in \Gamma$ represents the length of the shortest geodesic in the same free homotopy class as γ.

Let us now consider the case when S has punctures.

Example. $\Gamma = SL(2, Z)$, the modular group.

We can construct eigenfunctions of D on $\Gamma \backslash H$ as follows: We start wtih the function $a_s(\tau) = y^s$ which is an eigenfunction of D with eigenvalue $-s(1-s)$. We want to make a_s invariant under Γ by summing over $\gamma \in \Gamma$. Since a_s is already invariant under the subgroup

$$\Gamma_0 = \{\begin{pmatrix} 1 & n \\ 0 & 1 \end{pmatrix}\} \tag{9}$$

of Γ, we need sum only over $\gamma \in \Gamma/\Gamma_0$, that is, we define

$$E(\tau, s) = \sum_{\gamma \in \Gamma/\Gamma_0} a_s(\gamma\tau) . \tag{10}$$

A simple computation shows that, if $\gamma = \begin{pmatrix} k & \ell \\ m & n \end{pmatrix}$ then

$$a_s(\gamma\tau) = \frac{y^s}{|m\tau + n|^{2s}} , \tag{11}$$

thus,

$$E(\tau, s) = \sum \frac{y^s}{|m\tau + n|^{2s}} . \tag{12}$$

It is easily seen that the sum is over all pairs m, n of relatively prime integers.

The series in (12) converges for $\text{Re } s > 1$. The function $E(\tau, s)$ is known as the *Eisenstein series*. It has a meromorphic continuation to the whole complex s plane. Moreover $E(\tau, s)$ satisfies the functional equation

$$E(\tau, 1-s) = \chi(s)\, E(\tau, s) . \tag{13}$$

Here $\chi(s)$ has a simple expression in terms of the Γ function.

What is important for us here is that $E(\tau, s)$ gives the continuous part of the spectrum of D on $\Gamma \backslash H$. More precisely,

THEOREM 1 (Selberg, Roelke). *The spectral representation of a function* f *on* $\Gamma\backslash H$ *takes the form*

$$f(\tau) = \int_{Rs=\frac{1}{2}} F(s)E(\tau,s)\,ds + \sum F_j\, u(\tau, s_j) \;. \tag{14}$$

Here $u(\tau, s_j)$ *belong to* $L_2(\Gamma\backslash H)$ *and*

$$F(s) = \int_{\Gamma\backslash H} f(\tau)\bar{E}(\tau, s)\, \frac{dx\,dy}{y^2} \tag{15}$$

Problem 2. What are the values s_j which occur in (14)?

Any information on Problem 2 is probably very important for number theory.

For the general group Γ with h cusps in the fundamental domain, we can construct h Eisenstein series by transforming the cusps successively to infinity and then applying the same construction as for the modular group. Selberg has shown that these Eisenstein series have analytic continuations and give the continuous spectrum on $\Gamma\backslash H$.

Part II. *The three dimensional picture.* We now regard G as the three dimensional proper Lorentz group SL(1,2), that is, the group of 3 by 3 real matrices which leave invariant the quadratic form $t^2 - x^2 - y^2$ and have determinant + 1.

There are three types of orbits of G on R^3.

A. *One sheet of hyperboloid of two sheets.* As in the case of the upper half-plane H, we can identify this sheet with the quotient of G by the subgroup leaving a point fixed. In particular we see easily that the subgroup leaving (1, 0, 0) fixed is just the group of rotations about the t axis which is exactly the group K.

B. *The part of the light cone with* $t > 0$ *or* $t < 0$. This can be identified with G/B where

$$B = \{\begin{pmatrix} 1 & b \\ 0 & 1 \end{pmatrix}\} ,$$

C. *Hyperboloid of one sheet.* This can be identified with G/A where

$$A = \{\begin{pmatrix} a & 0 \\ 0 & a^{-1} \end{pmatrix}\} .$$

The differential operator which is invariant under G is the wave operator

$$\Box = \frac{\partial^2}{\partial t^2} - \frac{\partial^2}{\partial x^2} - \frac{\partial^2}{\partial y^2} .$$

The relation between $\Box$ and D is separation of variables. The general theory of separation of variables is rather complicated (see [1]). For our purposes we can think of it as follows: Let h be a solution of the wave equation

(16) $$\Box h = 0$$

in the interior of the forward light cone. Suppose that h is also homogeneous, that is

(17) $$h(rt, rx, ry) = r^s h(t, x, y) .$$

Thus h can be thought of as a function $\tilde{h}$ on the sheet of hyperbola through (1, 0, 0), that is, $\tilde{h}$ is a function on G/K which is the upper half-plane H.

(18) $$D\tilde{h} = -s(1-s)\tilde{h} .$$

The relation just described between $\Box$ and D is the analog of the classical relation between homogeneous harmonic polynomials in R^n and spherical harmonics on the unit sphere in R^n.

It is important to note the following: If h satisfies (16) in the forward light cone and is small at infinity (in a suitable sense) then we can decompose h into homogeneous parts by use of Mellin transform, that is,

(19) $$h = \int h_s ds$$

where the h_s satisfy (17) and are given by

$$h_s(t, x, y) = \int h(rt, rx, ry) r^{-s} \frac{dr}{r} \quad . \tag{20}$$

If $\Box h = 0$ then $\Box h_s = 0$ since $\Box$ commutes with scalar multiplication. We can think of h as a *generating function* for the h_s.

Now, let Γ be a discrete group as in Part I. If h is Γ invariant then so are all the functions h_s because Γ commutes with scalar multiplication. We wish to show how to construct Γ invariant solutions of the wave equation.

Let us take the case where $\Gamma = SL(2, Z)$ is the modular group. When we think of G as $SL(1, 2)$, Γ becomes a group of 3 by 3 matrices whose entries are integers. We denote this subgroup of $SL(1,2)$ again by Γ. The orbit under Γ of the point $(1, 1, 0)$ on the light cone is a discrete set of points $\Gamma(1,1,0)$ on the light cone, which is contained in the set of integer points. [$\Gamma(1,1,0)$ is the set of all lattice points on the forward light cone which satisfy a certain "relatively prime" condition.] We can also write $\Gamma(1,0,0)$ in the form

$$\Gamma(1,1,0) = \{\gamma(1,1,0)\}_{\gamma \epsilon \Gamma/\Gamma_0} \quad .$$

Here Γ_0 is the subgroup of Γ leaving the point $(1,1,0)$ fixed; Γ_0 is, as in Part I, the group

$$\Gamma_0 = \{\begin{pmatrix} 1 & n \\ 0 & 1 \end{pmatrix}\} \quad .$$

We denote by $d = \Sigma_{\gamma \epsilon \Gamma/\Gamma_0} \delta_{\gamma(1,1,0)}$ the measure formed by unit masses at each of the points $\gamma(1,1,0)$. We can use d to construct Γ invariant solutions h of $\Box h = 0$ in two ways.

Method a. Regard d as a measure in R^3 and form its 3-dimensional Fourier transform. More precisely, form

$$\hat{d}(t, x, y) = \Sigma \exp[(t, x, y) \cdot \gamma(1,1,0)] \quad . \tag{21}$$

Since d is Γ invariant, so is $\hat{d}$. Since the support of d is contained in the light cone, $\square \hat{d} = 0$.

Method b. Regard d as a measure on the forward light cone and solve the Dirichlet problem (see [3]) for $\square$ in the interior of the forward light cone with Dirichlet data equal to d. We thus construct a function d_1 satisfying

$$\square d_1 = 0 \qquad \text{in forward light cone} \tag{22}$$

$$d_1 = d \qquad \text{on boundary.} \tag{23}$$

Of course, (23) has to be taken in a suitable limit sense.

Since Γ commutes with $\square$, it is easily seen that d_1 is γ invariant. We can now state

THEOREM 2. *(Poisson summation formula).* $d_1 = \hat{d}$.

The reason for calling Theorem 2 a Poisson summation formula is as follows: the classical Poisson summation formula states that the sum of unit masses at *all* lattice points is its own Fourier transform. On the other hand, $\hat{d}$ is the Fourier transform of the sum of unit masses placed at *some* lattice points. Certainly $\hat{d} \neq d$ because $\square \hat{d} = 0$ so the support of $\hat{d}$ cannot be such a small set as a set of lattice points. Instead of $\hat{d} = d$ we have $\hat{d} = d_1$ which means that d is the Dirichlet data of $\hat{d}$.

We can use the explicit solution to the Dirichlet problem to find an explicit formula for d_1. We can then construct the functions $d_{1,s}$ by means of (20). This was carried out by the author and Mautner. The result is

$$\tilde{d}_{1,s}(\tau) = E(\tau, s) \;. \tag{24}$$

We can also form $\hat{\tilde{d}}_s(\tau)$. We find

$$\hat{\tilde{d}}_s(\tau) = [\chi(s)]^{-1} E(\tau, 1-s) \;. \tag{25}$$

Thus our Poisson summation formula gives the functional equation (13) for the Eisenstein series just as the ordinary Poisson summation formula gives the functional equation for the Riemann zeta function.

An analog of Theorem 2 holds for all Γ which are the fundamental groups of Riemann surfaces with $\Gamma \backslash G$ of finite measure. There are also analogs for general semi-simple Lie groups.

The details of Part II will appear in a book [1] being prepared by the author.

BIBLIOGRAPHY

[1] L. Ehrenpreis, *Representations of Semi-Simple Lie Groups*, in preparation.

[2] I. M. Gelfand, "Automorphic functions and the theory of representations," *International Congress*, Stockholm, 1962.

[3] J. Hadamard, *Lectures on Cauchy's Problem*, Dover, New York, 1952.

[4] A. Selberg, "Harmonic analysis and discontinuous groups," *Jour. Indian Math. Soc.*, Vol. 20 (1956), pp. 47-87.

RELATIONS BETWEEN QUADRATIC DIFFERENTIALS

Hershel M. Farkas*

Introduction

If S is a compact Riemann surface of genus g, $g \geq 2$, then the vector space of holomorphic quadratic differentials on S is a $3g-3$ dimensional space. We shall denote this space by A_2. If S is not hyperelliptic, then by Noether's theorem [6], one can choose a basis for A_2 from the $g(g+1)/2$ quadratic differentials $\zeta_i \zeta_j$, $i \geq j = 1, \ldots, g$ where $\zeta_1, \ldots, \zeta_g$ is a basis for the holomorphic abelian differentials on S. Choosing such a basis we can then express each $\zeta_i \zeta_j$ as a linear combination of the basis elements. Our problem is to determine the constants in these expressions.

In the case $g = 2$ $\dim A_2 = 3$ and $\zeta_1^2, \zeta_1\zeta_2, \zeta_2^2$ is a basis for A_2. In the case $g = 3$ $\dim A_2 = 6$ and there are two cases to consider: case (i) S is not hyperelliptic. In this case $\zeta_1^2, \zeta_1\zeta_2, \zeta_1\zeta_3, \zeta_2^2, \zeta_2\zeta_3, \zeta_3^2$ is a basis for A_2. Case (ii) S is hyperelliptic. In this case $\zeta_1^2, \zeta_1\zeta_2, \zeta_1\zeta_3, \zeta_2^2, \zeta_2\zeta_3, \zeta_3^2$ is not a basis for A_2 and we can exhibit a linear relation between these six quantities with the coefficients of the relation appearing explicitly as derivatives of a certain theta null.

In this paper we shall be primarily concerned with the case $g = 4$ and consider the two situations of no even theta null vanishing on S, and precisely one even theta null vanishing on S. The methods used however are quite general and apply to any genus $g \geq 2$. We consider the particular case $g = 4$ in order to show how the Schottky relation arises from the relations obtained.

In part I of this paper I shall briefly describe the key result which is

*Research partially supported by NSF Grant GP-12467.

needed to establish the desired relations. Details will be appearing elsewhere [2]. The results have already for the most part been announced in [1,3]. The main result is that one can attach 2 sets of theta functions to a compact Riemann surface of genus g and that these two sets of functions are related in a very special way. In part II, I shall show how this result enables one to write down many relations between quadratic differentials.

I. Let (S, Γ, Δ) denote a compact Riemann surface of genus $g \geq 2$ together with a cannonical homology basis $\Gamma = \gamma_1, \ldots, \gamma_g$ $\Delta = \delta_1, \ldots, \delta_g$ and let $\zeta_1, \ldots, \zeta_g$ denote the basis for the space of holomorphic abelian differentials dual to the cannonical homology basis Γ, Δ. Then $\int_{\gamma_j} \zeta_i = \delta_{ij}$ and $\int_{\delta_j} \zeta_i = \Pi_{ij}$. It is well known that the $g \times g$ matrix $\Pi = (\Pi_{ij})$ is a complex symmetric matrix with positive definite imaginary part. We shall call the $g \times 2g$ matrix $(I\ \Pi)$ the period matrix for (S, Γ, Δ) where I denotes the $g \times g$ identity matrix.

We define a map $\phi : S \to C^g$ as $p \to (\phi_1(p), \ldots, \phi_g(p))$ where $\phi_j(p) = \int_{p_0}^{p} \zeta_j$ and p_0 is a fixed point on S. ϕ_j is a finite regular multivalued function on S, the multivaluedness arising from the dependence on the path of integration. In order to make the map ϕ single valued we simply identify all possible images of a point p. Since any two images of p can differ only by an integral linear combination of the columns of the $g \times 2g$ period matrix for (S, Γ, Δ) we identify all such points in C^g. C^g under this identification is a compact commutative complex Lie group of complex dimension g called the Jacobi variety of S denoted by J(S).

Definition 1. A meromorphic multivalued differential V on (S, Γ, Δ) is said to be multiplicative (or a Prym-differential) with characteristic $\left[\begin{smallmatrix}\varepsilon \\ \varepsilon'\end{smallmatrix}\right] = \left[\begin{smallmatrix}\varepsilon_1, \ldots, \varepsilon_g \\ \varepsilon_1', \ldots, \varepsilon_g'\end{smallmatrix}\right]$ providing continuation of V along $\gamma_j(\delta_j)$ carries V into $(-1)^{\varepsilon_j} V ((-1)^{\varepsilon_j'} V)$ where the ε_i and ε_i' are elements of $Z/(2)$.

PROPOSITION 1. *Let* $\left[\begin{smallmatrix}\varepsilon \\ \varepsilon'\end{smallmatrix}\right]$ *be any non-zero characteristic. Then the set of everywhere finite multiplicative differentials with characteristic*

$[{\varepsilon \atop \varepsilon'}]$ *on* (S, Γ, Δ) is a $(g-1)$ *dimensional vector space.*

In this discussion we shall be concerned only with the vector space of everywhere finite multiplicative differentials with characteristic $\psi = [{\varepsilon \atop \varepsilon'}]$, $\varepsilon = (0,\dots,0)$ $\varepsilon' = (1,0,\dots,0)$. If $V_1,\dots,V_{g-1}$ is any basis for this space we can form the $(g-1)\times 2g$ period matrix (AB) where

$$A = (A_{ij})\, A_{ij} = \int_{\gamma_j} V_i \quad {i = 1,\dots,g-1 \atop j = 0,\dots,g-1}$$

and

$$B = (B_{ij}) \text{ where } B_{ij} = \int_{\delta_j} V_i \quad {i = 1,\dots,g-1 \atop j = 0,\dots,g-1} \quad .$$

It is then possible to choose a new basis $W_1,\dots,W_{g-1}$ so the period matrix will have the form

$$\begin{matrix} 1 & 0 & \cdots & 0 & \tau_{10} & \tau_{11} & \cdots & \tau_{1g-1} \\ 0 & 1 & & & & \tau_{21} & & \\ \vdots & & \ddots & & & \vdots & & \\ 0 & \cdots & 0 & 1 & \tau_{g-10} & \tau_{g-12} & \cdots & \tau_{g-1\,g-1} \end{matrix}$$

We shall refer to $W_1,\dots,W_{g-1}$ as the normal basis of everywhere finite multiplicative differentials with characteristic ψ on (S, Γ, Δ).

PROPOSITION 2. *The matrix* $\tau = (\tau_{ij})$ $i, j = 1,\dots,g-1$ *is a complex symmetric matrix and has positive definite imaginary part.*

We recall the definition of $\mathfrak{S}_g$, the Siegel upper half plane of degree g. It is the set of $g\times g$ complex symmetric matrices with positive definite imaginary aprt. Hence what we have shown till now is that we can attach to each (S, Γ, Δ) an element of $\mathfrak{S}_g$ and an element of $\mathfrak{S}_{g-1}$.

Definition 2. Let $Z = (Z_1, \ldots, Z_g)$ be a complex g-vector and $T = (t_{ij})$ an element of $\mathfrak{S}_g$. The first order theta function with g-characteristic $\left[{\varepsilon \atop \varepsilon'}\right] = \left[{\varepsilon_1 \cdots \varepsilon_g \atop \varepsilon'_1 \cdots \varepsilon'_g}\right]$ where ε_i and ε'_i are elements of $\mathbf{Z}$, $\theta\left[{\varepsilon \atop \varepsilon'}\right](Z, T)$, is defined by the following series which converges absolutely and uniformly on compact subsets of $C^g \times \mathfrak{S}_g$.

$$\theta\left[{\varepsilon \atop \varepsilon'}\right](Z, T) = \sum_N \exp 2\pi i \left\{\tfrac{1}{2}\left(N + \tfrac{\varepsilon}{2}\right) T \left(N + \tfrac{\varepsilon}{2}\right) + \left(N + \tfrac{\varepsilon}{2}\right)\left(Z + \tfrac{\varepsilon'}{2}\right)\right\}$$

where N runs over all integer vectors in C^g and all operations are matrix products. The theta constant or theta null with g-characteristic $\left[{\varepsilon \atop \varepsilon'}\right]$ at T is $\theta\left[{\varepsilon \atop \varepsilon'}\right] = \theta\left[{\varepsilon \atop \varepsilon'}\right](0, T)$.

It can be shown quite easily that the theta function $\theta\left[{\varepsilon \atop \varepsilon'}\right](Z, T)$ is an even or odd function of Z depending on whether $\varepsilon \cdot \varepsilon'$ is even or odd. If $\varepsilon \cdot \varepsilon'$ is even (odd) $\theta\left[{\varepsilon \atop \varepsilon'}\right](Z, T)$ is an even (odd) function of Z. Hence we see immediately that all odd theta constants vanish.

Definition 3. The Riemann theta function with characteristic $\left[{\varepsilon \atop \varepsilon'}\right]$ associated with (S, Γ, Δ) is $\theta\left[{\varepsilon \atop \varepsilon'}\right](Z, \Pi)$ and the Schottky theta function associated with (S, Γ, Δ) is $\eta\left[{\varepsilon \atop \varepsilon'}\right](Z, \tau)$, that is the theta function with $T = \tau$.
Note: The Z, Π variables in the definition of Riemann theta function are an element of $C^g \times \mathfrak{S}_g$ while the Z, τ variables in the definition of Schottky theta function are an element of $C^{g-1} \times \mathfrak{S}_{g-1}$.

For a fixed Π the Riemann theta function is a function on C^g. It is not a well defined function on $J(S)$ since $Z_1 \equiv Z_2$ does not imply that $\theta\left[{\varepsilon \atop \varepsilon'}\right](Z_1, \Pi) = \theta\left[{\varepsilon \atop \varepsilon'}\right](Z_2, \Pi)$; however it is true that $\theta\left[{\varepsilon \atop \varepsilon'}\right](Z_1, \pi) = E\theta\left[{\varepsilon \atop \varepsilon'}\right](Z_2, \Pi)$ where E is some exponential multiplier. Therefore, the zeros of the theta function are well defined on $J(S)$. Utilizing the map $\phi : S \to J(S)$ described earlier we can view the theta function $\phi\left[{\varepsilon \atop \varepsilon'}\right](\phi(p), \Pi)$ as a multivalued function on S with well defined zeros. The properties of this function are discussed in [5].

Returning now to definition and proposition 1 we can construct an $(\hat{S}, \hat{\Gamma}, \hat{\Delta})$ where $\hat{S}$ is a smooth two sheeted cover of S and where $\hat{\Gamma}, \hat{\Delta}$ is a cannonical homology basis for S constructed in a natural way from Γ, Δ. The properties of $\hat{S}$ are summarized in the following proposition.

PROPOSITION 3. *$\hat{S}$ is a smooth two sheeted cover of S, therefore of genus $2g-1$, on which all multiplicative differentials with characteristic ψ are single valued. If $\zeta_1, \dots, \zeta_g$; $W_1, \dots, W_{g-1}$ are the respective bases of holomorphic differentials on S dual to Γ, Δ and normal basis of everywhere finite multiplicative differentials with characteristic ψ on (S, Γ, Δ) then denoting the lift of ζ_i by $\hat{\zeta}_i$ and the left of W_i by $\hat{W}_i$, $\hat{\zeta}_1, \dots, \hat{\zeta}_g$, $\hat{W}_1 \dots \hat{W}_{g-1}$ is a basis for the vector space of holomorphic abelian differentials on $\hat{S}$. $\hat{S}$ permits a conformal fixed point free involution T, the sheet interchange, and one can choose a canonical homology basis $\hat{\Gamma}, \hat{\Delta}$ for $\hat{S}$ in such a way that $T(\hat{\gamma}_1)$ is homologous to $\hat{\gamma}_1$, $T(\hat{\delta}_1) = \hat{\delta}_1$. $T(\hat{\gamma}_i) = \hat{\gamma}_{g+i-1}$ $i = 2 \dots g$ and $T(\hat{\delta}_i) = \hat{\delta}_{g+i-1}$ $i = 2 \dots g$. The dual basis to $\hat{\Gamma}, \hat{\Delta}$ is $\hat{u}_1 = \hat{\zeta}_1$*

$$\hat{u}_i = \frac{\hat{\zeta}_i + \hat{W}_{i-1}}{2} \quad i = 2, \dots, g \qquad \hat{u}_{g+i-1} = \frac{\hat{\zeta}_i - \hat{W}_{i-1}}{2} \quad i = 2, \dots, g .$$

$(I, \hat{\Pi})$ is the period matrix for $(\hat{S}, \hat{\Gamma}, \hat{\Delta})$ and if we denote the Riemann theta function associated with $(\hat{S}, \hat{\Gamma}, \hat{\Delta})$ by $\hat{\theta}$ then at least $2^{g-2}(2^{g-1}-1)$ even theta constants $\hat{\theta}[{\varepsilon \atop \varepsilon'}]$ vanish.

One of the nice features of $\hat{S}$ is that aside from its theta function $\hat{\theta}$ we may also view the Riemann and Schottky theta functions associated with (S, Γ, Δ) as multivalued functions on $\hat{S}$. This is done via the maps

$$\phi_\zeta : p \to \left(\int_{p_0}^{p} \hat{\zeta}_1, \dots, \int_{p_0}^{p} \hat{\zeta}_g \right) \text{ and } \phi_W : p \to \left(\int_{p_0}^{p} \hat{W}_1, \dots, \int_{p_0}^{p} \hat{W}_{g-1} \right) .$$

The natural question to ask is what is the relation between $\hat{\theta}, \theta$ and η? This is partially answered by the following proposition.

PROPOSITION 4.

$$\theta\left[\begin{smallmatrix} g_1 & \delta & \delta \\ h_1 & \delta' & \delta' \end{smallmatrix}\right](Z_1, Z_2, \ldots, Z_g, Z_{g+1}, \ldots, Z_{2g-1}, \hat{\Pi}) =$$

$$\sum_{p \in F^{g-1}} (-1)^{(\delta - p)\cdot\delta'} \eta\left[\begin{smallmatrix} p \\ 0 \end{smallmatrix}\right](t_1, \ldots, t_{g-1}, 2\tau)\,\theta\left[\begin{smallmatrix} g_1 & \delta - p \\ h_1 & 0 \end{smallmatrix}\right](s_1, \ldots, s_g, 2\Pi)$$

where $F = Z/(2)$, $\left[\begin{smallmatrix} g_1 \\ h_1 \end{smallmatrix}\right]$ is a 1-characteristic $\left[\begin{smallmatrix} \delta \\ \delta' \end{smallmatrix}\right]$ is a $g-1$ characteristic and

$$Z_1 = s_1 \qquad Z_2 = \frac{s_2 + t_1}{2}, \ldots, Z_g = \frac{s_g + t_{g-1}}{2},$$

$$Z_{g+1} = \frac{s_2 - t_2}{2}, \ldots, Z_{2g-1} = \frac{s_g - t_{g-1}}{2}$$

In particular as functions on $\hat{S}$ we have $\hat{\theta}\left[\begin{smallmatrix} g_1 & \delta & \delta \\ h_1 & \delta' & \delta' \end{smallmatrix}\right] =$

$$\sum_{p \in F^{g-1}} (-1)^{(\delta - p)\cdot\delta'} \eta\left[\begin{smallmatrix} p \\ 0 \end{smallmatrix}\right](\phi_W, 2\tau)\,\theta\left[\begin{smallmatrix} g_1 & \delta - p \\ h_1 & 0 \end{smallmatrix}\right](\phi_\zeta, 2\Pi)\,.$$

We have already mentioned in proposition 3 that at least $2^{g-2}(2^{g-1}-1)$ even theta constants associated with $(\hat{S}, \hat{\Gamma}, \hat{\Delta})$ vanish. The next proposition tells us which theta constants vanish. The proof of the next proposition was obtained jointly with H. E. Rauch [3].

PROPOSITION 5. $\hat{\theta}\left[\begin{smallmatrix} 0 & \varepsilon & \varepsilon \\ 1 & \varepsilon' & \varepsilon' \end{smallmatrix}\right](\phi, \Pi) \equiv 0$ *on* $(\hat{S}, \hat{\Gamma}, \hat{\Delta})$ *for all choices of base point* p_0 *where* $\left[\begin{smallmatrix} \varepsilon \\ \varepsilon' \end{smallmatrix}\right]$ *ranges over all the* $2^{g-2}(2^{g-1}-1)$ *odd* $g-1$ *characteristics.*

From propositions 4 and 5 and a little bit of algebra we can deduce

PROPOSITION 6.

$$\frac{\eta\left[{\varepsilon \atop 0}\right](\phi_W, 2\tau)}{\theta\left[{0\ \varepsilon \atop 1\ 0}\right](\phi_\zeta, 2\Pi)} \equiv \frac{\eta\left[{\delta \atop 0}\right](\phi_W, 2\tau)}{\theta\left[{0\ \delta \atop 1\ 0}\right](\phi_\zeta, 2\Pi)}$$

for any vectors $\varepsilon, \delta \in F^{g-1}$ *and any base point* p_0 *with* $F = Z/(2)$.

It now follows from proposition 6 and known formulae [4 p. 233] relating theta functions with parameters Π and τ to theta functions with parameters 2Π and 2τ that

PROPOSITION 7.

$$\frac{\eta\left[{\varepsilon \atop \varepsilon'}\right]\eta\left[{\varepsilon \atop \varepsilon'}\right](\phi_W, \tau)}{\theta\left[{0\ \varepsilon \atop 0\ \varepsilon'}\right]\theta\left[{0\ \varepsilon \atop 1\ \varepsilon'}\right](\phi_\zeta, \Pi) + \theta\left[{0\ \varepsilon \atop 1\ \varepsilon'}\right]\theta\left[{0\ \varepsilon \atop 0\ \varepsilon'}\right](\phi_\zeta, \Pi)}$$

$$= \frac{\eta\left[{\delta \atop \delta'}\right]\eta\left[{\delta \atop \delta'}\right](\phi_W, \tau)}{\theta\left[{0\ \delta \atop 0\ \delta'}\right]\theta\left[{0\ \delta \atop 1\ \delta'}\right](\phi_\zeta + \Pi) + \theta\left[{0\ \delta \atop 1\ \delta'}\right](\phi_\zeta, \Pi)}$$

for any choice of base point p_0 *on* $\hat{S}$ *and any even* $g-1$ *characteristics* $\left[{\varepsilon \atop \varepsilon'}\right]$, $\left[{\delta \atop \delta'}\right]$.

The preceding formula is the key to our method of obtaining relations between quadratic differentials on S.

II.

We now consider the following expression:

$$\frac{\eta\left[{\varepsilon \atop \varepsilon'}\right]\eta\left[{\varepsilon \atop \varepsilon'}\right](\phi_W, \tau)}{\eta\left[{\delta \atop \delta'}\right]\eta\left[{\delta \atop \delta'}\right](\phi_W, \tau)}$$

$$-\frac{\theta[{}^{0\,\varepsilon}_{0\,\varepsilon'}]\theta[{}^{0\,\varepsilon}_{1\,\varepsilon'}](\phi_\zeta,\Pi)+\theta[{}^{0\,\varepsilon}_{1\,\varepsilon'}]\theta[{}^{0\,\varepsilon}_{0\,\varepsilon'}](\phi_\zeta,\Pi)}{\theta[{}^{0\,\delta}_{0\,\delta'}]\theta[{}^{0\,\delta}_{1\,\delta'}](\phi_\zeta,\Pi)+\theta[{}^{0\,\delta}_{1\,\delta'}]\theta[{}^{0\,\delta}_{0\,\delta'}](\phi_\zeta,\Pi)} = Q(P)$$

and observe that in a neighborhood of the point p_0, the base point of the maps ϕ_ζ and ϕ_W, this defines a holomorphic function on $\hat{S}$.(1) Letting Z be a local parameter about p_0 we can write out the Taylor expansion of Q. By virtue of proposition 7 however, all the coefficients must be zero since $Q = 0$ on $\hat{S}$. We shall here be concerned only with the first three coefficients of the Taylor expansion.

Our first lemma is the key to deriving the Schottky relation and Schottky type relations for $g > 4$ in [2,3].

Lemma 1. *For any two even* $g-1$ *characteristics* $[{}^{\varepsilon}_{\varepsilon'}]$, $[{}^{\delta}_{\delta'}]$ *we have*

$$\frac{\eta^2[{}^{\varepsilon}_{\varepsilon'}]}{\eta^2[{}^{\delta}_{\delta'}]} = \frac{\theta[{}^{0\,\varepsilon}_{0\,\varepsilon'}]\theta[{}^{0\,\varepsilon}_{1\,\varepsilon'})}{\theta[{}^{0\,\delta}_{0\,\delta'}]\theta[{}^{0\,\delta}_{1\,\delta'}]} .$$

Proof: This is simply the statement that the constant term in the expansion of Q is zero.

We have defined $\phi_\zeta(p) = (\int_{p_0}^{p} \hat{\zeta}_1, \ldots, \int_{p_0}^{p} \hat{\zeta}_g)$ and $\phi_W(p) = (\int_{p_0}^{p} \hat{W}_1, \ldots, \int_{p_0}^{p} \hat{W}_{g-1})$. We shall denote $\int_{p_0}^{p} \hat{\zeta}_i$ by s_i and $\int_{p_0}^{p} \hat{W}_i$ by t_i. Computing now $\frac{d}{dz}Q(p)$ we find

$$\frac{d}{dz}Q(z(p)) = \sum_{k=1}^{g-1} \frac{\eta[{}^{\varepsilon}_{\varepsilon'}]}{\eta[{}^{\delta}_{\delta'}]} \frac{\partial}{\partial t_k} \frac{\eta[{}^{\varepsilon}_{\varepsilon'}](\phi_W,\tau)}{\eta[{}^{\delta}_{\delta'}](\phi_W,\tau)} \frac{dt_k}{dz}(p)$$

$$-\sum_{i=1}^{g} \frac{\partial}{\partial s_i} \frac{\theta[{}^{0\,\varepsilon}_{0\,\varepsilon'}]\theta[{}^{0\,\varepsilon}_{1\,\varepsilon'}](\phi_\zeta,\Pi)+\theta[{}^{0\,\varepsilon}_{1\,\varepsilon'}](\phi_\zeta,\Pi)}{\theta[{}^{0\,\delta}_{0\,\delta'}]\theta[{}^{0\,\delta}_{1\,\delta'}](\phi_\zeta,\Pi)+\theta[{}^{0\,\delta}_{0\,\delta'}](\phi_\zeta,\Pi)} \frac{ds_i}{dz}(p)$$

(1) We assume that none of the quantities in the expression for Q(p) vanish identically for every choice of base point p_0 on $\hat{S}$.

Setting $z = 0$ or $p = p_0$ we gain no new information since each term in the sum vanishes separately. Computing now d^2/dz^2 we find

$$\frac{d^2}{dz^2} Q(Z(p))\Big|_{z=0} = \sum_{\ell, k=1}^{g-1} \frac{\eta[{\varepsilon \atop \varepsilon'}]}{\eta[{\delta \atop \delta'}]} \frac{\partial^2}{\partial t_\ell \, \partial t_k} \frac{\eta[{\varepsilon \atop \varepsilon'}](\phi_W, \tau)}{\eta[{\delta \atop \delta'}](\phi_W, \tau)}\Bigg|_{p=p_0} \frac{dt_k}{dZ} \frac{dt_\ell}{dZ} (p_0)$$

$$- \sum_{i,j=1}^{g} \frac{\partial^2}{\partial s_i \partial s_j} \frac{\theta[{0\ \varepsilon \atop 0\ \varepsilon'}]\theta[{0\ \varepsilon \atop 1\ \varepsilon'}](\phi_\zeta, \Pi) + \theta[{0\ \varepsilon \atop 0\ \varepsilon'}](\phi_\zeta, \Pi)}{\theta[{0\ \delta \atop 0\ \delta'}]\theta[{0\ \delta \atop 0\ \delta'}](\phi_\zeta, \Pi) + \theta[{0\ \delta \atop 0\ \delta'}](\phi_\zeta, \Pi)}\Bigg|_{p=p_0} \frac{ds_i}{dZ} \frac{ds_j}{dZ} (p_0)$$

Evaluating the partial derivatives at p_0 in the above expression, using the heat equations

$$\pi i(1 + \delta_{k\ell}) \frac{\partial \eta[{\varepsilon \atop \varepsilon'}](t, \tau)}{\partial \tau_{k\ell}} = \frac{\partial^2 \eta[{\varepsilon \atop \varepsilon'}](t, \tau)}{\partial t_k \partial t_\ell},$$

$$\pi i(1+\delta_{ij}) \frac{\partial \theta[{\varepsilon \atop \varepsilon'}](s, \pi)}{\partial \Pi_{ij}} = \frac{\partial^2 \theta[{\varepsilon \atop \varepsilon'}](s, \pi)}{\partial s_i \partial s_j}$$

lemma 1, and the fact that $ds_i = \hat{\zeta}_i \quad dt_k = \hat{W}_k$, we obtain

LEMMA 2.

$$\sum_{k \le \ell = 1}^{g-1} \frac{\partial}{\partial \tau_{k\ell}} \left(\frac{\eta[{\varepsilon \atop \varepsilon'}]}{\eta[{\delta \atop \delta'}]} \right) W_k W_\ell$$

$$= \sum_{i \le j = 1}^{g-1} \frac{\partial}{\partial \Pi_{ij}} \sqrt{\frac{\theta[{0\ \varepsilon \atop 0\ \varepsilon'}]\theta[{0\ \varepsilon \atop 0\ \varepsilon'}]}{\theta[{0\ \delta \atop 0\ \delta'}]\theta[{0\ \delta \atop 0\ \delta'}]}} \zeta_i \zeta_j$$

for any two even (g − 1) *characteristics* $[{\varepsilon \atop \varepsilon'}]$ $[{\delta \atop \delta}]$.

Observe that lemma 2 gives us many relations between quadratic differentials on S. In particular it expresses linear combinations of $W_k W_\ell$'s which are holomorphic quadratic differentials on S in terms of linear combinations of $\zeta_i \zeta_j$'s. Hence in order to obtain the desired relations we simply have to eliminate the $W_k W_\ell$'s from these relations.

THEOREM 1. *Let* S *be a compact Riemann surface of genus* 2 *and let* Γ,Δ *be a cannonical homology basis for* S. *Let* ζ_1,ζ_2 *be the basis of the vector space of holomorphic abelian differentials dual to* Γ,Δ *and let* W *be the normal basis of everywhere finite multiplicative differentials with characteristic* $\left[\begin{smallmatrix} 0 & 0 \\ 1 & 0 \end{smallmatrix}\right]$. *Then*

$$W^2 = \frac{1}{\dfrac{\partial}{\partial \tau}\left(\dfrac{\eta\left[\begin{smallmatrix} 0 \\ 0 \end{smallmatrix}\right]}{\eta\left[\begin{smallmatrix} 0 \\ 1 \end{smallmatrix}\right]}\right)} \sum_{i\geq j=1}^{2} \frac{\partial}{\partial \Pi_{ij}}\left(\sqrt{\frac{\theta\left[\begin{smallmatrix} 0 & 0 \\ 0 & 0 \end{smallmatrix}\right]\theta\left[\begin{smallmatrix} 0 & 0 \\ 1 & 0 \end{smallmatrix}\right]}{\theta\left[\begin{smallmatrix} 0 & 0 \\ 0 & 1 \end{smallmatrix}\right]\theta\left[\begin{smallmatrix} 0 & 0 \\ 1 & 1 \end{smallmatrix}\right]}}\right)\zeta_i\zeta_j$$

Proof: The proof follows from simply putting the given data into lemma 2.

The above theorem is of course the best one we can hope to obtain in genus 2 since necessarily $\zeta_1{}^2,\zeta_1\zeta_2,\zeta_2{}^2$ form a linearly independent set. We can similarly obtain results of this type for $g = 3$; however, our intention is to obtain relations solely between the quadratic differentials $\zeta_i\zeta_j$ not involving the $W_k W_\ell$. We shall do this explicitly in the case $g = 4$. To this end we must first establish certain theta identities. We shall begin by assuming the well known genus 1 theta identity $\theta^4\left[\begin{smallmatrix} 0 \\ 0 \end{smallmatrix}\right] = \theta^4\left[\begin{smallmatrix} 0 \\ 1 \end{smallmatrix}\right] + \theta^4\left[\begin{smallmatrix} 1 \\ 0 \end{smallmatrix}\right]$.

LEMMA 3. *For any* $\Pi \in \mathfrak{S}_2$ *we have*

$$\theta^2\left[\begin{smallmatrix} 0 & 0 \\ 0 & 0 \end{smallmatrix}\right]\theta\left[\begin{smallmatrix} 0 & 0 \\ 1 & 0 \end{smallmatrix}\right] = \theta^2\left[\begin{smallmatrix} 0 & 0 \\ 0 & 1 \end{smallmatrix}\right]\theta^2\left[\begin{smallmatrix} 0 & 0 \\ 1 & 1 \end{smallmatrix}\right] + \theta^2\left[\begin{smallmatrix} 0 & 1 \\ 0 & 0 \end{smallmatrix}\right]\theta^2\left[\begin{smallmatrix} 0 & 1 \\ 1 & 0 \end{smallmatrix}\right]$$

and for any $\Pi \in \mathfrak{S}_3$ *we have*

$$\theta\begin{bmatrix}000\\000\end{bmatrix}\theta\begin{bmatrix}000\\100\end{bmatrix}\theta\begin{bmatrix}000\\010\end{bmatrix}\theta\begin{bmatrix}000\\110\end{bmatrix}$$

$$= \theta\begin{bmatrix}000\\001\end{bmatrix}\theta\begin{bmatrix}000\\101\end{bmatrix}\theta\begin{bmatrix}000\\011\end{bmatrix}\theta\begin{bmatrix}000\\111\end{bmatrix} + \theta\begin{bmatrix}001\\000\end{bmatrix}\theta\begin{bmatrix}001\\100\end{bmatrix}\theta\begin{bmatrix}001\\010\end{bmatrix}\theta\begin{bmatrix}001\\110\end{bmatrix} .$$

Proof: The first half of lemma 3 follows quite simply from lemma 1 at least in the case where Π is a period matrix. In this case the theta constants are Riemann theta constants and by lemma 1 they are proportional to the Schottky theta constants which are genus 1 thetas. Hence we get the first half of lemma 3. The second half, again in the case where $\Pi \in \mathfrak{S}_3$ is a period matrix, follows from the first half of lemma 3 and lemma 1. The cases where the Π's are period matrices in $\mathfrak{S}_2$ and $\mathfrak{S}_3$ have the dimension of $\mathfrak{S}_2, \mathfrak{S}_3$ respectively and hence the result must hold for the whole of S_2 and $\mathfrak{S}_3$.

We now rewrite the second identity in lemma 3 in the following form

$$\frac{\eta\begin{bmatrix}000\\001\end{bmatrix}\ \eta\begin{bmatrix}000\\101\end{bmatrix}\ \eta\begin{bmatrix}000\\011\end{bmatrix}\ \eta\begin{bmatrix}000\\111\end{bmatrix}}{\eta\begin{bmatrix}000\\000\end{bmatrix}\ \eta\begin{bmatrix}000\\100\end{bmatrix}\ \eta\begin{bmatrix}000\\010\end{bmatrix}\ \eta\begin{bmatrix}000\\110\end{bmatrix}} + \frac{\eta\begin{bmatrix}001\\000\end{bmatrix}\eta\begin{bmatrix}001\\100\end{bmatrix}\eta\begin{bmatrix}001\\010\end{bmatrix}\eta\begin{bmatrix}001\\110\end{bmatrix}}{\eta\begin{bmatrix}000\\110\end{bmatrix}\eta\begin{bmatrix}000\\000\end{bmatrix}\eta\begin{bmatrix}000\\010\end{bmatrix}\eta\begin{bmatrix}000\\100\end{bmatrix})}$$

$$= 1$$

and observe that since it is an identity in $\mathfrak{S}_3$ its gradient must vanish identically on $\mathfrak{S}_3$. In particular these remarks hold for Schottky theta constants on a surface of genus 4.

In the above formula we have eight quotients of Schottky theta constants. Denoting them by $q_1, \ldots, q_8$ and the corresponding square roots of quotients of products of Riemann theta constants via lemma 1 by $r_1, \ldots, r_8$ we clearly have

LEMMA 4. *For any* $\tau_{k\ell}$ $k\ell = 1,\ldots,3$

$$\frac{\partial q_1}{\partial \tau_{k\ell}} q_2q_3q_4 + \frac{\partial q_2}{\partial \tau_{k\ell}} q_1q_3q_4 + \cdots + \frac{\partial q_4}{\partial \tau_{k\ell}} q_1q_2q_3 + \frac{\partial q_5}{\partial \tau_{k\ell}} q_6q_7q_8$$

$$+ \cdots + \frac{\partial q_8}{\partial \tau_{k\ell}} q_5q_6q_7 = 0 .$$

Proof: This is simply the statement that the gradient of $q_1q_2q_3q_4 + q_5q_6q_7q_8 - 1$ vanishes on S_3.

THEOREM 2. *Let* S *be a compact Riemann surface of genus* 4 *together with a cannonical homology basis* Γ, Δ. *Let* $\zeta_1, \ldots, \zeta_4$ *be the basis of the holomorphic abelian differentials on* S *dual to* Γ, Δ. *Then*

$$\sum_{i \le j = 1}^{4} \left(\frac{\partial r_1}{\partial \Pi_{ij}} r_2r_3r_4 + \cdots + \frac{\partial r_4}{\partial \Pi_{ij}} r_1r_2r_3 \right.$$

$$\left. + \frac{\partial r_5}{\partial \Pi_{ij}} r_6r_7r_8 + \cdots + \frac{\partial r_8}{\partial \Pi_{ij}} r_5r_6r_7 \right) \zeta_i\zeta_j = 0 .$$

Proof: By virtue of lemma 2 we have the eight relations

$$\sum_{k \le \ell = 1}^{3} \frac{\partial q_1}{\partial \tau_{k\ell}} W_k W_\ell = \sum_{i \le j = 1}^{4} \frac{\partial r_1}{\partial \Pi_{ij}} \zeta_i \zeta_j$$

$$\sum_{k \le \ell = 1}^{3} \frac{\partial q_8}{\partial \tau_{k\ell}} W_k W_\ell = \sum_{i \le j = 1}^{4} \frac{\partial r_8}{\partial \Pi_{ij}} \zeta_i \zeta_j$$

Multiplying the left hand side of the first relation by $q_2q_3q_4$ and the right hand side by $r_2r_3r_4$ which is clearly permissable by lemma 1, and continuing in this fashion finally multiplying the left hand side of the eighth relation by $q_5 q_6 q_7$ and the right side by $r_5 r_6 r_7$ and summing the eight relations we clearly obtain the statement of the theorem.

The only thing that one should check is that the coefficients in the relation obtained do not all vanish. The vector of coefficients however, is precisely the gradient of the Schottky relation and this is known to be non-zero. As a matter of fact if we replace $\zeta_i \zeta_j$ by $d\Pi_{ij}$, Theorem 2 gives the differential form of the Schottky relation. In any case, the relation in theorem 2 is the Noether relation in genus 4.

The following question now presents itself. Suppose the Riemann surface of genus 4 is such that precisely one Riemann theta constant say $\theta[{}^{\varepsilon}_{\varepsilon'}]$ vanishes. It can be shown then that this leads to a relation of the form

$$\sum_{i\leq j=1}^{4} \frac{\partial \theta[{}^{\varepsilon}_{\varepsilon'}]}{\partial \Pi_{ij}} \zeta_i \zeta_j = 0 .$$

Hence for such a surface in addition to the relation obtained in theorem 2 we have also another relation between quadratic differentials. This relation must be a multiple of the relation obtained in theorem 2 for if not we can deduce from Noether's theorem that the surface is hyperelliptic and then ten Riemann theta constants must vanish.

Appendix.

In the Nachträge to the collected works of Riemann [7, p. 108] the following 2 formulae appear without proof:

(1) $$\frac{\Omega(2\mu_1, \dots, 2\mu_g, 0 \cdots 0)}{\theta(\mu_1, \dots, \mu_g)\theta(\mu_1 \cdots \mu_{g-1}, \mu_g + \frac{\pi i}{2}}$$ is independent of μ

(2) $$\frac{\Omega(0, \dots, 0, 2\nu_1, \dots, 2\nu_{g-1})}{\theta^2(\nu_1, \dots, \nu_{g-1})}$$ is independent of ν .

These two formulae were mentioned by Mumford in his talk at the conference and constitute a special case of the theorem in Accola's talk at the conference. It therefore seems to be of interest to show how they can be proved from proposition 4 and 5 of this paper.

THEOREM 3. *For any* (g−1) *characteristic* $\left[{\varepsilon \atop \varepsilon'}\right]$ *we have*

$$\frac{\hat{\theta}\left[{0\ \varepsilon\ \varepsilon \atop 1\ \varepsilon'\varepsilon'}\right](2s_1, s_2, \ldots, s_g, s_2, \ldots, s_g;\ \hat{\Pi})}{\theta\left[{0\ \varepsilon \atop 1\ \varepsilon'}\right](s_1, \ldots, s_g;\ \Pi)\,\theta\left[{0\ \varepsilon \atop 0\ \varepsilon'}\right](s_1, \ldots, s_g;\ \Pi)} \quad \text{is independent of } s_1, \ldots, s_g$$

and

$$\frac{\hat{\theta}\left[{0\ \varepsilon\ \varepsilon \atop 1\ \varepsilon'\varepsilon'}\right](0, t_1, \ldots, t_{g-1}, -t_1, \ldots, -t_{g-1};\ \hat{\Pi})}{\eta^2\left[{\varepsilon \atop \varepsilon'}\right](\nu_1, \ldots, \nu_{g-1};\ \tau)} \quad \text{is independent of } t_1, \ldots, t_{g-1}\ .$$

Proof: It follows immediately from proposition 4 that

$$\hat{\theta}\left[{0\ \varepsilon\ \varepsilon \atop 1\ \varepsilon'\varepsilon'}\right](2s_1, s_2, \ldots, s_g s_2, \ldots, s_g)$$

$$= \sum_p (-1)^{(\varepsilon - p)\cdot\varepsilon}\,\eta\left[{p \atop 0}\right](0, 2\tau)\,\theta\left[{0\ \varepsilon - p \atop 1\ \ 0}\right](2s, 2\Pi)$$

and that

$$\hat{\theta}\left[{0\ \varepsilon\ \varepsilon \atop 1\ \varepsilon'\varepsilon'}\right](0, t_1, \ldots, t_{g-1}, -t_1, \ldots, -t_{g-1};\ \hat{\Pi})$$

$$= \sum_p (-1)^{(\varepsilon - p)\cdot\varepsilon'}\,\eta\left[{p \atop 0}\right](2t, 2\tau)\,\theta\left[{0\ \varepsilon - p \atop 1\ \ 0}\right](0, 2\Pi)\ .$$

Utilizing now proposition 6 and letting

$$\frac{\eta\left[{0 \atop 0}\right](0, 2\tau)}{\theta\left[{0\,0 \atop 1\,0}\right](0, 2\Pi)} = K(\Pi, \tau)$$

we can rewrite the above as

$$\hat{\theta}\left[\begin{smallmatrix}0 & \varepsilon & \varepsilon \\ 1 & \varepsilon' & \varepsilon'\end{smallmatrix}\right](2s_1, s_2, \ldots, s_g, s_2, \ldots, s_g, \hat{\Pi})$$

$$= K(\Pi,\tau) \sum_p (-1)^{(\varepsilon-p)\cdot\varepsilon'} \theta\left[\begin{smallmatrix}0 & p \\ 1 & 0\end{smallmatrix}\right](0,2\Pi)\,\theta\left[\begin{smallmatrix}0 & \varepsilon-p \\ 1 & 0\end{smallmatrix}\right](2s, 2\Pi)$$

and

$$\hat{\theta}\left[\begin{smallmatrix}0 & \varepsilon & \varepsilon \\ 1 & \varepsilon' & \varepsilon'\end{smallmatrix}\right](0, t_1, \ldots, t_{g-1}, -t_1, \ldots, -t_{g-1}\,;\, \hat{\Pi})$$

$$= \frac{1}{K(\Pi,\tau)} \sum_p (-1)^{(\varepsilon-p)\cdot\varepsilon'} \eta\left[\begin{smallmatrix}\varepsilon-p \\ 0\end{smallmatrix}\right](0,2\tau)\,\eta\left[\begin{smallmatrix}p \\ 0\end{smallmatrix}\right](2t, 2\tau)$$

Finally using the formula in [4, p 233] relating the theta functions with parameters Π and τ to theta functions with parameters 2Π and 2τ we obtain

$$\hat{\theta}\left[\begin{smallmatrix}0 & \varepsilon & \varepsilon \\ 1 & \varepsilon' & \varepsilon'\end{smallmatrix}\right](2s_1, s_2, \ldots, s_g, s_2, \ldots, s_g\,;\, \hat{\Pi})$$

$$= K(\Pi,\tau)\,\theta\left[\begin{smallmatrix}0 & \varepsilon \\ 1 & \varepsilon'\end{smallmatrix}\right](s_1, \ldots, s_g; \Pi)\,\theta\left[\begin{smallmatrix}0 & \varepsilon \\ 0 & \varepsilon'\end{smallmatrix}\right](s_1, \ldots, s_g\,;\, \Pi)$$

$$\hat{\theta}\left[\begin{smallmatrix}0 & \varepsilon & \varepsilon \\ 1 & \varepsilon' & \varepsilon'\end{smallmatrix}\right](0, t_1, \ldots, t_{g-1}, -t_1, \ldots, -t_{g-1}\,;\, \hat{\Pi})$$

$$= \frac{1}{K(\Pi,\tau)}\,\eta^2\left[\begin{smallmatrix}\varepsilon \\ \varepsilon\end{smallmatrix}\right](t_1, \ldots, t_{g-1}\,;\, \tau)$$

Having now proved Riemann's theorem in this form it should be observed that our lemma 1 is now a corollary of this theorem. Setting the s and t variables equal to zero and dividing one formula by the other we obtain our lemma 1 i.e.:

$$\frac{\theta^2\left[\begin{smallmatrix}\varepsilon \\ \varepsilon'\end{smallmatrix}\right]}{\theta\left[\begin{smallmatrix}0 & \varepsilon \\ 0 & \varepsilon'\end{smallmatrix}\right)\theta\left[\begin{smallmatrix}0 & \varepsilon \\ 1 & \varepsilon'\end{smallmatrix}\right]} = K^2(\Pi,\tau) \text{ for any } \left[\begin{smallmatrix}\varepsilon \\ \varepsilon'\end{smallmatrix}\right].$$

State University of New York at Stony Brook
Stony Brook, New York

REFERENCES

[1] Farkas, H. M., "Theta Constants, Moduli and Compact Riemann Surfaces," *Proc. of the Nat'l. Acad. of Sci. U.S.A.*, 62 (1969), pp. 320-325.

[2] ———, On the Schottky Relation and Its Generalization to Arbitrary Genus," *Annals of Math.*, July 1970.

[3] ——— and Rauch, H. E., "Two Kinds of Theta Constants and Period Relations on a Riemann Surface," *Proc. of the Nat'l. Acad. of Sci.*, 62, (1969), pp. 679-686.

[4] Igusa, J. I., "On the Graded Ring of Theta Constants," *Amer. Journal of Math.*, 86 (1964), pp. 219-246.

[5] Lewittes, J., "Riemann Surfaces and the Theta Function," *Acta Mathematica,* 111 (1964), pp. 37-61.

[6] Martens, H., "Varieties of Special Divisors on a Curve II," *Jnl. für die Reine und Ang. Math.*, 233, (1968), pp. 89-100.

[7] Riemann, B., "Collected Works," Nachträge, Dover 1953.

DEFORMATIONS OF EMBEDDINGS OF RIEMANN SURFACES IN PROJECTIVE SPACE

Frederick Gardiner*

Let Γ be a covering group of a compact Riemann surface of genus g where Γ operates on U, the upper half plane. Let $B_q(\Gamma, U)$ be the space of holomorphic functions ϕ in U which satisfy the relation $\phi(\gamma(z))\gamma'(z)^q = \phi(z)$ for all $\gamma \in \Gamma$. $B_q(\Gamma, U)$ is called the space of holomorphic q-differentials and we shall always take q to be an integer ≥ 1. The Riemann surface $S = U/\Gamma$, is mapped into $P(B_q(\Gamma, U)^*)$, the projective space of the dual of $B_q(\Gamma, U)$, in a natural way. Namely, given $p \in U/\Gamma$, if p is the image under the covering map of a point $z \in U$, we send p into the linear functional ℓ_p, where $\ell_p(\phi) = \phi(z)$. Then ℓ_p depends on the choice of z but only up to a multiplicative constant, so we have a well-defined element of $P(B_q(\Gamma, U)^*)$. Let us call this map Φ_q. In most cases Φ_q is an injection and everywhere holomorphic with non-vanishing derivative.

Likewise, for any Beltrami coefficient μ we can look at the same mapping for a covering group Γ^μ which covers the surface S^μ which is the same as S with a new complex structure determined by μ. Γ^μ will operate on some simply connected domain which we call U^μ. Just as before, we obtain a mapping $\Phi_q^\mu : S^\mu \to P(B_q(\Gamma^\mu, U^\mu)^*)$. The theory of moduli for Riemann surfaces supplies a natural non-singular complex linear map

---

* This research was supported by the U.S. Army Office of Research (Durham).
The author wishes to express his deepest appreciation for the encouragement, advice, and inspiration given him by Lipman Bers.

N_μ^q: $B_q(\Gamma,U) \to B_q(\Gamma^\mu,U^\mu)$ and, hence, we obtain a mapping of each Riemann surface S^μ into the same projective space $P(B_q(\Gamma,U)^*)$. It will turn out that the mapping N_μ^q will depend only on the Teichmüller class of μ.

A similar mapping Q_μ^q is also supplied by the theory of moduli. This mapping will be in some sense natural, but will depend on μ and not just the Teichmüller class of μ.

This paper investigates how the embeddings induced by N_μ^2 and Q_μ^2 vary as μ varies. In Section 1 we review well-known facts about the mapping Φ_q.

In Section 2 we define the mapping N_μ^q and, following Bers [5], we define a fibre space F over Teichmüller space, each of whose fibres is a surface of fixed genus with a complex structure determined by the image point in Teichmüller space. Then we show how N_μ^q induces a natural mapping Ψ_q from F to $P(B_q(\Gamma,U)^*)$.

In Section 3, we show that $\Psi_2 : F \to P(B_2(\Gamma,U)^*)$ maps onto an open set in $P(B_2(\Gamma,U)^*)$.

In Section 4, we define the map Q_μ^q. As far as we know, this mapping really depends on μ and not just the Teichmüller class of μ. The only excuse we give for trying to give an analysis of it is that the analysis itself is interesting.

In Section 5 we show that Q_μ^2 induces embeddings of the surfaces S_μ in such a way that they fill out an open set containing the image of S in $P(B_2(\Gamma,U)^*)$.

Many of the computational devices employed here are taken from Ahlfors' paper [1] and the basic mappings used are modifications of ones first explicitly written down by Bers [4, page 129].

§1. *Facts about the map* Φ. In this section we present some facts about the map $\Phi_q : S \to P(B_q(\Gamma,U)^*)$ where $S = U/\Gamma$. All of them are classical and we do not give proofs. First of all, note that the fact that there always is a non-vanishing q-differential (for $q \geq 1$) at any point $p \in U/\Gamma$ means that Φ is defined everywhere on U/Γ.

Pick a point $p \in S = U/\Gamma$ and a local parameter z in a neighborhood of p such that $z(p) = 0$. $\{\phi_1, \ldots, \phi_n\}$ is called a basis of $B_q(\Gamma, U)$ adapted to p if it is a basis of $B_q(\Gamma, U)$ and the power series developments of the ϕ_i's have the form $\phi_i(z) = z^{\nu_i} + a_{1i} z^{\nu_i + 1} + \cdots$ for $1 \leq i \leq n$ and where $0 = \nu_1 < \nu_2 < \cdots < \nu_n$. The numbers ν_i are functions of p and do not depend on the local parameter z.

DEFINITION. *Unless each* $\nu_i(p) = i-1$ *for* $1 \leq i \leq n$, p *is called a* q-*Weierstrass point*.

THEOREM *(Weierstrass). Let* $W(p) = \Sigma_{i=1}^{n}(\nu_i(p)-(i-1))$. *Let the genus of* $S = g$. *Then*

$$\sum_{p \in S} W(p) = (2q-1)^2(g-1)^2 g \quad \textit{if } q > 1$$

and

$$\sum_{p \in S} W(p) = (g+1)g(g-1) \quad \textit{if } q = 1 .$$

In particular, the number of q*-Weierstrass points for fixed* q *is finite.*

DEFINITION. S *is hyperelliptic if for* $q = 1$ *there is a point* $p \in S$ *such that* $\nu_2(p) > 1$:

LEMMA. *If* S *is not hyperelliptic, then for all* $q \geq 1$ *and for all* $p \in S$, $\nu_2(p) = 1$.

THEOREM. *If* S *is not hyperelliptic, then the map* $\Phi_q : S \to P(B_q(\Gamma, U)^*)$ *has non-vanishing derivative for all* $q \geq 1$ *and all* $p \in S$. *In fact,* Φ_q' *vanishes precisely at those points where* $\nu_2(p) > 1$.

THEOREM. *The map* $\Phi_1 : S \to P(B_1(\Gamma, U)^*)$ *is an injection unless* S *is hyperelliptic and in this case* Φ_1 *is two-to-one.*
The map $\Phi_q : S \to P(B_q(\Gamma, U)^*)$ *for* $q > 1$ *is an injection unless* $q = g = 2$ *and, in this case, all surfaces are hyperelliptic and the mapping is two-to-one.*

§2. *The fibre space of Teichmüller space and the natural linear map* N_μ^q. In all of this section we are summarizing notation and results given in Bers' Zurich notes [5].

First, let us define Teichmüller space. Let Γ be a Fuchsian group operating on the upper half plane U. Much of what we say will apply in this setting but, of course, if we want the results of Section 1 to apply as stated, then we must assume $S = U/\Gamma$ is a compact Riemann surface of genus g. Let $M(\Gamma) = \{\mu \mid \mu$ is bounded measurable complex valued function on C, $\|\mu\|_\infty < 1$, $\mu(\bar{z}) = \overline{\mu(z)}$, $\mu(\gamma)\overline{\gamma'(z)}\gamma'(z)^{-1} = \mu(z)$ for all $\gamma \in \Gamma\}$. Let $\Sigma^*(\Gamma) = \{w_\mu \mid \mu \in M(\Gamma),\ \bar{\partial} w_\mu = \mu \partial w_\mu$ and $w_\mu(0) = 0$, $w_\mu(1) = 1$, $w_\mu(\infty) = \infty\}$. The existence and uniqueness theorem for Beltrami equation tells us that $M(\Gamma) \ni \mu \mapsto w_\mu \in \Sigma^*(\Gamma)$ is a bijection. Define a normal subgroup $\Sigma_*(\Gamma)$ by $\Sigma_*(\Gamma) = \{w_\mu \in \Sigma^*(\Gamma) \mid w_\mu(x) = x$ for all $x \in R\}$.

Teichmüller space is defined to be $T(\Gamma) = \Sigma^*(\Gamma)/\Sigma_*(\Gamma)$. It is well known that if we take $\tilde{\Sigma}^*(\Gamma)$ (respectively, $\tilde{\Sigma}_*(\Gamma)$) to be those elements of $\Sigma^*(\Gamma)$ (respectively, $\Sigma_*(\Gamma)$) which are smooth, (C^k), in the interior of U then $T(\Gamma)$ is also given by $\tilde{\Sigma}^*(\Gamma)/\tilde{\Sigma}_*(\Gamma)$, at least in the case when U/Γ is compact. We will need this fact later.

Define w^μ to be the unique quasiconformal homeomorphism of C which is normalized by $w^\mu(0) = 0$, $w^\mu(1) = 1$, $w^\mu(\infty) = \infty$ and which satisfies $\bar{\partial} w^\mu = \hat{\mu} \partial w^\mu$ where

$$\hat{\mu}(z) = \begin{cases} \mu(z) & \text{for } z \in U \\ 0 & \text{for } z \in L \end{cases}.$$

Fix the following notations: $U^\mu = w^\mu(U)$, $L^\mu = w^\mu(L)$, $C^\mu = w^\mu(R)$, and if $\mu \in M(\Gamma)$, $\Gamma^\mu = w^\mu \Gamma w^{\mu-1}$. It is well known (see [2]) that U^μ, C^μ, and L^μ depend only on the class of μ in $\Sigma^*(\Gamma)/\Sigma_*(\Gamma)$.

In [5], Bers has shown that a system of complex analytic local charts for $T(\Gamma)$ can be obtained in the following way. Let $B_2(\Gamma^\mu, L^\mu)$ = the holomorphic quadratic differentials on L^μ with respect to Γ^μ. Suppose $[w_{\mu_0}] \in \Sigma^*(\Gamma)/\Sigma_*(\Gamma)$ and $\sigma \in M(\Gamma)$. Define ν by the relation $w^\sigma = w^\nu \circ w^{\mu_0}$.

Clearly ν will have support in $U^{\mu_0} \cup C^{\mu_0}$ and w^{μ} will be holomorphic in L^{μ_0}. Let $<w^{\nu}>$ be the Schwarzian derivative of w^{ν} in L^{μ_0}. Then for a sufficiently small neighborhood of $[w_{\mu_0}]$ in $T(\Gamma)$, the mapping $[w_{\sigma}] \to <w^{\nu}> \epsilon B_2(\Gamma^{\mu_0}, L^{\mu_0})$ is a chart. And, in fact, the tangent space to $T(\Gamma)$ at the point $[w_{\mu_0}]$ can be identified with $B_2(\Gamma^{\mu_0}, L^{\mu_0})$.

Now we can define a fibre bundle F over $T(\Gamma)$ in such a way that the fibre over each point $[w_{\mu}]$ is a Riemann surface. $S^{\mu} = U^{\mu}/\Gamma^{\mu}$. Let $\tilde{F} = \{(z, [w_{\mu}]) \mid z \epsilon U^{\mu}$ and $[w_{\mu}] \epsilon T(\Gamma)\}$. To define charts for $\tilde{F}$ we proceed as follows. Let $(z_0, [\mu_0]) \epsilon \tilde{F}$. Let the distance from z_0 to the boundary of U^{μ_0} be 2ϵ. Then because $w^{\sigma}(z)$ is continuous in σ, there exists a $\delta > 0$ such that if $\|\nu\|_{\infty} < \delta$ then $w^{\nu}(B_{\varepsilon}(z_0)) \subseteq U^{\mu}$. Of course, $B_{\varepsilon}(z_0) = \{z \epsilon C \mid |z - z_0| < \epsilon\}$. Let $B_{\delta}([w_{\mu_0}]) = \{[w_{\sigma}] \mid w^{\sigma} = w^{\nu} \cdot w^{\mu_0}$, $w_{\sigma} \epsilon \Sigma^*(\Gamma)$, $\|\nu\| < \delta\}$.

Consider the neighborhood of $(z_0, [w_{\mu_0}])$ in $\tilde{F}$ given by $B_{\varepsilon}(z_0) \times B_{\delta}([w_{\mu_0}])$. In this neighborhood a chart is given by $(z, [w_{\sigma}]) \to (z, <w^{\nu}>)$. The choice of δ and ε that we have made assures that if $z \epsilon B_{\varepsilon}(z_0)$, then $z \epsilon U^{\sigma}$. The mapping $\tilde{\pi}: \tilde{F} \to T(\Gamma)$ is the obvious one: $\tilde{\pi}(z, [w_{\mu}]) = [w_{\mu}]$. There is a complex analytic action of Γ on $\tilde{F}$ with respect to which the above system of charts are automorphic. This action is given by $\gamma(z, [w_{\mu}]) = (\gamma^{\mu}(z), [w_{\mu}])$, where $\gamma^{\mu} = w^{\mu} \gamma w^{\mu-1} \epsilon \Gamma^{\mu}$. Hence we can form $F = \tilde{F}/\Gamma$ and we obtain a fibre bundle $\pi: F \to T(\Gamma)$ which has holomorphic structure.

Now it is time to define the mappings $N_{\mu}^{q}: B_q(\Gamma, U) \to B_q(\Gamma^{\mu}, U^{\mu})$. We will simply write down the formula and from the fact that the behaviour of w^{μ} in L depends only on the Teichmüller class of μ it will follow that N_{μ}^{q} depends only on the Teichmüller class of μ.

$$(N_{\mu}^{q}\phi)(z) = \int_L \frac{\phi(\zeta)\lambda^{2-2q}(\zeta)\, w^{\mu\prime}(\zeta)^q}{(w^{\mu}(\zeta) - z)^{2q}}\, d\zeta d\bar{\zeta} \ .$$

Here z is an element of U^{μ} and λ is the Poincaré metric for $L: \lambda(\zeta) = 1/|\zeta-\bar{\zeta}|$. By Theorem 2 of [6] one can see that in the case $q = 2$ this formula is nearly the same as the derivative of right translation by $[w_{\mu}^{-1}]$ at the point $[w_{\mu}]$ if representations of the fibre of the tangent space to $T(\Gamma)$ at $[w_{\mu}]$ and 0 are chosen properly. We do not need this result for what follows so we shall not attempt to explain it. From Bers [5] we know that $N_{\mu}^{q}: B_q(\Gamma,U) \to B_q(\Gamma^{\mu},U^{\mu})$ is a bijective linear mapping.

We can now proceed to study an induced family of mappings of the surfaces S^{μ} in $P(B_q(\Gamma, U)^*)$. These mappings will be embeddings in almost all cases. Section 1 describes the exceptions when U/Γ is compact of genus g. The family of mappings will vary holomorphically. Define $\Psi_q: F \to P(B_q(\Gamma, U)^*)$ by $\Psi_q(z, [w_{\mu}])$ = the projective class of the non-trivial linear functional $\ell_{z,\mu}$ where $\ell_{z,\mu}(\phi) = (N_{\mu}^{q}\phi)(z)$. The projective class of $\ell_{z,\mu}$ depends only on $[w_{\mu}]$ and the images of z in U^{μ}/Γ^{μ} so the mapping Ψ_q is well defined on F.

It is clear that Ψ_q restricted to $\pi^{-1}([w_{\mu}]) = U^{\mu}/\Gamma^{\mu}$ has all the same properties as the Φ_q described in Section 1 because it differs from it only by composition with the complex linear bijective map N_{μ}^{q}.

§3. *The variation of the map* $\Psi_q: F \to P(B_q(\Gamma,U)^*)$. From the definitions made in Section 2 it is clear that Ψ_q lifts to a mapping $\tilde{\Psi}_q: \tilde{F} \to B_q(\Gamma,U)^* - \{0\}$ if we define $\tilde{\Psi}_q(z,[w_{\mu}]) = \ell_{z,\mu}$. Here $\ell_{z,\mu}$ is the linear functional on $B_q(\Gamma,U)$ defined by $\ell_{z,\mu}(\phi) = (N_{\mu}^{q}\phi)(z)$.

Let $(z_0, [w_{\mu_0}]) \in \tilde{F}$ and let $B_{\varepsilon}(z_0) \times B_{\delta}([w_{\mu_0}])$ be a neighborhood of this point in $\tilde{F}$. Recall that from Section 2

$$B_{\delta}([w_{\mu_0}]) = \{[w_{\sigma}] \in T(\Gamma) \mid w^{\sigma} = w^{\nu} \circ w^{\mu_0} \text{ and } \|\nu\|_{\infty} < \delta\} .$$

LEMMA 1. *The map* $\tilde{\Psi}_q(z,[w_{\mu}])$ *is holomorphic on* $\tilde{F}$.

Proof: Because of our definition of the charts of $\tilde{F}$, we must show that the map which sends (z, w^{σ}) into $(N_{\sigma}^{q}\phi)(z)$ is a holomorphic function of z and ν where $w^{\sigma} = w^{\nu} \circ w^{\mu_0}$ and $|z-z_0| < \varepsilon$ and $\|\nu\|_{\infty} < \delta$. This amounts to saying that

$$\int_L \frac{\phi(\zeta)\lambda^{2-2q}(\zeta)w^{\sigma\prime}(\zeta)^q d\zeta \overline{d\zeta}}{(w^\nu(\zeta)-z)^{2q}}$$

$$= \int_{L^{\mu_0}} \frac{\overline{(w^{\mu_0 - 1})_q^* j\phi(\zeta)}\lambda_{\mu_0}^{2-2q}(\zeta)\, w^{\nu\prime}(\zeta)^q d\zeta\overline{d\zeta}}{(w^\nu(\zeta)-z)^{2q}}$$

is a holomorphic function of z and ν. In the latter integral we have used the following notations:

(1) $$j : B_q(\Gamma, U) \to B_q(\Gamma, L) \text{ by } (j\phi)(z) = \overline{\phi(\bar{z})} \quad ,$$

(2) $$(w^{\mu_0 -1})_q^* : B_q(\Gamma, L) \to B_q(\Gamma^{\mu_0}, L^{\mu_0})$$

by

$$(w^{\mu_0 -1})_q^* \phi(z) = \phi(w^{\mu_0 -1}(z))\left(\frac{d}{dz} w^{\mu_0 -1}(z)\right)^q \quad ,$$

and

(3) $\lambda_{\mu_0}(\zeta)$ is the Poincaré metric for L^{μ_0} .

We know from the theory of the Beltrami equation and Bers [4] that the value of the latter integral of the above equation varies holomorphically with ν and z so long as $z \in U^{\mu_0}$. But we have ensured that $z \in U^{\mu_0}$ by our choice choice of ϵ . Q.E.D.

In the following theorem let $P = P(B_2(\Gamma, U)^*)$, $M = M(\Gamma)$, $T = T(\Gamma)$ and α be the natural projection of M onto T.

THEOREM. *At each point* $\mu_0 \in M$ *and* $z_0 \in U^{\mu_0}/\Gamma^{\mu_0}$ *there is a section* S *defined on a neighborhood* N *of* μ_0 *such that* $S(\mu_0) = (z_0, [w_{\mu_0}])$ *and such that*

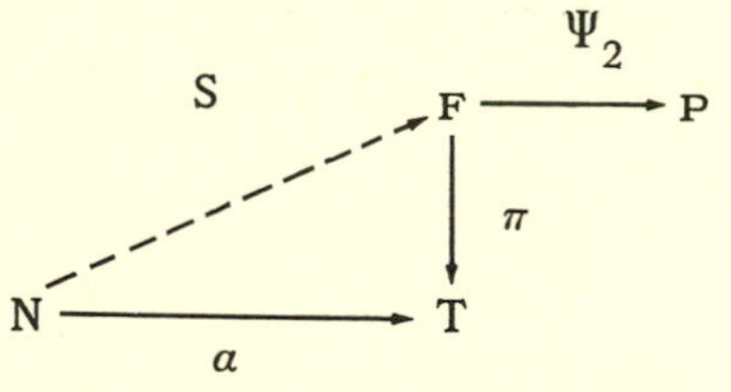

commutes and the mapping $\Psi_2 \circ S$ *has surjective derivative at* μ_0. *In particular,* $\Psi_2 : F \to P$ *is an open mapping.*

The proof will involve several steps and lemmas. Given $\sigma \in M$ which satisfies $w^\sigma = w^\nu \circ w^{\mu_0}$ we define $S(\sigma) = (w^\nu(z_0), [w_\sigma])$. Note that it suffices to prove that the map $\nu \to$ (projective class of $\ell_{w^\nu(z_0),\sigma}$) has a surjective derivative at $\nu = 0$. We do not change the projective class of $\ell_{w^\nu(z_0),\sigma}$ if we multiply it by $\partial w^\nu(z_0)^2$ and we can assume that $\partial w^\nu(z_0)$ is some well-defined number because we can assume that μ is smooth in U^{μ_0}.

By an obvious manipulation of the formula used in Lemma 1, we see that the above problem will be solved if we show the following lemma.

LEMMA 2. *Let* h_{ν,z_0} *be the conjugate complex linear map on* $B_2(\Gamma^{\mu_0}, L^{\mu_0})$ *defined by*

$$h_{\nu,z_0}(\psi) = \int_{L^{\mu_0}} \frac{\overline{\psi(\zeta)}\lambda_{\mu_0}(\zeta)\partial w^\nu(\zeta)^2 \partial w^\nu(z_0)^2}{(w^\nu(\zeta) - w^\nu(z_0))^4} d\zeta\, d\zeta \quad .$$

Then as ν *varies in* $\{\nu \mid \|\nu\|_\infty < \delta$, $\operatorname{supp}\nu \subseteq U^{\mu_0} \cup C^{\mu_0}$ *and*

$\nu(\gamma^\mu(z))\overline{\gamma^{\mu'}(z)} = \gamma^{\mu'}(z)\nu(z)$ *for all* $\gamma^\mu \in \Gamma^\mu\}$, h_{ν,z_0} *fills out a neighborhood of* h_{0,z_0} *in the conjugate dual space to* $B_2(\Gamma^{\mu_0}, L^{\mu_0})$.

To prove Lemma 2 we will use several ideas of Ahlfors in [1]. Let $K(z,\zeta) = 1/(z-\zeta)^2$ and $K^\mu(z,\zeta) = K(w^\mu(z), w^\mu(\zeta))\partial w^\mu(z)\partial w^\mu(\zeta)$. Then

$$h_{\mu,z_0}(\psi) = \int_{L^{\mu_0}} \overline{\psi(\zeta)}\lambda_{\mu_0}^{-2}(\zeta) K^\mu(\zeta, z_0)^2 \, d\zeta\, d\overline{\zeta} \quad .$$

LEMMA 3. *Let* $\operatorname{supp}\mu \subseteq U^{\mu_0} \cup C^{\mu_0}$. *Let* $\zeta \in L^{\mu_0}$ *and* $z_0 \in U^{\mu_0}$. *Then* $K^{\mu}(\zeta, z_0)$ *is a complex holomorphic function of* μ *and*

$$\frac{d}{dt} K^{t\mu}(\zeta, z_0)\bigg|_{t=0} = \frac{1}{2\pi i} \int_{U^{\mu_0}} K(\zeta,\eta)\, K(\eta, z_0)\, \mu(\eta)\, d\eta\, \overline{d\eta} \quad ,$$

where t *is a complex variable.*

This formula can be viewed as a consequence of one derived by Ahlfors in [1, page 170, (7.3)]. Here we shall derive it using the power series technique. There is no loss of generality in assuming that μ has compact support because the formula is invariant under application of linear fractional mappings A if we establish the convention that $K(A(\zeta), A(\eta))\, A'(\zeta)\, A'(\eta) = K_D(\zeta,\eta)$ where $A(D) = U$. And we can arrange for D to be the interior of the unit circle by picking A to be defined by $A(z) = -i(z+i)/(z-i)$.

Define (as in [2]) for ν with compact support

$$P\nu(z) = \frac{1}{2\pi i} \int_C \frac{\nu(\zeta)\, d\zeta\, \overline{d\zeta}}{(\zeta - z)}$$

and

$$T\nu(z) = \frac{1}{2\pi i}\ \text{p.v.} \int_C \frac{\nu(\zeta)\, d\zeta\, \overline{d\zeta}}{(\zeta - z)^2}$$

Then it is proved in [3] that

$$w^{\nu}(z) = \frac{z + P\nu(z) + P\nu\, T\nu(z) + \cdots}{1 + P\nu\, 1 + P\nu\, 1 + \cdots} \quad .$$

It is easy to see that the denominator in this expression will cancel when we differentiate $K^{t\nu}$ at $t = 0$. In fact,

$$K^{t\nu}(z,\zeta) - K(z,\zeta) = \frac{(1 + tT\nu(z) + o(t)(1 + T\nu(\zeta) + o(t)}{(z-\zeta + t(P\nu(z) - P\nu(\zeta)) + o(t))^2} - \frac{1}{(z-\zeta)^2}$$

$$= \frac{1}{(z-\zeta)^2}\left[T\nu z + T\nu\zeta - 2\frac{P\nu z - P\nu\zeta}{z-\zeta} + o(1)\right] .$$

Hence,

$$\frac{d}{dt}K^{t\nu}(z\zeta)\Big|_{t=0} = \frac{1}{(z-\zeta)^2}\left[T\nu z + T\nu\zeta - 2\frac{P\nu z - P\nu\zeta}{z-\zeta}\right] .$$

If we collect the integrals P and T all into one integral in this expression, a simple calculation shows that the resulting kernel is

$$\frac{1}{(z-\zeta)^2}\,\frac{1}{(\eta-z)^2} + \frac{1}{(\eta-\zeta)^2} - \frac{z}{(z-\zeta)}\left(\frac{1}{\eta-z}\,\frac{1}{\eta-\zeta}\right) = \frac{1}{(\eta-z)^2}\frac{1}{(\eta-\zeta)^2}$$

This shows that

$$\frac{d}{dt}K^{t\nu}(z\zeta)\Big|_{t=0} = \frac{1}{2\pi i}\int_C K(z\eta)\,K(\eta\zeta)\,\nu(\eta)\,d\eta\,\overline{d\eta}$$

and completes the proof of Lemma 3 because $\operatorname{supp}\nu \subseteq U^{\mu_0} \cup C^{\mu_0}$.

To prove Lemma 2 we start by differentiating the mapping $\nu \to h_{\nu,z_0}$ at $\nu = 0$. One finds that

$$\frac{d}{dt}h_{t\nu,z_0}\Big|_{t=0}(\psi)$$

$$= \frac{1}{\pi i}\int_{L^{\mu_0}} K(\zeta,z_0)\overline{\psi(\zeta)}\lambda_{\mu_0}^{-2}(\zeta)\int_{U^{\mu_0}} K(\zeta,\eta)\,K(\eta,z_0)\,\nu(\eta)\,d\eta\,\overline{d\eta}\,d\zeta\,\overline{d\zeta} .$$

In this expression η varies in U^{μ_0}, ζ varies in L^{μ_0}, and z_0 is a fixed point in U^{μ_0}. What we want to show is that as ν varies in $L_\infty(\Gamma^{\mu_0}, U^{\mu_0}) = \{\rho \mid \operatorname{supp}\rho \subset U^{\mu_0} \cup C^{\mu_0}, \rho(\gamma^{\mu}(z))\overline{\gamma^{\mu'}(z)} = \gamma^{\mu'}(z)\rho(z), \|\rho\|_\infty < \infty\}$, then

$$\frac{d}{dt} h_{t\nu, z_0}\Big|_{t=0}$$

yields all conjugate linear functionals on $B_2(\Gamma^{\mu_0}, L^{\mu_0})$. Hence, we must show that if

$$\frac{d}{dt} h_{t\nu, z_0}\Big|_{t=0} (\psi) = 0$$

for all such ν then $\psi = 0$. After changing the order of integration (which we will justify later) one sees that this reduces to showing that if

$$(*) \qquad \int_{L^{\mu_0}} K(\zeta, z_0)\, K(\zeta, \nu)\, \lambda_{\mu_0}^{-2}(\zeta)\, \overline{\psi(\zeta)}\, d\zeta\, \overline{d\zeta} = 0 \quad \text{for all } \eta \in U^{\mu_0}$$

then $\psi = 0$.

Define a mapping $\mathfrak{L}: B_2(\Gamma^{\mu_0}, L^{\mu_0}) \to B_2(\Gamma^{\mu_0}, U^{\mu_0})$ by

$$\mathfrak{L}\psi(z) = \int_{L^{\mu_0}} K(\zeta, z)^2 \lambda_{\mu_0}^{-2}(\zeta)\, \overline{\psi(\zeta)}\, d\zeta\, \overline{d\zeta} \ .$$

We know from [4] and [5] that $\mathfrak{L}$ is a bijection. Hence, to show $\psi = 0$ it suffices to show $\mathfrak{L}\psi \equiv 0$. The formula (*) evaluated at $\eta = z_0$ tells us that $\mathfrak{L}\psi(z_0) = 0$. If we differentiate (*) k times with respect to η and evaluate at $\eta = z_0$ we find that (*) tells us that

$$\left(\frac{d}{dt}\right)^k \mathfrak{L}\psi(z_0) = 0 \ .$$

Hence, $\mathfrak{L}\psi \equiv 0$ in U^{μ_0}.

To justify the change in order of integration and the application of duality between L_1 and L_∞ we restrict attention to those elements ν of $L_\infty(\Gamma^{\mu_0}, U^{\mu_0})$ such that $\operatorname{supp} \nu \subseteq U^{\mu_0} \cup C^{\mu_0} - B_a(z_0)$ where $a > 0$. Then the whole integrand is absolutely integrable as a function of ζ and η over $L^{\mu_0} \times U^{\mu_0}$. Since a is arbitrary we easily see that (*) is valid.

This completes the proof of Lemma 2 because of the inverse function theorem. Therefore, Theorem 1 also is proved.

Summary. The image of $\Psi_2 : F \to P(B_2(\Gamma, U)^*)$ is an open set. Recall that $\pi : F \to T(\Gamma)$. If genus $(U/\Gamma) > 2$ then Ψ_2 restricted to $\pi^{-1}([w_{\mu_0}])$ is an embedding for every $[w_{\mu_0}] \in T(\Gamma)$. Moreover, Ψ_2 is a holomorphic mapping.

§4. *The mapping* Q^q_μ. Just as with N^q_μ, the idea for constructing Q^q_μ is motivated by computing the derivative of the right translation map between $T(\Gamma)$ and $T(\Gamma_\mu)$.

Let Γ be a covering group for a compact surface S of genus g operating on L. So $S = L/\Gamma$. Let $[w_\mu] \in T(\Gamma)$ and $\Gamma_\mu = w_\mu \Gamma w_\mu^{-1}$. Then there is a translation map $R^\mu : T(\Gamma_\mu) \to T(\Gamma)$ defined by $R^\mu([w_\sigma]) = [w_\sigma \circ w_\mu]$. If we let μ be of the form $\lambda^{-2}(\zeta)\psi(\overline{\zeta})$ for some quadratic differential ψ, then it is an easy computation to show that

$$(dR^{\mu-1})_\mu(\phi)(z) = \int_U \frac{\partial w_\mu(\zeta)^2 \phi(\overline{\zeta})\, d\zeta\, \overline{d\zeta}}{(w_\mu(\zeta) - z)^4} \quad .$$

Clearly this formula still has meaning even if we allow μ to be an arbitrary element of $M(\Gamma)$.

We define $Q^q_\mu : B_q(\Gamma, L) \to B_q(\Gamma_\mu, L)$ by

$$Q^q_\mu(\phi)(z) = \int_U \frac{\lambda^{2-2q}(\zeta)\phi(\overline{\zeta})\partial w_\mu(\zeta)^q d\zeta\, \overline{d\zeta}}{(w_\mu(\zeta) - z)^{2q}} \quad , \text{ where } q \geq 2.$$

It is a simple matter to check that $\mu \in M(\Gamma)$ and $\phi \in B_q(\Gamma, L)$ implies $Q^q_\mu \phi \in B_q(\Gamma_\mu, L)$. Therefore, for each μ one obtains an embedding of L/Γ_μ into $P(B_q(\Gamma, L)^*)$. Define $\Xi^q : L/\Gamma \times M(\Gamma) \to P(B_q(\Gamma, L)^*)$ by saying $\Xi^q(z, \mu)$ is equal to the projective class of $\ell_{\mu,z}$ where $\ell_{\mu,z}(\phi) =$ (evaluation at $w_\mu(z)$ of $Q^q_\mu \phi$). Then

$$\ell_{\mu,z}(\phi) = \int_U \frac{\lambda^{2-2q}(\zeta)\phi(\overline{\zeta})\partial w_\mu(\zeta)^q}{(w_\mu(\zeta) - w_\mu(z))^{2q}}\, d\zeta\, \overline{d\zeta} \ .$$

We do not change the projective class of $\ell_{\mu,z}$ if we multiply it by $\partial w_\mu(z)^q$ and for computational purposes it will be convenient to do this.

§5. *Variation of* $\Xi^2 : L/\Gamma \times M(\Gamma) \to P(B_2(\Gamma,L)^*)$. For technical reasons in this section $M(\Gamma) = \{\mu : L \to \mathbf{C} \mid \mu$ has continuous first derivatives in the interior of L, $\|\mu\|_\infty < 1$, and $\mu(\gamma(z))\overline{\gamma'(z)} = \gamma'(z)^{-1}\mu(z)$ for all $\gamma \in \Gamma\}$.

THEOREM. (a) *Let* Ξ^2_μ *be the restriction of* Ξ^2 *to* $L/\Gamma \times \{\mu\}$. *Then if* $g > 2$, Ξ^2_μ *induces a quasiconformal embedding of* L/Γ *into* $P(B_2(\Gamma,L)^*)$ *and a holomorphic embedding of* $L/w_\mu \Gamma w_\mu^{-1}$.

(b) *The map* Ξ^2 *is continuously differentiable on* $L/\Gamma \times M(\Gamma)$.

(c) *The image of* Ξ^2 *in* $P(B_2(\Gamma,L)^*)$ *contains an open set in* $P(B_2(\Gamma,L)^*)$ *which contains the image of* Ξ^2_0.

Remark: The image of Ξ^2_0 is just the image of the canonical map of L/Γ into $P(B_2(\Gamma,L)^*)$.

Part (a) of this theorem is obvious because $\Xi^2_\mu \circ w_\mu^{-1}$ is just the canonical map of $L/w_\mu \Gamma w_\mu^{-1}$ into $P(B_2(\Gamma_\mu,L)^*)$ composed with the map Q^2_μ which is linear.

Part (b) is obvious as soon as one observes that $\tilde{\Xi} : L \times M(\Gamma) \to B_2(\Gamma,L)^*$ defined by

$$\tilde{\Xi}(z,\mu) = \left(\phi \to \int_U \frac{\lambda^2(\zeta)\,\phi(\overline{\zeta})\,\partial w_\mu(\zeta)^2 \partial w_\mu(z)^2}{(w_\mu(\zeta) - w_\mu(z))^4}\, d\zeta\, \overline{d\zeta}\right)$$

is continuously differentiable as a function of (z,μ). This is true because of well-known properties of the Beltrami equation and because we have chosen μ to be smooth.

To prove part (c) we will compute

$$\left.\frac{d}{dt}\tilde{\Xi}(z_0, t\mu)\right|_{t=0}$$

where t is a real variable. Instead of doing this directly it is easier to find the complex linear and conjugate complex linear parts of this derivative. In general, if F is a function of μ, set

$$\dot{F}[\mu] = \frac{d}{dt} F(t\mu)\bigg|_{t=0} \quad \text{where } t \in \mathbf{R}$$

and

$$DF[\mu] = \tfrac{1}{2}(\dot{F}[\mu] - i\dot{F}[i\mu])$$

$$\bar{D}F[\mu] = \tfrac{1}{2}(\dot{F}[\mu] + i\dot{F}[i\mu]) \ .$$

The first step involved in finding these derivatives is to find the derivative of $K_\mu(\zeta,z) = K(w_\mu(\zeta), w_\mu(z))\partial w_\mu(\zeta)\partial w_\mu(z)$ where $K(\zeta,z) = 1/(\zeta - z)^2$. Then Ahlfors in [1, page 170, (7.3)] has shown that

$$DK(z,\zeta)[\mu] = \frac{1}{2\pi i}\int_L K(z,\eta)K(\eta,\zeta)\mu(\eta)\,d\eta\,\overline{d\eta}$$

and

$$\bar{D}K(z,\zeta)[\mu] = \frac{1}{2\pi i}\int_U K(z,\eta)K(\eta,\zeta)\overline{\mu(\eta)}\,d\eta\,\overline{d\eta} \ .$$

These formulas are valid if interpreted as Cauchy principal values. With the help of these results one finds that

$$D\tilde{\Xi}(z_0,0)[\nu] = \Big(\phi \mapsto \frac{1}{\pi i}\int_L \nu(\eta)K(\eta,z_0) \tag{1}$$

$$\cdot \int_U K(\eta,\zeta)K(\zeta,z_0)\lambda^{-2}(\zeta)\phi(\zeta)\,d\zeta\,\overline{d\zeta}\,d\eta\,\overline{d\eta}\Big) \ .$$

$$D\tilde{\Xi}(z_0,0)[\nu] = \Big(\phi \mapsto \frac{1}{\pi i}\int_U \overline{\nu(\bar{\eta})}K(\eta,z_0) \tag{2}$$

$$\cdot \int_U K(\eta,\zeta)K(\zeta,z_0)\lambda^{-2}(\zeta)\phi(\zeta)\,d\zeta\,\overline{d\zeta}\,d\eta\,\overline{d\eta}\Big) \ .$$

To arrive at these expressions one must change the order of integration. Justification for this step is subtle and involves use of the fact that the kernel K gives the Hilbert transform which is an isometry. For a similar argument see Ahlfors [2, page 171]; we will not give the argument here.

One simple trick we can employ to avoid difficulties with equation (1) is to let $\nu \equiv 0$ in an ε-neighborhood of z_0 and then, at the end of the argument, allow ε to approach 0.

It is clear that we will show

$$\frac{d}{dt}\tilde{\Xi}(z_0, t\nu)\Big|_{t=0}$$

is surjective if we prove two facts:

(1) $$D\tilde{\Xi}(z_0, 0)[\nu](\phi) = 0 \quad \text{for all } \nu \implies \phi = 0$$

and

(II) $$D\tilde{\Xi}(z_0, 0)[\nu](\phi) = 0 \quad \text{for all } \nu \implies \phi = 0 .$$

The proof of statement II is most interesting. By duality between L_1 and L_∞ and Equation (2) what we must show is that if

$$\int_U K(\eta,\zeta)\,K(\zeta,z_0)\,\lambda^{-2}(\zeta)\,\delta(\overline{\zeta})\,d\zeta\,d\overline{\zeta} = 0 \quad \text{for all } \eta \in U \text{ then } \phi \equiv 0.$$

To this end, let

$$h(\zeta) = \begin{cases} \dfrac{\lambda^{-2}(\zeta)\,\phi(\overline{\zeta})}{(\zeta - z_0)^2} & \text{for } \zeta \in U \\[2ex] 0 & \text{for } \zeta \in L . \end{cases}$$

Let P and T be the singular integral operators defined on page 10. For sufficiently well behaved functions h they are related by

$$\frac{\partial\, P\,h(\eta)}{\partial\eta} = Th(\eta) \quad \text{and} \quad \frac{\partial\, P\,h(\eta)}{\partial\overline{\eta}} = h(\eta) .$$

Hence, the hypothesis to II tells us that

$$\frac{\partial P h(\eta)}{\partial \eta} = 0 \quad \text{for } \eta \in U,$$

and by definition $h(\eta) = 0$ for $\eta \in L$ so

$$\frac{\partial}{\partial \bar{\eta}} P h(\eta) = 0 \qquad \text{for} \quad \eta \in L.$$

Therefore $Ph(\eta)$ is holomorphic in L and antiholomorphic in U. Moreover, $Ph(\eta)$ is continuous everywhere because it satisfies a Hölder condition, (see [2]). If $\eta \in L$, $\bar{\eta} \in U$ and $Ph(\bar{\eta})$ is holomorphic in L. So is $Ph(\eta)$ and $Ph(\eta) = Ph(\bar{\eta})$ for $\eta \in R$. We conclude that $Ph(\eta) = Ph(\bar{\eta})$ everywhere. Therefore, for $\eta \in U$,

$$h(\eta) = \frac{\partial}{\partial \bar{\eta}} Ph(\eta) = \frac{\partial}{\partial \bar{\eta}} Ph(\bar{\eta}) = \overline{\frac{\partial}{\partial \eta} Ph(\eta)} = \overline{Th(\eta) = 0} \quad .$$

Hence $\phi = 0$.

To prove I, we must show that if

$$\int_U K(\eta,\zeta)\, K(\zeta,z_0)\, \lambda^{-2}(\zeta) \phi(\overline{\zeta})\, d\zeta\, d\bar{\zeta} = 0 \qquad \text{for all } \eta \in L$$
$$\text{then } \phi \equiv 0 \,.$$

By letting $\eta = z_0$ we find immediately by use of the reproducing formula that $\phi(z_0) = 0$, and by looking at derivatives of order k, we find

$$\frac{\partial^k}{\partial z^k} \phi(z_0) = 0 \quad \text{for all } k \geq 0 \,.$$

Thus $\phi \equiv 0$ and the theorem is proved.

BIBLIOGRAPHY

[1] Ahlfors, L., Curvature properties of Teichmüller's space, *Journal d'Analyse*, vol. 9 (1961), pp. 161-176.

[2] ———, *Lectures on Quasiconformal Mapping*, Van Nostrand, Mathematical Studies no. 10, Princeton, N.J., 1966.

[3] Ahlfors, L. and Bers. L., Riemann's mapping theorem for variable metrics, *Annals of Mathematics*, vol. 72 (1960), pp. 385-404.

[4] Bers, L., A non-standard integral equation with applications to quasiconformal mappings, *Acta Mathematica*, vol. 116 (1966), pp. 113-134.

[5] ———, On moduli of Riemann surfaces, *Notes from Eidgenossiche Technische Hochschule*, Zurich, summer term, 1964.

[6] Gardiner, F., An analysis of the group operation in universal Teichmüller space, *Trans. of Am. Math. Soc.*, vol. 132, number 2 (1968), pp. 471-486.

LIPSCHITZ MAPPINGS AND THE p-CAPACITY OF RINGS IN n-SPACE

F. W. Gehring[1]

1. Introduction. Given a domain D in euclidean n-space R^n and given $1 \leq p < \infty$, we let $N_p(D)$ denote the collection of all continuous complex-valued functions u defined on D which are absolutely continuous in the sense of Tonelli or ACT in D with

$$\|u\|_p = \sup_D |u| + \left(\int_D |\nabla u|^p \, dm_n\right)^{1/p} < \infty .$$

Then $N_p(D)$ is a Banach algebra under pointwise addition and multiplication with $\| \ \|_p$ as norm. We call it the *Royden* p-*algebra* of D.

Suppose next that D and D′ are domains in R^n. In 1960, M. Nakai [13] showed that when $n = 2$, the algebras $N_n(D)$ and $N_n(D')$ are isomorphic if and only if there exists a quasiconformal mapping of D onto D′. Recently L. G. Lewis [9] obtained the same result for the case where $n > 2$. Lewis also proved that when $n \geq 2$ and $1 < p < n$, $N_p(D)$ and $N_p(D')$ are isomorphic if and only if there exists a bi-Lipschitz homeomorphism of D onto D′. The proof of this last result appears to depend crucially on the fact that when $p \neq n$, a homeomorphism which, together with its inverse, does not increase the p-capacity of spherical rings by more than a fixed factor is bi-Lipschitzian.

In the present paper, we study homeomorphisms which do not increase the p-capacity of spherical rings by more than a fixed factor in order to supply a proof of the above mentioned fact in Theorem 2.

---

[1] This research was supported in part by the National Science Foundation, Contract GP-7234.

For an investigation of a related class of homeomorphisms, see [14].

2. **Notation.** We consider sets in euclidean n-space R^n, $n \geq 2$, and in its one point compactification $\overline{R}^n$ obtained by adding the point ∞. Points in R^n are designated by the capital letter P or by small letters x and y. In the latter case, x_i will always denote the i-th coordinate of x. Points in R^n are treated as vectors, and $|P|$ and $|x|$ will denote the norms of P and x.

For each set E we let ∂E, $\overline{E}$, and C(E) denote respectively the boundary, closure, and complement of E in $\overline{R}^n$. Next for $1 \leq k \leq n$, we let m_k denote the k-dimensional Hausdorff measure in R^n; in particular, m_n will denote Lebesgue measure in R^n. We also let

$$\omega_n = m_{n-1}(\partial B^n), \qquad \tau_n = m_n(B^n) ,$$

where B^n denotes the unit ball $\{x: |x| < 1\}$ in R^n. Thus $\omega_n = n\tau_n$.

A domain $R \subset R^n$ is said to be a *ring* if C(R) has exactly two components C_0 and C_1. For convenience of notation we shall always assume that $\infty \in C_1$. Then for $1 \leq p < \infty$ we define the p-capacity of R as

$$(1) \qquad \text{cap}_p(R) = \inf_u \int_R |\nabla u|^p dm_n , \qquad \nabla u = \left(\frac{\partial u}{\partial x_1}, \ldots, \frac{\partial u}{\partial x_n} \right) ,$$

where the infimum is taken over all functions u which are continuous in $\overline{R}^n$ and ACT in R^n with $u = 0$ on C_0 and $u = 1$ on C_1. We call each such u an *admissible function* for R. We say that u is a *simple admissible function* if, in addition, $0 \leq u \leq 1$ in R and the set where $0 < u < 1$ is contained in the finite union of closed n-simplices $T_i \subset R$, in each of which u is linear. Then arguing as in [5] or [12], it is not difficult to see that (1) still holds when the infimum is taken over the subclass of all simple admissible functions u.

A ring R is said to be a spherical ring if it is bounded by concentric spheres, that is, if $R = \{x: a < |x - P| < b\}$ where $0 < a < b < \infty$ and $P \in R^n$. In this case it is easy to verify that

(2) $$\mathrm{cap}_p(R) = \begin{cases} \omega_n(\int_a^b r^{q-1}\,dr)^{1-p}, & q = \frac{p-n}{p-1}, \text{ if } p > 1, \\ \omega_n\, a^{n-1} & \text{if } p = 1 . \end{cases}$$

Suppose next that D and D′ are domains in R^n and that f is a homeomorphism of D onto D′. Then f maps each ring $R \subset D$ onto a ring $f(R) \subset D'$. We say that $f \in Q_p(K)$, $0 < K < \infty$, if

(3) $$\mathrm{cap}_p(f(R)) \leq K\ \mathrm{cap}_p(R)$$

for all spherical rings R with $\bar{R} \subset D$. When $p = n$, a homeomorphism is in $Q_p(K)$ for some K if and only if it is a quasiconformal mapping. (See [6], [12], or [15].) Next for each $P \in D$ we set

$$L(P,f) = \limsup_{r \to 0} \frac{|f(x)-f(P)|}{r}, \qquad \ell(P,f) = \liminf_{r \to 0} \frac{|f(x)-f(P)|}{r}$$

$$H(P,f) = \limsup_{r \to 0} \frac{|f(x)-f(P)|}{|f(y)-f(P)|}, \qquad J(P,f) = \limsup_{r \to 0} \frac{m_n(f(U))}{m_n(U)},$$

where $|x-P| = |y-P| = r$ and $U = \{x : |x-P| < r\}$. We say that $f \in \mathrm{Lip}(K)$, $0 < K < \infty$, if $L(P,f) \leq K$ in D.

The purpose of this paper is to establish some relations between the classes $Q_p(K)$ and Lip(K). These are given in Theorems 2 and 3 in section 5.

3. Bounds for the p-capacity. We collect in this section several different estimates for the p-capacity of rings.

We begin with some extremal length bounds. Given a ring R with non-degenerate complementary components C_0 and C_1, we set

$$L(R) = \mathrm{dist}(C_0, C_1), \qquad A(R) = \inf_{\Sigma} m_{n-1}(\Sigma), \qquad V(R) = m_n(R),$$

where the infimum is taken over all polyhedral surfaces $\Sigma \subset R$ which separate C_0 and C_1.

LEMMA 1. *If* R *is a ring with nondegenerate complementary components, then*

$$\frac{A(R)^p}{V(R)^{p-1}} \leq \mathrm{cap}_p(R) \leq \frac{V(R)}{L(R)^p} \quad . \tag{4}$$

Proof: Let u be an arbitrary simple admissible function for R. Then for all but a finite set of $t \in (0, 1)$, $u^{-1}(t)$ is a polyhedral surface in R which separates C_0 and C_1, whence

$$A(R) \leq \int_{u^{-1}(t)} dm_{n-1} \quad .$$

The co-area formula (see Theorem 3.1 in [4]) and Hölder's inequality imply that

$$A(R) \leq \int_0^1 \left(\int_{u^{-1}(t)} dm_{n-1} \right) dt = \int_R |\nabla u| \, dm_n \leq \left(\int_R |\nabla u|^p dm_n \right)^{1/p} V(R)^{\frac{p-1}{p}}$$

and taking the infimum over all such u yields the first part of (4). For the second part, let

$$u(x) = \min\left(1, \frac{\mathrm{dist}(x, C_0)}{L(R)}\right), \qquad u(\infty) = 1 \ .$$

Then u is admissible for R, $|\nabla u| \leq \frac{1}{L(R)}$ a.e. in R, and hence

$$\mathrm{cap}_p(R) \leq \int_R |\nabla u|^p dm_n \leq \frac{V(R)}{L(R)^p} \quad .$$

The following result is the analogue for p-capacity of a well known inequality due to Carleman [2].

LEMMA 2. *If* R *is a ring with complementary components* C_0 *and* C_1, *if* R^* *is a spherical ring with complementary components* C_0^* *and* C_1^*, *and if* $m_n(C_0^*) = m_n(C_0)$ *and* $m_n(C_0^* \cup R^*) = m_n(C_0 \cup R)$, *then*

$$\text{cap}_p(R) \geq \text{cap}_p(R^*) \ .$$

Proof: This inequality was obtained in [5] for the special case where $n = p = 3$ by means of point symmetrization. Moreover this argument can be easily modified to cover the case where $n \geq 2$ and $p \geq 1$, since it depends neither on the dimension n nor, except for an application of Hölder's inequality, on the exponent p. An alternative proof, which does not involve symmetrization, for an even more general inequality is given in [8].

We have another quite different lower bound for the p-capacity of a ring when $p > n-1$.

LEMMA 3. *If* R *is a ring whose complementary components* C_0 *and* C_1 *both meet a sphere with radius* $r > 0$ *and center* $P \in C_0$ *and if* $p > n-1$, *then*

$$\text{cap}_p(R) \geq c\, r^{n-p} \,, \tag{5}$$

where c *is a positive constant which depends only on* n *and* p.

Proof: By performing a preliminary translation, we may assume that $P = 0$. Next for $0 < s < \infty$, let $R^*(s)$ denote the ring bounded by the segment $\{x: -s \leq x_1 \leq 0,\ x_2 = \cdots = x_n = 0\}$ and by the ray

$$\{x:\ s \leq x_1 < \infty,\ x_2 = \cdots = x_n = 0\} \cup \{\infty\} \ .$$

Then by means of spherical symmetrization one can show that

$$\text{cap}_p(R) \geq \text{cap}_p(R^*(r)) \ . \tag{6}$$

This inequality was obtained in [5] for the case where $n = p = 3$, and the argument given there can be modified easily to cover the case where $n \geq 2$ and $p \geq 1$. Details for the case where $n = p \geq 2$ were recently given in [1] and [12].

Next suppose that $v(x) = u(rx)$. Then u is an admissible function for $R^*(r)$ if and only if v is an admissible function for $R^*(1)$, in which case

$$\int_{R^*(r)} |\nabla u|^p dm_n = r^{n-p} \int_{R^*(1)} |\nabla v|^p dm_n .$$

From this we conclude that

$$\operatorname{cap}_p(R^*(r)) = r^{n-p} \operatorname{cap}_p(R^*(1)) . \tag{7}$$

Finally the fact that $p > n-1$ implies that the p-capacity of $R^*(1)$ is positive. This can be proved directly by adapting the argument in [10] where the case $n = p \geq 2$ is handled. When $p \leq n$, it follows from the main theorem in [17]. Hence (5) follows from (6) and (7) with $c = \operatorname{cap}_p(R^*(1))$.

LEMMA 4. *If* R *is a ring and if* $n < p < \infty$, *then*

$$\operatorname{cap}_n(R)^p \leq \operatorname{cap}_p(R)^n V(R)^{p-n} . \tag{8}$$

If, in addition, R *is the spherical ring* $\{x: a < |x-P| < b\}$, *then*

$$\operatorname{cap}_p(R)^n V(R)^{p-n} \leq \left(\frac{b}{a}\right)^{n(p-n)} \operatorname{cap}_n(R)^p . \tag{9}$$

Proof: If u is an admissible function for R, then Hölder's inequality implies that

$$\operatorname{cap}_n(R) \leq \int_R |\nabla u|^n dm_n \leq \left(\int_R |\nabla u|^p dm_n\right)^{n/p} V(R)^{\frac{p-n}{p}} ,$$

and taking the infimum over all such u yields (8). Inequality (9) follows directly from (2).

Finally we shall need a sharpened form of the first part of inequality (4) for the case where $p = n$. Given a ring R and a function g which is nonnegative and Borel measurable in R, we set

$$A(g,R) = \inf_{\Sigma} \int_{\Sigma} g^{n-1} dm_{n-1} , \quad V(g,R) = \int_R g^n dm_n ,$$

where the infimum is taken over all polyhedral surfaces $\Sigma \subset R$ which separate the complementary components C_0 and C_1. Next we say that R is *approximable from outside* if there exists a sequence of rings R_k containing $\bar{R}$ such that $\bar{R}$ separates the complementary components of each R_k and

(10) $$\operatorname{cap}_n(R) = \lim_{k \to \infty} \operatorname{cap}_n(R_k) .$$

LEMMA 5. *If* R *is a ring which is approximable from outside, then*

(11) $$\operatorname{cap}_n(R) = \sup_g \frac{A(g,R)^n}{V(g,R)^{n-1}} ,$$

where the supremum is taken over all functions g *which are nonnegative and continuous in* R *with* $A(g,R) > 0$ *and* $V(g,R) < \infty$.

Proof: From [7] and [18] we see that (11) holds for an *arbitrary* ring R if the supremum is taken over the larger class of functions g which are Borel measurable in R with $A(g,R)$ and $V(g, R)$ not simultaneously 0 or ∞. Hence in order to establish Lemma 5, it is sufficient to show that when R is approximable from outside, for each $\varepsilon > 0$ there exists a function g which is nonnegative and continuous in R with $A(g,R) > 0$ and $V(g,R) < \infty$ such that

(12) $$\operatorname{cap}_n(R) \leq \frac{A(g,R)^n}{V(g,R)^{n-1}} + \varepsilon .$$

Fix $\varepsilon > 0$. Then (10) implies that there exists a ring R' containing $\bar{R}$ such that $\bar{R}$ separates the components of $C(R')$ and

(13) $$\operatorname{cap}_n(R) \leq \operatorname{cap}_n(R') + \varepsilon .$$

Since $\bar{R} \subset R'$, the components of $C(R)$ are nondegenerate, $\operatorname{cap}_n(R) > 0$ by [10], and thus we may assume also that $\operatorname{cap}_n(R') > 0$. This implies that the components of $C(R')$ are nondegenerate and hence that there exists a unique extremal admissible function v for R' such that

(14) $$\mathrm{cap}_n(R') = \int_{R'} |\nabla v|^n \, dm_n \, .$$

(See [6] or [12].) Choose r so that $0 < r < \mathrm{dist}(\bar{R}, C(R'))$, and let

$$g(x) = \left(\frac{1}{m_n(U)} \int_U |\nabla v(x+y)|^{n-1} dm_n(y) \right)^{\frac{1}{n-1}} ,$$

where $U = \{y: |y| < r\}$. Then g is continuous and nonnegative in R, and from Minkowski's inequality and (14) it follows that

(15) $$V(g, R) \leq \mathrm{cap}_n(R') < \infty \, .$$

Next arguing as in [7] or [18], we see that

$$\int_\Sigma g^{n-1} dm_{n-1} \geq \mathrm{cap}_n(R')$$

for each polyhedral surface $\Sigma \subset R$ which separates the components of $C(R)$. Thus

(16) $$A(g, R) \geq \mathrm{cap}_n(R') > 0,$$

and (12) follows from (13), (15), and (16).

4. Volume and area distortion properties. We derive next some analytic properties of mappings in $Q_p(K)$.

LEMMA 6. *Suppose that* $f \in Q_p(K)$ *and that* $P \in D$. *If* $1 \leq p < n$, *then*

$$J(P, f) \leq K^{\frac{n}{n-p}} \, .$$

If $n < p < \infty$, *then*

$$J(P, f) \geq K^{\frac{n}{n-p}} \, .$$

Proof: For $0 < a < b < \mathrm{dist}(P, C(D))$, let R and R^* denote respectively the spherical rings $\{x: a < |x - P| < b\}$ and $\{x: a^* < |x| < b^*\}$ with complementary components C_0, C_1 and C_0^*, C_1^*, where a^* and b^* are chosen so that

$$m_n(C_0^*) = m_n(f(C_0)) , \qquad m_n(C_0^* \cup R^*) = m_n(f(C_0 \cup R)) .$$

Then Lemma 2 implies that

$$\text{(17)} \qquad \operatorname{cap}_p(R^*) \le \operatorname{cap}_p(f(R)) \le K \operatorname{cap}_p(R) .$$

Now suppose that $p > 1$. Then (2) and (17) imply that

$$\left(\frac{(b^*)^q - (a^*)^q}{q}\right)^{1-p} \le K\left(\frac{b^q - a^q}{q}\right)^{1-p} ,$$

where $q = (p-n)/(p-1)$. If $1 < p < n$, then $q < 0$,

$$(a^*)^{n-p} \le K a^{n-p}(1-(\tfrac{b}{a})^q)^{1-p} ,$$

and holding b fixed we obtain

$$J(P,f) = \limsup_{a \to 0} (\tfrac{a^*}{a})^n \le K^{\frac{n}{n-p}} .$$

Next if $n < p < \infty$, then $q > 0$,

$$(b^*)^{n-p} \le K b^{n-p}(1-(\tfrac{a}{b})^q)^{1-p} ,$$

and after letting $a \to 0$ we obtain

$$J(P,f) = \limsup_{b \to 0} (\tfrac{b^*}{b})^n \ge K^{\frac{n}{n-p}} .$$

The case where $p = 1$ follows similarly.

LEMMA 7. *Suppose that* $f \in Q_p(K)$ *and that* E *is a Borel set in* D. *If* $1 \le p < n$, *then*

$$\text{(18)} \qquad m_n(f(E)) \le K^{\frac{n}{n-p}} m_n(E) .$$

If $n < p < \infty$, *then*

$$m_n(f(E)) \geq K^{\frac{n}{n-p}} m_n(E) \quad .$$

Proof: Suppose that $1 \leq p < n$. Since (18) follows trivially when $m_n(E) = \infty$, we may assume that $m_n(E) < \infty$. Then for each $\varepsilon > 0$ there exists an open set $G \supset E$ with

$$m_n(G) \leq m_n(E) + \varepsilon \quad . \tag{19}$$

Next by Lemma 6, for each $\delta > 0$ and each $P \in D$ there exists a ball $U = \{x: |x-P| < r\}$ with $0 < r < \delta$ and $U \subset D \cap G$ such that

$$m_n(f(U)) \leq (K^{\frac{n}{n-p}} + \varepsilon)\, m_n(U) \quad . \tag{20}$$

If $m_n(E) = 0$, then by a covering theorem due to Morse [11], there exists a countable subcollection of balls $U_k = \{x: |x-P_k| < r_k\}$ covering E such that the balls $\{x: |x-P_k| < r_k/5\}$ are disjoint. With (20) we obtain

$$m_n(f(E)) \leq (K^{\frac{n}{n-p}} + \varepsilon) \sum_k m_n(U_k) \leq (K^{\frac{n}{n-p}} + \varepsilon)\, 5^n m_n(G) \quad ,$$

and since ε is arbitrary, (19) implies that $m_n(f(E)) = 0$ and hence that (18) holds. Next if $m_n(E) > 0$, then by the Vitali covering theorem, there exists a countable subcollection of disjoint balls U_k such that $m_n(H) = 0$, where $H = E - \cup_k U_k$. Then $m_n(f(H)) = 0$,

$$m_n(f(E)) \leq (K^{\frac{n}{n-p}} + \varepsilon) \sum_k m_n(U_k) \leq (K^{\frac{n}{n-p}} + \varepsilon)\, m_n(G) \quad ,$$

and (18) follows by letting $\varepsilon \to 0$.

The case where $n < p < \infty$ is handled by a similar argument.

Lemma 7 shows that the n-dimensional measure or volume of Borel sets is not increased (decreased) by more than the factor $K^{n/(n-p)}$ under mappings in $Q_p(K)$ when $1 \leq p < n$ ($n < p < \infty$). We require an analogous estimate for the distortion of the (n–1)-dimensional measure or area. The following result shows that the de Giorgi perimeter [3] of spheres is not increased by more than the factor $K^{(n-1)/(n-p)}$ under mappings in $Q_p(K)$ when $1 \leq p < n$. All these bounds are sharp.

LEMMA 8. *Suppose that* $f \in Q_p(K)$ *where* $1 \leq p < n$, *and that* U *is an open ball with* $\overline{U} \subset D$. *Then for each* $\varepsilon > 0$ *there exists an open polyhedron* $G' \supset f(U)$ *such that each point of* G' *lies within distance* ε *of* $f(U)$ *and*

$$m_{n-1}(\partial G') \leq K^{\frac{n-1}{n-p}}(1+\varepsilon)\, m_{n-1}(\partial U) .$$

Proof: Suppose that $U = \{x: |x-P| < a\}$ and let $R = \{x: a < |x-P| < b\}$, where b is chosen so that $\overline{R} \subset D$, so that each point of $f(R)$ lies within distance ε of $f(U)$, and so that

$$V(R) < (1+\varepsilon)\, A(R) L(R) .$$

Then Lemma 1 implies that

$$\operatorname{cap}_p(R) < \frac{A(R)^p}{V(R)^{p-1}}(1+\varepsilon)^p ,$$

Lemmas 1 and 7 imply that

$$\frac{A(f(R))^p}{V(R)^{p-1}} \leq K^{\frac{n(p-1)}{n-p}} \operatorname{cap}_p(f(R)) ,$$

and hence we obtain

$$A(f(R)) < K^{\frac{n-1}{n-p}}(1+\varepsilon)\, A(R) .$$

Since $A(R) = m_{n-1}(\partial U) > 0$, there exists a polyhedral surface $\Sigma' \subset f(R)$ which separates the components of $C(f(R))$ such that

$$m_{n-1}(\Sigma') \leq K^{\frac{n-1}{n-p}}(1+\varepsilon)\, m_{n-1}(\partial U) .$$

If we let G' denote the component of $C(\Sigma')$ which contains $f(U)$, then G' has all of the desired properties.

LEMMA 9. *Suppose that* $f \in Q_p(K)$ *where* $1 \leq p < n$, *and that* E *is a compact set in* D *with* $m_{n-1}(E) > 0$. *Then for each* $\varepsilon > 0$ *there exists an open polyhedron* $G' \supset f(E)$ *such that each point of* G' *lies within distance* ε *of* $f(E)$ *and*

$$m_{n-1}(\partial G') \leq c K^{\frac{n-1}{n-p}} m_{n-1}(E) ,$$

where c *is a constant which depends only on* n.

Proof: Choose δ so that $0 < \delta < \varepsilon$ and $(1+\delta)^2 \leq 2$. Since E is compact with $m_{n-1}(E) > 0$, there exists a finite collection of open balls U_k covering E, with $\overline{U}_k \subset D$ and $\operatorname{dia} f(U_k) < \varepsilon - \delta$ for all k, such that

$$\sum_k \tau_{n-1} \left(\frac{\operatorname{dia} U_k}{4} \right)^{n-1} \leq (1+\delta)\, m_{n-1}(E) .$$

Next for each k, let G_k' be the open polyhedron of Lemma 8 corresponding to U_k and δ. Then $G' = \bigcup_k G_k'$ satisfies the above requirements with

$$c = 2^n \frac{\omega_n}{\tau_{n-1}} .$$

5. **Main results.** We now combine various lemmas of the last two sections to establish the main theorems of this paper. We begin with a result relating the classes $Q_p(K)$.

THEOREM 1. *If* $f, f^{-1} \in Q_p(K)$ *where* $p \neq n$, *then* $f, f^{-1} \in Q_n(K_0)$ *where* K_0 *depends only on K, n, and p.*

Proof: By symmetry, it is sufficient to show that for each spherical ring $R = \{x : a < |x-P| < b\}$ with $\overline{R} \subset D$

$$\operatorname{cap}_n(f(R)) \leq K_0 \operatorname{cap}_n(R) , \tag{21}$$

where K_0 depends only on K, n, and p.

Suppose that $n < p < \infty$, fix an integer k, and for $j = 1, \dots, k$, let

$$R_j = \{x : a(\tfrac{b}{a})^{\frac{j-1}{k}} < |x-P| < a(\tfrac{b}{a})^{\frac{j}{k}}\} .$$

Then the R_j are disjoint spherical rings which separate the components of C(R) and

$$\operatorname{cap}_n(R_j) = k^{n-1} \operatorname{cap}_n(R) \tag{22}$$

by (2). Lemmas 4 and 7 imply that

$$\operatorname{cap}_n(f(R_j))^p \le K^{2n} \operatorname{cap}_p(R_j)^n V(R_j)^{p-n} \le K^{2n} (\tfrac{b}{a})^{\frac{n(p-n)}{k}} \operatorname{cap}_n(R_j)^p \tag{23}$$

for each j, and combining (22) and (23) with Lemma 2 of [5] or Lemma 7.2 of [12] yields

$$\operatorname{cap}_n(f(R)) \le \left(\sum_j \operatorname{cap}_n(f(R_j))^{\frac{1}{1-n}} \right)^{1-n} \le K^{\frac{2n}{p}} (\tfrac{b}{a})^{\frac{n(p-n)}{pk}} \operatorname{cap}_n(R) .$$

We then obtain (21) with $K_0 = K^{2n/p}$ by letting $k \to \infty$.

Suppose next that $1 \le p < n$. Since R is a spherical ring with $\bar{R} \subset D$, it is not difficult to see from Lemma 6 in [6] or Theorem 6.1 in [12] that f(R) is approximable from outside. Hence by Lemma 5,

$$\operatorname{cap}_n(f(R)) = \sup_h \frac{A(h, f(R))^n}{V(h, f(R))^{n-1}} , \tag{24}$$

where the supremum is taken over all functions h which are nonnegative and continuous in f(R) with $A(h, f(R)) > 0$ and $V(h, f(R)) < \infty$. We show that given any such h,

$$A(h, f(R)) \le c K^{\frac{n-1}{n-p}} A(g, R) , \quad V(g, R) \le K^{\frac{n}{n-p}} V(h, f(R)) , \tag{25}$$

where $g = h \circ f$ and c is the constant in Lemma 9.

For the first part of (25), fix $\varepsilon > 0$ and let Σ be any polyhedral surface in R which separates the components of C(R). Since g is continuous in R, we can express Σ as the finite union of compact sets E_k, where $m_{n-1}(E_k) > 0$ and

$$\operatorname*{osc}_{E_k} g^{n-1} < \varepsilon \tag{26}$$

for all k and $m_{n-1}(E_j \cap E_k) = 0$ for all j and k, $j \neq k$. Next using Lemma 9 and (26), we can choose for each k an open polyhedron $G_k' \supset f(E_k)$ such that $\overline{G}_k' \subset f(R)$,

$$\operatorname*{osc}_{\overline{G}_k'} h^{n-1} < \varepsilon \, , \qquad m_{n-1}(\partial G_k') \leq c K^{\frac{n-1}{n-p}} m_{n-1}(E_k) \tag{27}$$

for all k. Then $\cup_k \partial G_k'$ contains a polyhedral surface $\Sigma' \subset f(R)$ which separates the components of $C(f(R))$, and (26) and (27) imply that

$$\begin{aligned} A(h, f(R)) &\leq \sum_k \int_{\partial G_k'} h^{n-1} \, dm_{n-1} \leq \sum_k (g(P_k)^{n-1} + \varepsilon) \, m_{n-1}(\partial G_k') \\ &\leq c K^{\frac{n-1}{n-p}} \sum_k \int_{E_k} (g^{n-1} + 2\varepsilon) \, dm_{n-1} \\ &= c K^{\frac{n-1}{n-p}} \int_{\Sigma} (g^{n-1} + 2\varepsilon) \, dm_{n-1} \, , \end{aligned}$$

where $P_k \in E_k$. Since ϵ and Σ are arbitrary, we obtain the first part of (25). The proof for the second part is similar; we decompose R into disjoint Borel sets on which the oscillation of g^n is small and then apply Lemma 7 to f^{-1}.

Finally (24), (25), and Lemma 5 imply (21) with $K_0 = c^n K^{\frac{2n(n-1)}{n-p}}$.

THEOREM 2. *If* $f, f^{-1} \in Q_p(K)$ *where* $p \neq n$, *then* $f, f^{-1} \in \operatorname{Lip}(K_0)$ *where* K_0 *depends only on* K, n, *and* p.

Proof: Theorem 1 implies that $f \in Q_n(K_1)$, where K_1 depends only on K, n, and p. Next by Lemma 8 in [6] or Theorem 9.2 in [12],

$$\sup_D H(P,f) \leq K_2 < \infty, \tag{28}$$

where K_2 depends only on K_1 and n, while from Lemmas 6 and 7 it follows that

$$\sup_D J(P, f) \leq K^{\frac{n}{|n-p|}}. \tag{29}$$

Finally since $L(P, f)^n \leq H(P, f)^n J(P, f)$, we conclude from (28), (29), and the above argument applied to f^{-1} that $f, f^{-1} \in \mathrm{Lip}(K_0)$ where $K_0 = K_2 K^{1/|n-p|}$.

Theorem 2 shows that $f, f^{-1} \in \mathrm{Lip}(K_0)$ whenever $f, f^{-1} \in Q_p(K)$. It is natural to ask what can be said if we know only that $f \in Q_p(K)$.

THEOREM 3. *If* $f \in Q_p(K)$ *where* $n-1 < p < n$, *then* $f \in \mathrm{Lip}(K_0)$. *If* $f \in Q_p(K)$ *where* $n < p < \infty$, *then* $f^{-1} \in \mathrm{Lip}(K_0)$. *In both cases* K_0 *depends only on* K, n, *and* p.

Proof: For the first part, fix $P \in D$ and for $0 < a' < \mathrm{dist}(f(P), C(D'))$ let $V' = \{x: |x - f(P)| < a'\}$, $V = f^{-1}(V')$,

$$a = \min_{x \in \partial V} |x - P|, \qquad b = \max_{x \in \partial V} |x - P| .$$

Then a and b are continuous in a' and

$$L(P,f) = \limsup_{a' \to 0} \frac{a'}{a}. \tag{30}$$

Fix a' so that $\{x: |x - P| \leq b\} \subset D$. If $b \leq 2a$, then

$$a' = \left(\frac{m_n(V')}{\tau_n}\right)^{1/n} \leq K^{\frac{1}{n-p}} \left(\frac{m_n(V)}{\tau_n}\right)^{1/n} \leq 2K^{\frac{1}{n-p}} a \tag{31}$$

by Lemma 7. Next if $2a < b$, then $R = \{x: a < |x-P| < b\}$ is a spherical ring with $\overline{R} \subset D$, and

$$\operatorname{cap}_p(R) = \omega_n \left(\frac{b^q - a^q}{q} \right)^{1-p} \leq d\, a^{n-p}$$

by (2), where d depends only on p and n. Moreover both components of $C(f(R))$ meet $\partial V'$,

$$\operatorname{cap}_p(f(R)) \geq c\,(a')^{n-p}$$

by Lemma 3, and hence

$$a' \leq \left(\frac{Kd}{c}\right)^{\frac{1}{n-p}} a \; . \tag{32}$$

Finally (30), (31), and (32) imply that $f \in \operatorname{Lip}(K_0)$, where K_0 depends only on K, n, and p.

We omit the proof for the second part of Theorem 3, since it is quite similar to the argument given above.

6. **Examples.** We conclude this paper with a pair of examples which show that the ranges for the exponent p in Theorem 3 are almost best possible. They also show that the alternative short proof of Theorem 2, which Theorem 3 provides for the case where $n-1 < p < \infty$, $p \neq n$, cannot be extended to give the general case where $p \neq n$.

LEMMA 10. *If* f *is a diffeomorphism of a domain* $D \subset R^n$ *and if for some* K, $0 < K < \infty$,

$$J(P,f) \leq K\,\ell(P,f)^p$$

for all $P \in D$, *then* $f \in Q_p(K)$.

Proof: Let u be a simple admissible function for a ring R with $\overline{R} \subset D$, and set $v = u \circ f^{-1}$. Then v is admissible for $f(R)$,

$$\operatorname{cap}_p(f(R)) \leq \int_R |\nabla v(f(x))|^p J(x,f)\, dm_n(x) \leq K \int_R |\nabla u(x)|^p\, dm_n(x),$$

and taking the infimum over all such u yields

$$\mathrm{cap}_p(f(R)) \le K\,\mathrm{cap}_p(R) \ .$$

EXAMPLE 1. *For* $1 \le p < n-1$ *or* $n \le p < \infty$, *there exists* $f_1 \in Q_p(2^p)$ *such that* $f_1 \notin \mathrm{Lip}(K_0)$ *for all* K_0.

EXAMPLE 2. *For* $1 \le p \le n$, *there exists* $f_2 \in Q_p(2)$ *such that* $f_2^{-1} \notin \mathrm{Lip}(K_0)$ *for all* K_0.

Proofs: For Example 1, suppose first that $1 \le p < n-1$ and set

$$f_1(x) = \left(\frac{x_1}{x_n}, \ldots, \frac{x_{n-1}}{x_n}, \frac{x_n^{\,n-p}}{n-p}\right)$$

in $D_1 = \{x\colon x_1^2 + \cdots + x_{n-1}^2 < 1, \frac{8}{3} < x_n < \infty\}$. Then it is not difficult to verify that

$$J(x,f_1) = x_n^{-p} \le 2^p \ell(x,f_1)^p \ , \quad L(x,f_1) \ge x_n^{n-1-p}$$

in D_1. Hence $f_1 \in Q_p(2^p)$ by Lemma 10, while $f_1 \notin \mathrm{Lip}(K_0)$ for all K_0 since $L(x,f_1)$ is unbounded in D_1.

Suppose next that $n \le p < \infty$ and set $f_1(x) = |x|x$ in $D_1 = \{x\colon 1 < |x| < \infty\}$. Then

$$J(x, f_1) = 2|x|^n \le 2\ell(x,f_1)^p \ , \quad L(x,f_1) = 2|x|$$

in D_1, and again f_1 has the required properties.

Finally for Example 2, consider $f_2(x) = |x|x$ in $D_2 = \{x\colon |x| < 1\}$.

Since $f \in Q_n(K)$ implies that $f^{-1} \in Q_n(K_0)$ where K_0 depends only on K and n, the above examples also show that Theorem 2 does not hold for $p = n$.

REFERENCES

[1] P. Caraman, *Homeomorfisme cvasiconforme* n- *dimensionale*, Bucharest 1968.

[2] T. Carleman, Über ein Minimalproblem der mathematischen Physik, *Math. Zeit.* 1 (1918) pp. 208-212.

[3] E. de Giorgi, Su una teoria generale della misura (r – 1)-dimensionale in uno spazio ad r dimensioni, *Annali di Matematica* 36 (1954) pp. 191-213.

[4] H. Federer, Curvature measures, *Trans. Amer. Math. Soc.* 93 (1959) pp. 418-491.

[5] F. W. Gehring, Symmetrization of rings in space, *Trans. Amer. Math. Soc.* 101 (1961) pp. 499-519.

[6] _____, Rings and quasiconformal mappings in space, *Trans. Amer. Math. Soc.* 103 (1962) pp. 353-393.

[7] _____, Extremal length definitions for the conformal capacity of rings in space, *Mich. Math. J.* 9 (1962) pp. 137-150.

[8] _____, Inequalities for condensers, hyperbolic capacity, and extremal lengths, Mich. Math. J. 18 (1971) pp. 1-20.

[9] L. G. Lewis, Quasiconformal mappings and Royden algebras in space, *Trans. Amer. Math. Soc.* (to appear).

[10] C. Loewner, On the conformal capacity in space, *J. Math. Mech.* 8 (1959) pp. 411-414.

[11] A. P. Morse, A theory of covering and differentiation, *Trans. Amer. Math. Soc.* 55 (1944) pp. 205-235.

[12] G. D. Mostow, Quasi-conformal mappings in n-space and the rigidity of hyperbolic space forms, *Inst. Hautes Études Sci. Publ. Math.* 34 (1968) pp. 53-104.

[13] M. Nakai, Algebraic criterion on quasiconformal equivalence of Riemann surfaces, *Nagoya Math. J.* 16 (1960) pp. 157-184.

[14] H. M. Reimann, Über harmonische Kapazität und quasikonforme Abbildungen im Raum, *Comm. Math. Helv.* 44 (1969) pp. 284-307.

[15] J. Väisälä, On quasiconformal mappings in space, *Ann. Acad. Sci. Fenn.* 298 (1961) pp. 1-36.

[16] ________, *Lectures on* n*-dimensional quasiconformal mappings*, Van Nostrand Reinhold (to appear).

[17] H. Wallin, *A connection between* α*-capacity and* L^p*-classes of differentiable functions*, Ark. Mat. 5 (1963-1965) pp. 331-341.

[18] W. P. Ziemer, *Extremal length and conformal capacity*, Trans. Amer. Math. Soc. 126 (1967) pp. 460-473.

University of Michigan
Ann Arbor, Michigan

SPACES OF FUCHSIAN GROUPS AND TEICHMÜLLER THEORY *

by

W. J. Harvey

§1. Introduction

The primary aim of this article is to describe an approach to the study of moduli of Fuchsian groups. The basic ideas, which have roots in the work of Fricke, Nielsen, and others, are those of Macbeath as exhibited in his papers and lectures, and many of the results are due to him. Certain simplifications of Teichmüller theory appear, in particular the simple demonstration that the fixed set in Teichmüller space of an element of the modular group is homeomorphic to a Teichmüller space, a result originally obtained by Kravetz in the classical case.

§2. Preliminaries on Fuchsian Groups.

Let G be a Fuchsian group, that is, a discrete subgroup of the real linear fractional group $\mathfrak{L}$. If G is of the first kind and finitely generated, it is isomorphic to an abstract group Γ with presentation of the form

generators: $x_1, x_2, \ldots, x_k$; $a_1, b_1, \ldots, a_\gamma, b_\gamma$; $p_1, \ldots, p_m$;

relations: $x_j^{n_j} = 1, \quad j = 1, \ldots, k$;

---

* This is a slight elaboration of a talk presented at the Conference on Riemann Surfaces at Stony Brook, New York, July 1969.

This work has been partially supported by the National Science Foundation under Grants NSF–GP - 8735 and NSF–GP - 13693.

$$x_1 x_2 \cdots x_k \prod_{i=1}^{\gamma} [a_i, b_i] p_1 \cdots p_m = 1,$$

where $[a_i, b_i] = a_i b_i a_i^{-1} b_i^{-1}$.

We refer to Γ as the group of Fuchsian type with *signature* $(\gamma; n_1, \ldots, n_k; m)$, since this set of non-negative integers defines the isomorphism class of Γ if we regard the set $\{n_1, n_2, \ldots, n_k\}$ of periods as unordered [9]. Associated with such a group Γ there is a number $\mu(\Gamma)$ defined by

$$\mu(\Gamma) = 2\gamma - 2 + \sum_{j=1}^{k} (1 - \frac{1}{n_j}) + m .$$

A necessary and sufficient condition for the group Γ with signature $(\gamma; n_1, \ldots, n_k; m)$ to be of Fuchsian type is that $\mu(\Gamma)$ be greater than 0.

Let $\theta : \Gamma \to G$ be an isomorphism between Γ and G a Fuchsian group of the first kind such that the sets of elements $\{\theta(x_i)\ i = 1 \cdots k\}$, $\{\theta(a_i), \theta(b_i) \mid i = 1, \ldots, \gamma\}$ and $\{\theta(p_i) \mid i = 1, \ldots, m\}$ are respectively of elliptic, hyperbolic, and parabolic type. G then has the property that $\mathfrak{L}/G$ has finite volume and, equivalently, if U denotes the Poincaré upper half plane, that U/G has finite measure. If $m = 0$, then $\mathfrak{L}/G$ and U/G are compact. We say that such a group Γ is of compact Fuchsian type. If in addition $k = 0$, we obtain the important class of surface groups of genus γ, for $\gamma \geq 2$.

§3. Spaces of Fuchsian Groups

We denote by $R(\Gamma)$ the space of all representations r of a group Γ of Fuchsian type into $\mathfrak{L}$, with the topological structure inherited as a subspace of $\mathfrak{L}(\Gamma)$, The subspace of $R(\Gamma)$ consisting of those r which are faithful and type-preserving and such that $r(\Gamma)$ is discrete and $\mathfrak{L}/r(\Gamma)$ has finite volume, we denote by $R_0(\Gamma)$.

Note: 1. A theorem of Weil [11] implies that if Γ is of compact Fuchsian type then $R_0(\Gamma)$ is open in $R(\Gamma)$.

2. $R_0(\Gamma)$ is a real manifold.

Consider the set S of all discrete subgroups of $\mathfrak{L}$ (in the notation of Macbeath [7, 8], $S = S(\mathfrak{L})$) with the following topological structure. Let $\{G_n\}$ be a sequence of elements of S. We say that G_n tends to a group $G \in S$ if and only if

(i) for each $g \in G$ there is a sequence $\{g_n | g_n \in G_n, \ n = 1, 2, \ldots\}$ with $g_n \to g$ in $\mathfrak{L}$,

(ii) for each sequence $\{g_n | g_n \in G_n\}$ with $g_n \to g$ an element of $\mathfrak{L}$, it follows that $g \in G$.

We note that this gives the same structure as that defined in [7].

Let S_0 denote the subspace of S consisting of those discrete groups G with $\mathfrak{L}/G$ compact.

THEOREM 1. (Macbeath). *Let* $G \in S_0$. *Then there is a neighborhood* $N(G)$ *of* G *in* S *consisting entirely of groups isomorphic to* G.

Let Γ be a group of Fuchsian type. We denote by $S(\Gamma)$ the space of all groups $G \in S$ which are isomorphic to Γ and which have $\mathfrak{L}/G$ of finite volume. An immediate consequence of Theorem 1 is the result that if Γ is of compact type, then $S(\Gamma)$ is open and closed in S_0, and S_0 is an open subspace of S. We observe that S_0 is not closed in S; for example, as in [7], one may construct a sequence of Fuchsian groups G_n, with signature $(0; 2, 3; n; 0)$ for $n \geq 7$, which tend to the modular group $LF(2, Z)$ with signature $(0; 2, 3; 1)$.

§4. Group Actions on $R_0(\Gamma)$

Let $\mathfrak{A}(\Gamma)$ denote the group of automorphisms of Γ. $\mathfrak{A}(\Gamma)$ acts on $R_0(\Gamma)$ by the rule $r \mapsto r \cdot \alpha$, for $\alpha \in \mathfrak{A}(\Gamma)$. There are no fixed points, and it is easy to show that the action is properly discontinuous ([8]).
There is an obvious mapping of $R_0(\Gamma)$ onto $S(\Gamma)$ given by the rule $r \mapsto r(\Gamma)$, which we denote by j.

We have

THEOREM 2. (Macbeath [8]). *The mapping* $r \overset{j}{\longmapsto} r(\Gamma)$ *is a local homeomorphism if* Γ *is of compact Fuchsian type.*

Note: It follows that $S(\Gamma)$ is homeomorphic to $R_0(\Gamma)/\mathfrak{A}(\Gamma)$ and is a real manifold.

The group $\mathfrak{L}$ also acts on $R_0(\Gamma)$ by conjugation. If $t \in \mathfrak{L}$ we denote by t^* the self mapping $r \longrightarrow t^*(r)$ where $t^*(r): x \to t^{-1} r(x) t$ for each $x \in \Gamma$. This mapping is without fixed points since Γ is not cyclic.

The quotient space $R_0(\Gamma)/\mathfrak{L}^*$ is denoted by $T(\Gamma)$. It can be shown to be the union of two disjoint copies of the usual Teichmüller space, there being two possible orientations for each Riemann surface $U/r(\Gamma)$, $r \in R_0(\Gamma)$. One may avoid this slight discrepancy by working with the full hyperbolic group of conformal and anticonformal mappings of U on itself, but then one should consider all discrete subgroups of the hyperbolic group rather than solely the Fuchsian groups. This theory provides a unified treatment of moduli of orientable and non-orientable surfaces.

Note: We remark that $R_0(\Gamma)$ may be shown to be homeomorphic to the Cartesian product $T(\Gamma) \times \mathfrak{L}$, since the subspace of $R_0(\Gamma)$ consisting of the isomorphisms normalized appropriately with respect to the action of $\mathfrak{L}^*$ can be shown to be homeomorphic to $T(\Gamma)$ —see for example Bers [2]. This observation is due to Singerman [10].

§5. The Action of $\mathfrak{A}(\Gamma) \times \mathfrak{L}^*$

We first sketch the proof of a lemma on mappings between R_0-spaces.

Let $K \overset{i}{\longrightarrow} \Gamma$ be an injective homomorphism between two groups K and Γ of Fuchsian type such that $i(K)$ has finite index in Γ. Then there is an induced mapping $\hat{i}: R_0(\Gamma) \to R_0(K)$ given by $\rho \mapsto \rho \circ i$ for $\rho \in R_0(\Gamma)$.

LEMMA 3. $\hat{i}$ *is a real analytic homeomorphism onto a closed subspace of* $R_0(K)$.

Proof: If $x \in \Gamma - i(K)$ then, for some integer $n > 0$, $x^n \in i(K)$. Let

$r \in R_0(K)$; in order that r equal $\rho \circ i$ for some $\rho \in R_0(\Gamma)$, it is necessary to solve for $\rho(x)$ the equation

$$(\rho(x))^n = \rho(x^n) = r(x^n) \ .$$

This has essentially one solution in $\mathfrak{L}$ if $r(x^n)$ is hyperbolic or parabolic (in which case $\rho(x)$ is of the same type). But one need only deal with hyperbolic elements, since every Fuchsian group of our type may be generated by a finite number of hyperbolic elements. It follows that, since $i(K)$ has finite index in Γ, the correspondence $\rho \mapsto r$ consists in extraction of a finite number of nth roots, and consequently it is analytic and bicontinuous. The image set is clearly closed in $R_0(K)$.

There is an induced mapping $\hat{i}$ from $T(\Gamma)$ to $T(K)$ which makes a commutative diagram with the projections π_Γ, π_K:

$$\begin{array}{ccc} R_0(\Gamma) & \xrightarrow{\hat{i}} & R_0(K) \\ {\scriptstyle \pi_\Gamma}\downarrow & & \downarrow{\scriptstyle \pi_K} \\ T(\Gamma) & \xrightarrow{\hat{i}} & T(K) \end{array}$$

$\hat{i}$ is likewise a real-analytic homeomorphism.

We consider the transformation group $(\mathfrak{A}(K) \times \mathfrak{L}^*, R_0(K))$; noting that the actions of $\mathfrak{A}(K)$ and $\mathfrak{L}^*$ commute, we find that if the quotient space is denoted by $\mathfrak{R}(K)$, there is a commutative diagram

$$\begin{array}{ccc} R_0(K) & \xrightarrow{j} & S(K) \\ {\scriptstyle \pi_K}\downarrow & & \downarrow{\scriptstyle P_K} \\ T(K) & \longrightarrow & \mathfrak{R}(K) \end{array}$$

where the maps $S(K) \xmapsto{P_K} \mathfrak{R}(K)$, $T(K) \to \mathfrak{R}(K)$ represent quotient maps with respect to the actions induced by $\mathfrak{L}^*$ and $\mathfrak{A}(K)$ respectively. The action $\bar{t}$ of $t \in \mathfrak{L}$ on $S(K)$ is simply conjugation: if $G \in S(K)$, then $G \xrightarrow{\bar{t}} t^{-1}Gt$. Thus $\mathfrak{R}(K)$ is the space of $\mathfrak{L}$-conjugacy classes of

Fuchsian groups isomorphic to K. When K is a surface group of genus $\gamma \geq 2$, $\mathfrak{R}(K)$ is Riemann's space of moduli of closed surfaces of genus γ.

Note: We remark that if K is of compact type $\mathfrak{R}(K)$ appears as the quotient of the real manifold S(K) by the action of $\mathfrak{L}$. The fibre $P_K^{-1}([G])$ over a point $[G] \in \mathfrak{R}(K)$ is isomorphic to $\mathfrak{L}/\mathfrak{N}(G)$, where $\mathfrak{N}(G)$ denotes the normalizer in $\mathfrak{L}$ of G. It will follow from further discussion that the set of groups $G \in S(K)$ with $\mathfrak{N}(G) = G$ is dense in S(K) except in a few special cases.

§6. The Loci of Special Branching

Let $r \in R_0(K)$. We denote by Stab(r) the set of elements in $\mathfrak{A}(K) \times \mathfrak{L}^*$ which fix r. Thus $\mathrm{Stab}(r) \cong \mathfrak{N}(r(K))$, since $(\alpha, t^*)(r) = r$ if and only if $t \in \mathfrak{N}(r(K))$ and $\alpha = r^{-1} \circ (\mathfrak{t})^{-1} \circ r$. Let $L(K) = \{r \in R_0(K): \mathfrak{N}(r(K)) \neq r(K)\}$. For any element $r \in L(K)$ there is a group Γ of Fuchsian type with $\Gamma \cong \mathfrak{N}(r(()))$, such that for some $\rho \in R_0(\Gamma)$ there is an injection $i: K \to \Gamma$ with $\rho \circ i = r$. This defines a homeomorphism $\hat{i}: R_0(\Gamma) \to R_0(K)$ whose image lies entirely in L(K). Let $\mathfrak{F}(K)$ denote the (finite) family of all isomorphism classes of groups Γ of Fuchsian type which contain normal subgroups of finite index isomorphic to K. Then we have, from the above discussion

PROPOSITION 4. *L(K) is a union of real analytic subspaces of $R_0(K)$, each one homeomorphic to $R_0(\Gamma)$ for some $\Gamma \in \mathfrak{F}(K)$.*

More precisely, we observe that for each group, Γ, and each surjection $\phi: \Gamma \to H$ with H finite and $\ker \phi \cong H$ there is an injection $i_{\Gamma,\phi}: K \to \ker \phi$. If one takes equivalence classes of surjections modulo automorphisms of Γ and of H and isomorphism classes of groups Γ, there results a corresponding finite set of injections $\{i_{\Gamma,\phi}\}$ such that every subspace of L(K) homeomorphic to $R_0(\Gamma)$ in the above way is a translate of one of the image spaces $\{\mathrm{Im}\ \hat{i}_{\Gamma,\phi}\}$ by the action of $\mathfrak{A}(K)$. The action of $\mathfrak{L}^*$ takes any such image onto itself.

§7. The Teichmüller Theory

We now consider the implications of these results for the space T(K). One finds that since the subgroup $\mathfrak{J}$(K) (the inner automorphisms of K) of $\mathfrak{A}$(K) acts trivially on T(K), it is natural to consider $\mathfrak{R}$(K) as the quotient of T(K) under the action of the factor group M(K) = $\mathfrak{A}(K)/\mathfrak{J}(K)$, known usually as the *mapping-class group* (or *Teichmüller modular group*) of K. It is known that M(K) acts properly discontinuously on T(K). The locus of fixed points is $\pi_K(L(K)) = \Lambda(K)$, and consists therefore of the union of a finite number of homeomorphic images of Teichmüller spaces T(Γ) with $\Gamma \in \mathfrak{F}(K)$, together with translates under the action of M(K). One may obtain simply a description of the fixed point set of an element or subgroup of M(K).

THEOREM 5. *The fixed point set (assumed non-empty) of any subgroup* $H \subset M(K)$ *of finite order is homeomorphic to a Teichmüller space* T(Γ), *where* Γ *is a group of Fuchsian type with an exact sequence*

$$1 \to K \to \Gamma \to H' \to 1$$

where H′ *is anti-isomorphic to* H.

For the proof see [5].

For simplicity we produce the group Γ for H a cyclic group. Let [γ] denote the $\mathfrak{J}$(K)-equivalence class of $\gamma \in \mathfrak{A}(K)$. Suppose $[r] \in T(K)$ is fixed by [γ]. Then there is an element $t = t(\gamma) \in \mathfrak{L}$ such that $r \circ \gamma = t^* \circ r$. Consequently t normalizes r(K), and the group $\Gamma = \langle r(K), t\rangle$ is itself Fuchsian. The exponent of t in r(K) is the order of [γ], and there is a natural homomorphism onto the cyclic group Γ/r(K).

Note 1. It was shown by Kravetz [6] that when K is a surface group, every finite subgroup of M(K) has non-empty fixed point set. As yet, one does not know if this extends to general Teichmüller spaces.

Note 2. One has also a converse to the theorem ([5]): whenever there is an exact sequence

$$1 \to K \xrightarrow{i} \Gamma \longrightarrow H' \longrightarrow 1$$

with K and Γ of Fuchsian type and H' a finite group, there is a subgroup of $M(K)$ anti-isomorphic to H' with non-empty fixed point set $\hat{i}(T(\Gamma))$.

Note 3. In the context of the usual formulation of Teichmüller space (see Bers [2] or Kravetz [6]), one finds that when K is a surface group of genus γ, then Teichmüller classes $\{[r]\}$ lying in the fixed point set of a subgroup $H \subset M(K)$ correspond to surfaces $\{U/r(K) = X_r\}$, each admitting a conformal automorphism group H_r anti-isomorphic to H and such that between two surfaces X_r, X_s in any two classes there is a homeomorphism $\theta_{rs}: X_r \to X_s$ which preserves the "markings" and commutes with the actions of the groups, that is $H_r = \theta^{-1} H_s \theta$. In the case when $H \cong Z_2$ and Γ is the group with signature $(0; 2^{(2\gamma+2)}; 0)$ the conformal automorphisms are the hyperelliptic sheet-interchanges, and a fixed set of such a subgroup H is a hyperelliptic equivalence class in the sense of Ahlfors ([1] page 51).

An immediate corollary to Theorem 5 and Section 6 is obtained by comparing dimensions of $T(\Gamma)$ and $T(K)$ for groups K, Γ of Fuchsian type with $K \subset \Gamma$. Except for a finite number of special cases ([3]), $\Lambda(K)$ is an analytic set in $T(K)$ of positive codimension. The same statement is true for $L(K)$ as a subspace of $R_0(K)$, and one may deduce

PROPOSITION 6. *If* K *is of compact Fuchsian type, the set of groups* $G \in S(K)$ *with* $\mathfrak{N}(G) = G$ *is dense in* S(K) *and its complement is an analytic set.*

§8. An Example

We consider as illustration the situation where a cyclic group Z_p of automorphisms acts on a compact surface with genus $\gamma \geq 2$. The uniformizing group K corresponds to an exact sequence

$$(*) \qquad 1 \to K \xrightarrow{i} \Gamma \xrightarrow{\phi} Z_p \longrightarrow 1$$

as in note 7.2, which imposes restrictions on the signature of Γ (see

Harvey [4]). The surjection ϕ fixes the geometric nature of the action of the automorphism group in that it gives the rotation angles of of the automorphism group in that it gives the rotation angles of the group elements ataany points on the surface fixed under them. A con- corresponds to (that is provides the base surfaces for) an M(K)-orbit of points in T(K) and an $\mathfrak{A}(K) \times \mathfrak{L}^*$-orbit in $R_0(K)$.

To be more concrete, let X be a Riemann surface of genus 3 admitting an automorphism group Z_7. Then X is uniformized by a Fuchsian group arising from a sequence (*) with Γ having signature (0; 7, 7, 7; 0). When one examines surjections $\phi : \Gamma \to Z_7$, there are found to be only two classes modulo the automorphism groups of Γ and Z_7, namely $x_1, x_2, x_3 \mapsto 1, 1, 5$ and $x_1, x_2, x_3 \mapsto 1, 2, 4$ respectively, where x_1, x_2, x_3 are generators for Γ as in Section 2. One finds that in each case there are further automorphisms, the Fuchsian groups in question being subgroups of larger ones. The first class is in fact hyperelliptic: there is an extension Γ_1 of Γ which has signature (0; 2, 7, 14; 0) which admits a homomorphism $\phi_1 : \Gamma_1 \to Z_7 \oplus Z_2$ such that $\ker \phi_1 = K$, $\phi_1 | \Gamma = \phi$ and the subgroup $\phi_1^{-1}(Z_2)$ is a hyper-elliptic group. The other case corresponds therefore to Klein's surface with automorphism group LF(2, 7) of order 168,* which cannot be hyperelliptic since it has simple automorphism group. Here the Fuchsian group K is a normal subgroup of the (0; 2, 3, 7; 0)-group.

REFERENCES

[1] Ahlfors, L. V., "The Complex Analytic Structure of the Space of Closed Riemann Surfaces," *Analytic Functions*, pp. 45-66. Princeton, 1960.

[2] Bers, L., "Quasiconformal Mappings and Teichmüller's Theorem," *Analytic Functions*, pp. 89-119. Princeton, 1960.

*Note. This fact was recalled to me by R. Accola. One can in fact recover simple equations for these curves, as respectively $w^7 = z(z-1)$ and $w^7 = z(z-1)^2$, with automorphism group the sheet interchanges $w \to \eta w$, $z \to z$, $\eta^7 = 1$.

[3] Greenberg, L., "Maximal Fuchsian Groups," *Bull. A.M.S.* (1963), pp. 569-573.

[4] Harvey, W. J., "Cyclic Groups of Automorphisms of a Compact Riemann Surface," *Quarterly J. Math.*, 17 (1966), 86-97.

[5] ———, "On Branch Loci in Teichmüller Space," *Trans. A. M. S.* 151 (1970).

[6] Kravetz, S., "On the Geometry of Teichmüller Spaces and the Structure of Their Modular Groups," *Ann. Acad. Soc. Sci. Fenn.* (1959).

[7] Macbeath, A. M., and Swierczkowski, S., "Limits of Lattices in a Compactly Generated Group," *Can. J. Math.*, 12 (1960), pp. 426-436.

[8] ———, "Groups of Homeomorphisms of a Simply Connected Space," *Ann. Math.*, 79 (1964), pp. 473-488.

[9] ———, "The Classification of Non-Euclidean Plane Crystallographic Groups," *Can. J. Math.*, (1966), pp. 1192-1205.

[10] Singerman, D., Ph.D. Thesis. University of Birmingham, 1969.

[11] Weil, A., "On Discrete Subgroups of Lie Groups," *Ann. Math.*, 72 (1960), pp. 369-384.

ON FRICKE MODULI

Linda Keen*

1. Introduction

Let S be a closed Riemann surface of genus g from which m conformal disks have been removed. S has signature (g; m). The Teichmüller space T(S) is a topological manifold of real dimension $\tau = 6g - 6 + 3m$ and has a natural real analytic structure. In [10], Fricke describes a set of global real analytic coordinates for this space. They are the traces of certain matrices in SL(2, R), and simple real analytic functions of them; these determine the Fuchsian group G which represents S. In this paper a set of global real analytic coordinates for T(S) is defined. They are all traces of matrices of SL(2, R), and all the matrices involved correspond to elements of the Fuchsian group. Therefore, these moduli all have a geometric interpretation on the surface S as lengths of certain simple closed geodesics.

These moduli are essentially the same as the ones that Fricke describes. If $\mathcal{S}$ is a *standard set* of generators for G; the traces of all of the elements of $\mathcal{S}$ are moduli. Since $\mathcal{S}$ contains only $2g + m$ elements, more parameters are needed. There are various possibilities as Fricke indicates, and each has its advantages and disadvantages. It is possible to take traces of various products of the generators, as described in this paper. Taking these products symmetrically gives more parameters than necessary; however the relations among them can be explicitly written down. To see the cell structure most clearly it is necessary to take simple functions of the traces. Fricke describes one such set in [10]. Other sets are described in Fenchel-Nielsen [8], Keen [12] and Zieschang-Coldewey [7].

---

* This paper was partially supported by NSF Grant No. 8789.

All of the sets of moduli are obtained by decomposing the surface S into simpler subsurfaces. Groups are constructed corresponding to the subsurfaces and then combined as free products with amalgamation. Each of the constructions mentioned above is different and the proofs are independent. Each associates a particular region of discontinuity to the Fuchsian group G. Both Fricke [10] and Frenchel [8] use surfaces of signature (0;3) as basic building blocks. Zieschang-Coldewey [7] uses surfaces of signature (1 ; 1) to build upon. In that paper the proofs are much less group theoretic and depend more heavily on the Poincaré geometry of the disk. In this paper surfaces of signature (0 ; 3) and surfaces of signature (1 ; 1) are used. They are combined using the methods of [12]. The decomposition of the surface into building blocks is shown in Figures 1 and 2.

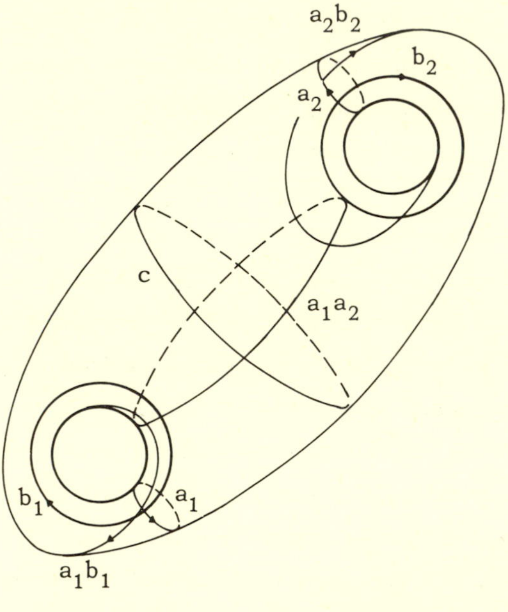

Figure 1. A surface of signature (2, 0). The curves are the geodesics whose lengths are moduli.

The surfaces of signature (0 ; 3) are constructed in Section III and methods of combining them are described in Sections IV and V. The sur-

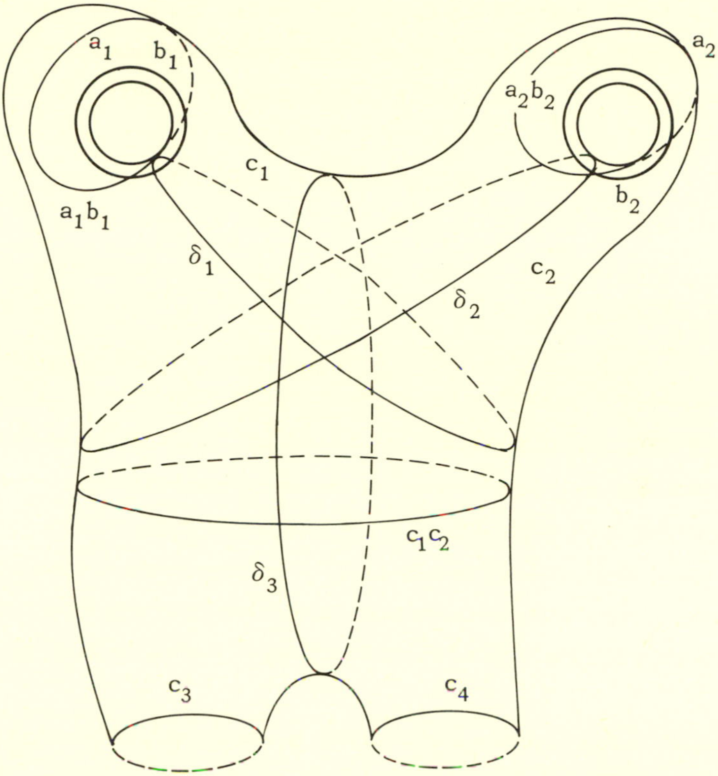

Figure 2. A surface of signature (2, 2). The curves are the geodesics whose lengths are moduli.

faces of signature (1 ; 1) are constructed in Section VI. Sections VIII and IX describe further methods of combination of various types of surfaces to obtain the most general hyperbolic surfaces. Section VII describes the special case (2 ; 0).

These constructions extend to all surfaces of finite type easily using the techniques of Section VI of [12].

II. Preliminaries

The points of Teichmüller space are equivalence classes of marked Riemann surfaces. That is, let S be a Riemann surface of signature (g ; m) and let α be a canonical homotopy basis for Π_S, the fundamental group of S with base point p. α consists of the simple closed curves a_i, b_i, c_j, $i = 1,\ldots,g$, $j = 1,\ldots,m$ with the intersection properties:

$$a_i * a_j = b_i * b_j = c_i * c_j = 0$$
$$a_i * b_j = \begin{cases} 0, & i \neq j \\ 1, & i = j \end{cases}$$

and the relation:

$$a_1 b_1 a_1^{-1} b_1^{-1} \cdots a_g b_g a_g^{-1} b_g^{-1} c_1 \cdots c_m \sim 0 .$$

The curves a_i and b_i go around the "handles" while the curves c_j go around the "holes." If α' is another canonical homotopy basis for S with base point q, α is equivalent to α' if and only if there exists a curve γ from p to q such that the curves $\gamma\alpha'\gamma^{-1}$ are respectively homotopic to those of α. (S,α) is a *marked surface* with marking α and $[S, \alpha]$ is the class of all surfaces with equivalent markings; a point in Teichmüller space T(S).

The universal covering surface of S is conformally equivalent to U, the upper half plane. The covering transformations are conformal homeomorphisms of U and hence linear fractional transformations $(az+b)/(cz+d)$ of the plane, with a, b, c, d real and $ad-bc = 1$. Π_S is thus canonically isomorphic to a properly discontinuous (Fuchsian) group G of elements $(az+b)/(cz+d)$ as above which is determined uniquely up to conjugation. The equivalence class of markings $[\alpha]$ determines a *standard set of generaotrs* $\mathcal{S}$ for G, so that a Fuchsian group with a given set of generators also determines a point in Teichmüller space. A more precise discussion of the correspondence between Fuchsian groups and Teichmüller space can be found in Bers [5].

Associate to each element $A(z) = (az+b)/(cz+d)$, the absolute value of its trace, $|\mathrm{tr}\, A| = |a+d|$. If $|a+d| > 2$, A(z) is called *hyperbolic*. It has two distinct fixed points p_A and q_A on the real axis R. $p_A = \lim_{n\to\infty} A^n(z)$ for all $z \in U$, and is called the *attracting fixed point of* A.

$q_A = \lim_{n\to\infty} A^{-n}(z)$ for all $z \in U$, and is called the *repelling fixed point* of A. The circle h_A orthogonal to R through p_A and q_A is called the *axis* of A. It is a geodesic in the Poincaré metric $\delta(z_1, z_2)$ on U. $d_A = \inf_{z\in U} \delta(z, A(z))$ is achieved if and only if z lies on h_A. d_A is called the *translation length* of A and is related to $|\mathrm{tr}\, A|$ by the formula

(1) $$|\mathrm{tr}\, A| = 2 \cosh d_A \ .$$

See Caratheodory [6] for a derivation of this formula. The Fuchsian groups representing surfaces of signature (g ; m) contain only hyperbolic elements and are the only ones we consider here. A proof of this fact can be found in Ford [9].

III. Construction of a group of signature (0 ; 3)

A Fuchsian group of signature (0 ; 3) is a free group on two generators. Given three real numbers, the problem is to write down generators C_1, C_2 for such a Fuchsian group. These numbers are to be the lengths of the geodesics in the free homotopy classes determined by C_1, C_2 and C_2C_1 on the surface corresponding to the group. If the group can be constructed from tr C_1, tr C_2 and tr C_2C_1, by (1) the problem will be solved.

Since a Fuchsian group with a standard set of generators $\mathcal{S}$ is determined only up to conjugation, one of signature (0 ; 3) with $\mathcal{S} = \{C_1, C_2, C_2C_1\}$ can be normalized so that the attracting and repelling fixed points of C_2C_1 are at 0 and ∞ respectively and so that the repelling fixed point of C_2 is at 1. When C_1, C_2 and C_2C_1 come from a marked surface they satisfy the following "geometric conditions." The axes of C_1 and C_2 do not intersect and the fixed points of C_2C_1 both lie between the attracting fixed point of C_2 and the repelling fixed point of C_1. This is proved in Keen [12]. This implies that the fixed points of C_1C_2 both lie between the attracting fixed point of C_1 and the repelling fixed point of C_2 and also that the fixed points of $C_2^2 C_1 C_2^{-1}$ which are $C_2(0)$ and $C_2(\infty)$ lie between 0 and the attracting fixed point of C_1. See Figure 3.

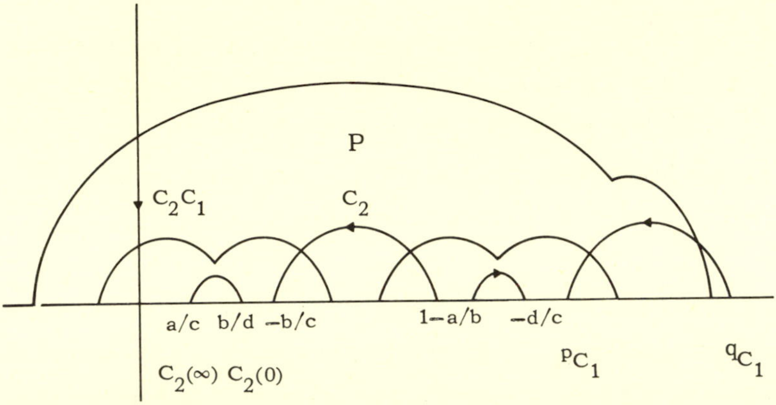

Figure 3. Canonical Modular polygon and the axes of some of the elements of a group of signature (0, 3).

Specifically, if C_1, C_2 and C_2C_1 are a standard set of generators for a group normalized as above they can be written as:

$$C_1 = \begin{pmatrix} d\lambda & -b/\lambda \\ -c\lambda & a/\lambda \end{pmatrix} \quad C_2 = \begin{pmatrix} a & b \\ c & d \end{pmatrix} \quad C_2C_1 = \begin{pmatrix} \lambda & 0 \\ 0 & 1/\lambda \end{pmatrix}$$

where

$$\left|\frac{a}{\lambda} + d\lambda\right| > 2, \quad |a+d| > 2 \quad \text{and} \quad |\lambda + 1/\lambda| > 2 \ .$$

The normalization conditions are:

$$a + b = c + d, \quad |\lambda| < 1 \quad \text{and} \quad 0 < -b/c < 1 \ ,$$

and the geometric conditions are

$$0 < \frac{a}{c} < \frac{b}{d} < -\frac{b}{c} < 1 < -\frac{a}{b} < -\frac{d}{c} < \frac{\lambda d - \frac{a}{\lambda} \pm 2\sqrt{(d\lambda + \frac{a}{\lambda})^2 - 4}}{-2c\lambda}$$

The signs of the traces of C_1 and C_2 are not determined; however, once they are chosen, the sign of the trace of C_2C_1 is determined.

LEMMA 1. *Assume that the traces of* C_2C_1 *and* C_2 *are both either positive or negative. Then the trace of* C_1 *is negative.*

Proof: It follows from (2) that either

I. $a > 0, b < 0, c > 0, d < 0$ or

II. $a < 0, b > 0, c < 0, d > 0$.

Since $a/c < -d/c$, case I implies $a + d < 0$ and case II implies $a + d > 0$.

Assume $a + d > 2$ and $\lambda > 0$. Since $-d/c < (\lambda d - a/\lambda)/(-2c\lambda)$ and $-2c\lambda > 0$, $2d\lambda > \lambda d - a/\lambda$ and $\lambda d + a/\lambda < 0$. Similarly assume $a+d < -2$ and $\lambda < 0$. Then $-2c\lambda > 0$ and from $-d/c < (\lambda d - a/\lambda)/(-2c\lambda)$ it follows that $2\lambda d < \lambda d - a/\lambda$ and $\lambda d + a/\lambda < 0$. q.e.d.

The geodesics on S corresponding to the generators C_1, C_2 and C_2C_1 go around the "holes" of S. In the following sections some of these holes will be "attached" to the holes of surfaces of signature (1; 1). The elements corresponding to the holes on each of the surfaces must agree since this "attaching" is achieved by taking the free product of the two groups and amalgamating over the cyclic subgroup generated by the element corresponding to the hole. For a surface of signature (1; 1) this element is the commutator of its generators; therefore the sign of its trace is independent of the signs of the traces of the generators. In fact, since the commutator is hyperbolic, formula (4) of Section VI shows that this sign is always negative. It is for this reason that the traces are always taken as negative in the construction of a (0; 3) group. Lemma 1 assures that this doesn't involve any loss of generality.

Let $k_1 < -1$, $k_2 < -1$ and $k_3 < -1$ be given; let $K_1 = \sqrt{k_1^2 - 1}$, $K_2 = \sqrt{k_2^2 - 1}$ and $K_3 = \sqrt{k_3^2 - 1}$. Consider the matrices:

$$C_2C_1 = \begin{pmatrix} k_3 + K_3 & 0 \\ 0 & k_3 - K_3 \end{pmatrix} \qquad C_2 = \begin{pmatrix} k_2 - J & L \\ M & k_2 + J \end{pmatrix}$$

where J, L and M are to be determined. The trace of $C_1 = C_2^{-1}(C_2C_1)$ is to be $2k_1$, hence $2k_1 = (k_3 + K_3)(k_2 + J) + (k_3 - K_3)(k_2 - J)$ and $J = (k_1 - k_2k_3)/K_3 < 0$. If C_2 is to have a fixed point at 1, $L = M + 2J$. If $\det C_2 = 1$, $k^2 - J^2 - LM = 1$ and $M^2 + 2JM + J^2 - K_2^2 = 0$; $M = -J \pm K_2$ and $L = J \pm K_2$. Since the group being constructed is to satisfy (2),

$M = -J \pm K_2 > 0, \;\; 0 < (-J \mp K_2)/(-J \pm K_2) < 1$ and $-J \mp K_2 < -J \pm K_2$;

therefore $M = -J + K_2$, $L = J + K_2$.

$$C_1 = \begin{pmatrix} k_1 + k_2 K_3 + k_3 J & -(K_2 + J)(k_3 - K_3) \\ (k_3 + K_3)(K_2 - J) & k_1 - k_2 K_3 - k_3 J \end{pmatrix} .$$

It is possible to check that all the conditions are fulfilled now. That is,

THEOREM 1. *The elements* C_1, C_2 *and* $C_2 C_1$ *constructed from* k_1, k_2 *and* k_3 *generate a Fuchsian group of signature* $(0; 3)$.

Proof: Since all the conditions (2) are fulfilled a canonical modular polygon can be constructed for these generators and therefore the group they generate is Fuchsian.

These are exactly the moduli Fricke obtains. In fact, he uses the same normalization and derives essentially these transformations as generators of the group. In order to perform his construction, Fricke needs to consider complicated inequalities involving a commutator of the generators.

IV. Construction of a surface of signature (0 ; 4)

If $\mathcal{S} = \{C_1, C_2, C_3, C_4\}$ is a standard set of generators for a surface of signature $(0; 4)$ the axes of the generators are mutually disjoint and the fixed points occur on the real axis as shown in Figure 4. The axis of C_2C_1 divides U so that the axes of C_3 and C_4 are separated from those of C_1 and C_2.

LEMMA 2. *Given a group of signature* $(0; 4)$ *and a standard set of generators* $\mathcal{S} = \{C_1, C_2, C_3, C_4\}$ *the axis of* $C_3C_2 = (C_1C_4)^{-1}$ *separates the axes of* C_2 *and* C_3 *from those of* C_1 *and* C_4 *and so intersects the axis of* C_2C_1.

Proof: Since $\{C_1, C_2, C_3, C_4\}$ is a standard set of generators the fixed points occur as shown in Figure 4. Consequently $C_2(\infty)$, $C_2(0)$, $C_1(\infty)$,

$C_1(0)$, $C_3(\infty)$, $C_3(0)$, $C_4(\infty)$ and $C_4(0)$ are also in the positions shown in the figure. Let x be a fixed point of C_3C_2. If $x \in (p_3, q_2)$, $C_2(x) \in (C_2(\infty), q_2)$ and $C_3C_2(x) \in (C_3(0), C_3(\infty))$. Therefore $x \notin (p_3, q_2)$. If $x \in (-\infty, q_4)$, $C_2(x) \in (C_2(\infty), C_2(0))$ and $C_3C_2(x) \in (C_3(0), C_3(\infty))$. Therefore $x \notin (-\infty, q_4)$. If $x \in (p_1, \infty)$, $C_2(x) < 0$ and $C_3C_2(x) < C_3(0) < p_1$. Therefore $x \notin (p_1, \infty)$. Consequently $x \in (q_4, p_3)$ or (q_2, p_1) or both. There is exactly one fixed point x in each of these intervals. Note that $C_3C_2(\infty) < C_3C_2(0)$ and hence that C_3C_2 moves 0 and ∞ in opposite directions. This is impossible unless only the attracting fixed point of C_3C_2 lies between them. Therefore $p_{32} \in (q_4, p_3)$ and $q_{32} \in (q_2, p_1)$. Now since the fixed points of C_3C_2 and C_2C_1 alternate, their axes must intersect.

The group G of signature (0; 4) which will be constructed will be a free product with amalgamation:

$$G = H * H' \operatorname{am}\{C_2C_1\}$$

where

(i) H is the group constructed in Section III,

(ii) $\{C_2C_1\}$ is the cyclic subgroup generated by C_2C_1 and

(ii) H' is generated by elements C_3 and C_4 such that $C_4C_3 = (C_2C_1)^{-1}$ and the fixed points of C_3 and C_4 occur as in Figure 4.

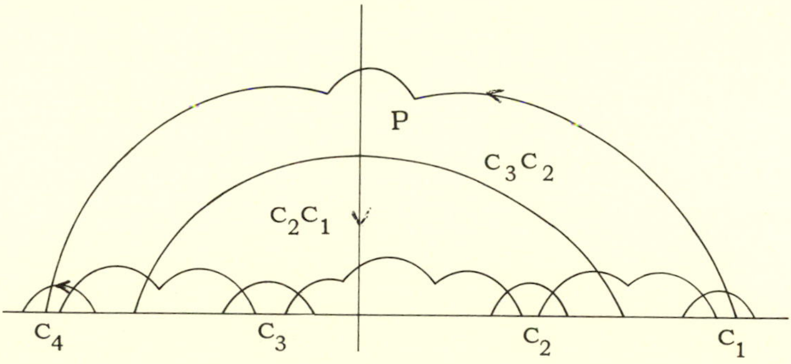

Figure 4. Surface of signature (0, 4) and a canonical modular polygon P for the group representing it.

Normalize so that C_2C_1 has attracting fixed point at 0, repelling fixed point at ∞ and C_2 has repelling fixed point at 1.

To construct H' proceed as follows. Given $k_3 < -1$, $k_4 < -1$, and $k_{34} = k_{12} < -1$, construct a group $\tilde{H}$ in the same manner as H was constructed. $\tilde{H}$ can be conjugated by a linear fractional transformation R which sends 0 to ∞, ∞ to 0 and the positive real axis to the negative real axis. R is not uniquely determined yet; it depends on one parameter. But $6g - 6 + 3m = 3 \cdot 4 - 6 = 6$ and only 5 parameters have been used so far. Fricke uses the trace of R as his sixth parameter. But, R doesn't belong to G.

Let the generators of $\tilde{H}$ be $\tilde{C}_3$ and $\tilde{C}_4$. Let $C_3' = R^{-1}\tilde{C}_3 R$ and $C_4' = R^{-1}\tilde{C}_4 R$. Since

$$(C_3' C_4')^{-1} = C_2C_1 = \begin{pmatrix} \lambda & 0 \\ 0 & 1/\lambda \end{pmatrix}$$

write

$$C_3' = \begin{pmatrix} \alpha & \beta \\ \gamma & \delta \end{pmatrix} \qquad C_4' = \begin{pmatrix} \lambda\delta & -\beta\lambda \\ -\gamma/\lambda & \alpha/\lambda \end{pmatrix}$$

where

$$\alpha + \delta = 2k_3, \quad \lambda\delta + \alpha/\lambda = 2k_4, \quad \alpha\delta - \beta\gamma = 1 . \tag{3}$$

$$C_3'C_2 = \begin{pmatrix} a\alpha + \beta c & \alpha b + \beta d \\ a\gamma + c\delta & b\gamma + d\delta \end{pmatrix} .$$

a, b, c, d are known constants (from Section III), α and δ can be obtained from (3) and β and $\gamma = (\alpha\delta - 1)/\beta$ depend analytically on the free parameter of R.

$$\operatorname{tr} C_3'C_2 = a\alpha + \beta c + b\gamma + d\delta = P\beta + \frac{Q}{\beta} + R, \quad P, Q \text{ and } R \text{ constant.}$$

If $(d \operatorname{tr} C_3'C_2)/d\beta = P - Q/\beta^2 = 0$, $\beta^2 = Q/P$, compute that the conditions on the fixed points described in Lemma 1 are not satisfied; the axes of $C_3'C_2$ and C_1 intersect. (The calculation is easier with a different normalization.) Therefore $k_{32} = (\operatorname{tr} C_3'C_2)/2 < -1$ can be taken as the sixth parameter and H' determined uniquely.

THEOREM 2: *Given* $k_1 < -1$, $k_2 < -1$, $k_3 < -1$, $k_4 < -1$, $k_{12} < -1$ *and* $k_{23} < -1$, *construct groups* $H = \langle C_1, C_2 \rangle$, $H' = \langle C_3, C_4 \rangle$ *and* $G = H * H' \text{ am} \{C_2C_1 = C_4C_3\}$. G *is a Fuchsian and represents a marked surface of signature* (0 ; 4). *Every such marked surface is so representable.*

Proof: H and H′ are Fuchsian by Theorem 1. Since the fixed points of their generators separate properly their canonical modular polygons can be combined and one obtained for G. The theorem now follows.

Figure 5 shows a surface of signature (0 ; 4) with the geodesics whose lengths are moduli.

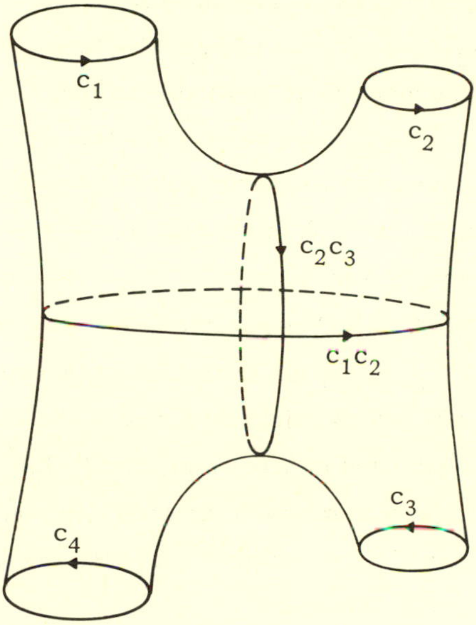

Figure 5. A surface of signature (0, 4) with geodesics whose lengths are moduli.

V. Construction of surfaces of signature (0 ; m)

Let S be a surface of signature (0 ; m) and let α be a canonical homotopy basis for S. α consists of m simple curves $c_1, c_2, \ldots, c_m$ with pairwise intersection number zero and such that each c_i goes around a different hole and $c_1 c_2 \cdots c_m \sim 1$. Consider the curves c_1c_2, $c_1c_2c_3, \ldots, c_1c_2c_3 \ldots c_m$ these are all simple dividing curves on the surface. Cutting the surface along them renders it into $m-2$ pieces, each of which is a surface of signature (0;3). See Figure 5. Let $\{C_1, \ldots, C_m\}$ be the standard set of generators of the group G representing S which comes from α. The fixed points of the generators and the fixed points of C_2C_1, $C_3C_2C_1, \ldots, C_{m-2} \ldots C_1$ occur as shown in Figure 6. This is proved in Keen [12].

Given $k_1 < -1$, $k_2 < -1$, $k_3 < -1$, $k_{123} < -1$, $k_{12} < -1$, and $k_{23} < -1$, construct the group G of Section IV where C is now $C_3C_2C_1$. Also let $H = H_1$ and $H' = H_2$. Conjugate G so that the attracting fixed point of $C_3C_2C_1$ is 0, the repelling fixed point of $C_3C_2C_1$ is ∞ and the repelling fixed point of C_3 is 1. A new group H_3 generated by $C_3C_2C_1$ and $\tilde{C}_4$ can be constructed given $k_{123} < -1$, $k_4 < -1$ and $k_{1\tilde{2}34} < -1$. The fixed points of $C_3C_2C_1$ are 0 and ∞, the fixed point of $\tilde{C}_4$ are negative and the fixed points of $\tilde{C}_4C_3C_2C_1$ are to the left of those of $\tilde{C}_4$. Again, H_3 depends on one parameter (which Fricke chooses as the fixed point of $\tilde{C}_4$) and again compute, as indicated in Section IV, that tr $\tilde{C}_4C_3$ is a one-to-one analytic function of the parameter, given the conditions on the fixed points. Therefore if $-1 > k_{34} = \operatorname{tr} C_4C_3/2$ is given, C_4, H_3 and $G_3 = G * H_3 \text{ am } \{C_3C_2C_1\}$ can be determined.

Construct $H_4, H_5, \ldots, H_{m-2}$ as above and let $G_i = G_{i-1} * H_i \text{ am } \{C_iC_{i-1} \cdots C_1\}$. $G = G_{m-2} = G_{m-3} * H_{m-2} \text{ am } \{C_{m-2} \cdots C_1\}$ is a free group on the generators $C_1, \ldots, C_{m-1}$ and is the desired group. This is summarized in

THEOREM 3. *The Teichmüller space of a marked surface of signature* (0 ; m) *is real analytically homeomorphic to a product of* $3m-6$ *intervals* $I = (-\infty, -1)$.

VI. Construction of a surface of signature (1 ; 1)

Let F be a group representing a surface of signature (1 ; 1). F is a free group on two generators A and B such that their axes intersect and their commutator C is hyperbolic. Normalize so that the attracting fixed point of A is at -1, the repelling fixed point of A is at 1, the attracting fixed point of B is at $-e^{i\alpha}$ and the repelling fixed point of B is at $e^{i\alpha}$. If $t > 1$, $s > 1$, $T = \sqrt{t^2-1}$, $S = \sqrt{s^2-1}$, A and B can be written as:

$$A = \begin{pmatrix} t & -T \\ -T & t \end{pmatrix} \qquad B = \begin{pmatrix} s & -Se^{i\alpha} \\ -Se^{-i\alpha} & s \end{pmatrix} .$$

Then $2r = \operatorname{tr} BA = 2(st + ST \cos \alpha)$ and $|r| > 1$.

(4) $2k = \operatorname{tr} C = 2(s^2 + t^2 - s^2t^2 + S^2T^2 \cos 2\alpha) = 4t^2 + 4s^2 + 4r^2 - 8rst - 2$.

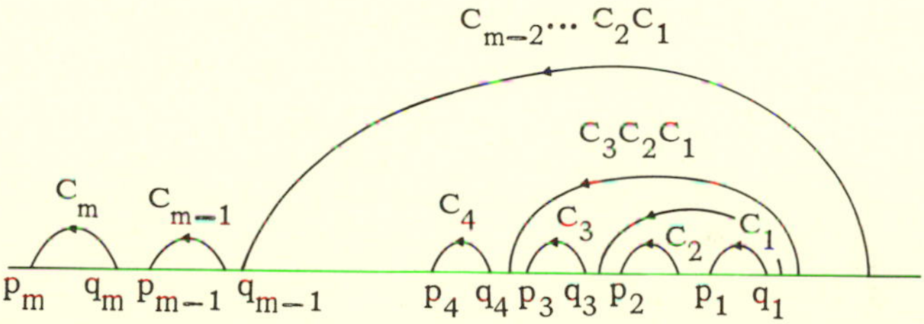

Figure 6. The axes of some elements of a group of signature (0, m).

If C is hyperbolic $|2k| > 2$ but in fact it follows from (4) that $2k < -2$.

In [12], t, α and k are the moduli. They can vary as follows: $1 < t < \infty$, $-1 > k > -\infty$ and $0 < \alpha < \pi$. A canonical modular polygon for G is constructed from them. Given $|r| > 1$, $|s| > 1$ and $|t| > 1$, it is possible to solve for α whenever the inequality

(5) $$r^2 + s^2 + t^2 - 2rst < 0 \text{ holds.}$$

However, this inequality is just the assertion that C is hyperbolic. Since the above mentioned construction can now be performed, $|r|$, $|s|$ and $|t|$

are half traces which can be chosen as moduli. If t and s are greater than 1, it is clear from (5) that r is also greater than 1. This is summarized in:

THEOREM 4: *Given* $r > 1$, $s > 1$ *and* $t > 1$ *such that* $r^2+s^2+t^2-2rst < 0$, *the group* $F = \langle A, B \rangle$ *with* A *and* B *as above and* $\alpha = \cos^{-1}(\frac{r-st}{ST})$ *is Fuchsian and represents a surface of signature* $(1;1)$. *Every such surface is so representable.*

These are precisely the moduli Fricke obtains although his constructios are completely different. He reduces the $(1;1)$ case to a $(0;4)$ case through the introduction of an elliptic transformation of order two in order to prove these are moduli.

VII. Construction of a surface of signature (2; 0)

Let S be a surface of signature $(2;0)$ and let α be a canonical homotopy basis on S. α consists of four simple closed curves a_1, b_1, a_2, b_2. The curve $c = a_1 b_1 a_1^{-1} b_1^{-1}$ is a simple dividing cycle on S and cutting the surface along c renders it into two pieces, each a surface of signature $(1;1)$. Given $r_1 > 1$, $s_1 > 1$ and $t_1 > 1$ such that (5) holds, construct A_1 and B_1 generating a group $\tilde{F}_1$ of signature $(1;1)$ as in Section VI. Given $r_2 > 1$, $s_2 > 1$ and $t_2 > 1$ such that

$$r_1^2 + s_1^2 + t_1^2 - 2r_1 s_1 t_1 = r_2^2 + s_2^2 + t_2^2 - 2r_2 s_2 t_2 < 0,$$

construct A_2 and B_2 generating a group $\tilde{F}_2$ of signature $(1;1)$ as in Section VI. Note that there are six parameters but that they must satisfy a relation.

Renormalize as follows. Let R_1 send the attracting fixed point of $C_1 = B_1^{-1} A_1^{-1} B_1 A_1$ to 0, the repelling fixed point of C_1 to ∞ and the attracting fixed point of A_1 to 1. Consider $R_1 \tilde{F}_1 R_1^{-1}$ and call it F_1. Consider $R_2(x)$ which sends the repelling fixed point of $C_2 = B_2^{-1} A_2^{-1} B_2 A_2$ to 0, the attracting fixed point of C_2 to ∞ and the attracting fixed point of A_2 to some negative number x. Let $R_2 \tilde{F}_2 R_2^{-1} = F_2(x)$. Define $G(x) = F_1 * F_2(x) \operatorname{am}\{C_1 = C_2^{-1}\}$. Since three points are already prescribed, G is fully normalized.

For any Fuchsian group of signature (2 ; 0) with a standard sequence of generators $\{A_1, B_1, A_2, B_2\}$. the fixed points of A_1, B_1, A_2, B_2 and $C = B_1^{-1}A_1^{-1}B_1A_1$ occur as shown in Figure 7.

LEMMA 3. *The fixed points of* A_2A_1 *occur as in Figure 7 and therefore the axis of* A_2A_1 *intersects the axis of* C.

Proof: Consider the canonical modular polygon P for the group G shown in Figure 7. $A_2(P)$ lies under the axis of B_2. Consequently

$$A_2((q_{A_2^{-1}B_2A_2}, \infty) \cup (-\infty, p_{A_2})) \subset (q_{B_2}, p_{A_2}) .$$

Since A_1 leaves $(1, q_{A_1})$ invariant, there can be no fixed point of A_2A_1 in $(1, q_A)$.

Looking at P again, $A_1(P)$ lies under the axis of B_1 and so

$$A_1((q_{A_1^{-1}B_1A_1}, \infty) \cup (-\infty, 1)) \subset (q_{B_1}, 1) .$$

$A_2(q_{B_1}, 1) < x$ and A_2A_1 cannot have a fixed point in (x, 1).

$$A_1((-\infty, q_{B_2}) \cup (q_{A_1^{-1}B_1A_1}, \infty)) \subset (q_{B_1}, 1)$$

and $A_2(q_{B_1}, 1) < p_{A_2}$.

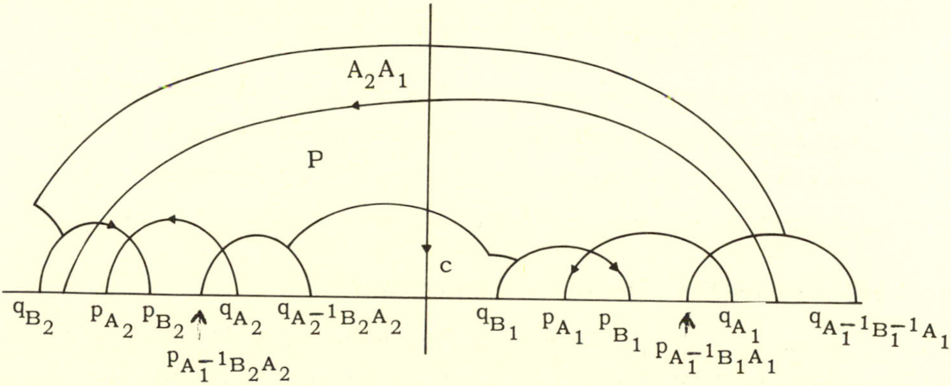

Figure 7. A group of signature (2, 0) and its canonical modular polygon.

so that A_2A_1 cannot have a fixed point in either of these intervals. The fixed points of A_2A_1 must lie in either (q_{B_1}, x) or $(q_{A_1}, q_{A_1^{-1}B_1A_1})$ or both.

$A_1(0) \subset (q_{B_1}, p_{A_1})$ and $A_2A_1(0) \subset (q_{B_2}, p_{A_2})$ so that A_2A_1 moves 0 to the left. $A_1(\infty) \subset (q_{B_1}, p_{A_1})$ and $A_2A_1(\infty) \subset (q_{B_1}, p_{A_1})$ so that A_2A_1 moves ∞ to the right. There can be only one fixed point in between, so the attracting fixed point of A_2A_1 lies in (q_{B_2}, p_{A_2}) and the repelling fixed point lies in $(q_{A_1}, q_{A_1^{-1}B_1A_1})$.

Let $\tau = (\mathrm{tr}\, A_2A_1)/2$; Lemma 1 implies τ is less than -1. Again compute that $d\tau/dx = 0$ implies the axis of A_2A_1 intersects the axis of A_2. By Lemma 3 this cannot happen and τ is a one-to-one analytic function of x.

THEOREM 5. *Given the seven real numbers* $r_1 > 1, s_1 > 1, t_1 > 1$; $r_2 > 1, s_2 > 1, t_2 > 1$ *and* $\tau < -1$ *such that*

$$r_1^2 + s_1^2 + t_1^2 - 2r_1s_1t_1 = r_2^2 + s_2^2 + t_2^2 - 2r_2s_2t_2 < 0,$$

construct a set of elements A_1, B_1, A_2, B_2 *such that*

$$B_2^{-1}A_2^{-1}B_2A_2B_1^{-1}A_1^{-1}B_1A_1 = 1$$

as above. The group they generate is Fuchsian and of signature (2 ; 0).

Proof: Construct the elements A_1, B_1, A_2, B_2 in the manner described in this section. Since the fixed points of the elements occur in the order given in Figure 7 a canonical modular polygon for the group they generate exists implying that the group is Fuchsian and of signature (2 ; 0).

VIII. Construction of a surface of signature (1; 2)

Given $r_1 > 1, s_1 > 1$ and $t_1 > 1$ such that (5) holds, construct $\tilde{F}$ generated by $\tilde{A}, \tilde{B}, \tilde{C} = \tilde{B}^{-1}\tilde{A}^{-1}\tilde{B}\tilde{A}$ as described in Section VI.

Determine $k_{12} = \text{trace } \tilde{\tilde{C}}/2 < -1$ from (4). Then given $k_1 < -1$ and $k_2 < -1$, construct the group H generated by C_1, C_2 and C_2C_1 as described in Section III. Conjugate $\tilde{\tilde{F}}$ by a transformation R which sends the unit circle onto the upper half plane and the upper semi-circle onto the upper left quadrant. There is a one parameter family of transformations R which will accomplish this. R can be written in terms of the image x, $x < 0$ of of the attracting fixed point of $B = R\tilde{\tilde{B}}R^{-1}$. Let $F = R\tilde{\tilde{F}}R^{-1}$. Figure 8 results.

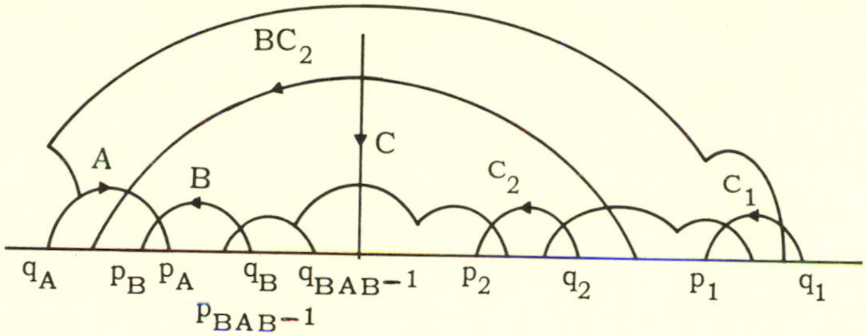

Figure 8.

LEMMA 3. *If* G *is a Fuchsian group of signature* (1 ; 2) *and* $\mathcal{S} = \{A, B, C, C_2^{-1}, C_1^{-1}\}$ *is a standard set of generators normalized as in Figure 8, then the axis of* BC_2 *is also as shown in the diagram.*

Proof: Since G is Fuchsian the canonical modular polygon drawn in Figure 8 exists. B moves all points to the right of $q_{BAB^{-1}}$ into the interval (q_A, x). Since C_2 sends $(q_{BAB^{-1}}, 1)$ into itself, BC_2 cannot have a fixed point in the interval $(q_{BAB^{-1}}, 1)$.

$C_2(-\infty, 0) \subset (0, p_2)$ and therefore $BC_2(x, q_{BAB^{-1}}) < p_B$ so that BC_2 cannot have a fixed point in $(x, q_{BAB^{-1}})$. Moreover $BC_2(-\infty, q_A) \subset (q_A, p_B)$ so that BC_2 cannot have a fixed point there.

$C_2(p, \infty) \subset (0, p_2)$ so that $BC_2(p, \infty) \subset (q_A, p_B)$ implying that the fixed points of BC_2 lie in either (q_A, p_B) or in $(1, p_1)$ or both. BC_2 moves

0 to the left and ∞ to the right so there must be one fixed point in each of these intervals.

Compute as before, that under the geometric conditions imposed on the fixed points, $\tau = (\mathrm{tr}\ BC_2)/2 < -1$ can be used as a sixth parameter to construct $G = F * H \ \mathrm{am}\{C\}$.

THEOREM 6. *Given the six real numbers* $r_1 > 1$, $s_1 > 1$, $t_1 > 1$, $k_1 < -1$, $k_2 < -1$, *and* $\tau < -1$, *such that* (5) *holds, determine* $k_{12} < -1$ *and construct elements* A, B, C, C_1 and C_2 *such that* $C_2C_1B^{-1}A^{-1}BA = 1$ *as above. The group they generate is Fuchsian and represents a surface of signature* (1; 2).

Proof: Perform the construction indicated in this section. The separation properties of the fixed points imply the construction of the canonical modular polygon for the group with these generators can be performed. Consequently the group is Fuchsian.

IX. Construction of a surface of signature (g; m)

Let α be a canonical homotopy basis on a surface S of signature (g; m). It consists of $2g + m$ simple closed curves a_i, b_i, c_j, $i = 1, \ldots, g$, $j = 1, \ldots, m$. The a_i's and the b_i's go around the handles and the c_j's go around the holes. Let $c_i = a_ib_ia_i^{-1}b_i^{-1}$. Cutting the surface S along the c_i's renders it into g pieces of signature (1; 1) and 1 piece of signature $(0; g + m - 2)$.

To construct a Fuchsian group of signature (g; m) proceed as follows. Given 3g numbers $r_i > 1$, $s_i > 1$ and $t_i > 1$, $i = 1, \ldots, g$ such that $r_i^2 + s_i^2 + t_i^2 - 2r_is_it_i < 0$ construct groups F_i of signature (1; 1) in the manner of Section VI. Let the generators of the constructed groups be A_i, B_i, $C_i = B_i^{-1}A_i^{-1}B_iA_i$. Determine $k_i = \mathrm{trace}\ C_i/2$. Given $2g + 3m - 6$ real numbers less than -1 and the g numbers k_i, construct a group of signature $(0; g + m - 2)$ such that the g numbers k_i are half traces of elements C_i which correspond to g of the holes, using the methods of

Section V. Given g more numbers $\tau_1, \ldots, \tau_g$ greater than 1 determine the free products

$$G_1 = F_1 * H \text{ am}\{C_1\}$$
$$G_2 = F_2 * G_1 \text{ am}\{C_2\}$$
$$G = G_g = F_g * G_{g-1} \text{ am}\{C_g\}$$

using the procedure developed in Section VII.

THEOREM 7. *Given* 3g *numbers* r_i, s_i, t_i, $i = 1, \ldots, g$ *greater than* 1, *such that* $r_i^2 + s_i^2 + t_i^2 - 2r_i s_i t_i < 0$, g *numbers* τ_j, $j = 1, \ldots, g$ *greater than* 1 *and* $2g + 3n - 6$ *numbers* k_ℓ, $\ell = 1, \ldots, 2g + 3m - 6$ *less than* -1 *construct the group* G *as above. It is Fuchsian and represents a surface* S *of signature* (g ; m). *The numbers* r_i, s_i, t_i *and* τ_i *are half traces of elements of* G *which correspond to nondividing (homologically non-trivial) cycles on the surface* S *and the numbers* k_ℓ *are half traces of elements of* G *which correspond to dividing (homologically trivial) cycles on* S. *The numbers* s_i, t_i, $i = 1, \ldots, g$ *and* k_ℓ, $\ell = 1, \ldots, g$ *are half traces of a standard set of generators for* G. *Every such group is so representable. (The case* (2 ; 0) *is an exception and is treated separately in Section VIII).*

Proof: All the previous constructions insure that the fixed points of the elements of the group separate in such a way that a canonical modular polygon can be constructed for the group. The group is therefore Fuchsian and represents a surface of signature (g ; m).

BIBLIOGRAPHY

[1] L. V. Ahlfors, Finitely generated Kleinian groups, *Amer. J. Math.*, LXXXVI, No. 2 (1964).

[2] ________ and L. Bers, Riemann's mapping theorem for variable metrics, *Ann. of Math.*, 72 (1960), 385-408.

[3] P. Appell and E. Goursat, *Théorie des Fonctions Algébraiques d'une Variable*, Vol. II, Fonctions Automorphes, P. Fatou, Gautier-Villars, Paris, 1930.

[4] L. Bers, Uniformization by Beltrami Equations, *Comm. Pure Appl. Math.*, XIV, No. 3.

[5] ________, Quasi-conformal mappings and Teichmüller's theorem, *Analytic Functions*, Princeton, 1960.

[6] C. Caratheodory, *Theory of Functions of a Complex Variable*, Vol. I, Chelsea, 1960.

[7] H. D. Coldewey and H. Zieschang, *Der Raum der markierten Riemannschen Flächen*, to appear.

[8] W. Fenchel and J. Nielsen, *Discontinuous Groups of Non-euclidean Motions*, unpublished manuscript.

[9] L. Ford, *Automorphic Functions*, Chelsea, 1929.

[10] R. Fricke and F. Klein, *Vorlesungen über die Theorie der Automorphen Funktionen*, Leipzig, 1926.

[11] L. Keen, Canonical polygons for finitely generated Fuchsian groups, *Acta Math.*, (115), 1966).

[12] ________ Intrinsic moduli on Riemann surfaces, *Ann. Math.*, 84 (1966).

Lehman College of
City University of N. Y.

EICHLER COHOMOLOGY AND THE STRUCTURE OF FINITELY GENERATED KLEINIAN GROUPS

by

Irwin Kra[1]

1. Introduction.

There are a number of theorems on the structure of finitely generated Kleinian groups whose proofs depend on the cohomology theory introduced by Eichler [9] for Fuchsian groups. Foremost among these are Ahlfors' finiteness theorem [1] and Bers' area inequalities [6]. Also, the structure of these cohomology groups (Ahlfors [3], Kra [13] and [14]) is of interest, since it may reveal further information about the Kleinian groups. In this note we present an outline of this cohomology theory as developed by the author in [13] and [14] and derive from it the finiteness theorem and area inequalities. Along the way we shall encounter several open problems, and we will state several natural conjectures. It should be pointed out that most of the results have previously appeared in the literature ([1], [6], [3], [13], [14]). Section 7 (the case $q = 1$) as well as most of section 8 is new. Previously, the Fuchsian case has been treated by Eichler [9], Gunning [11], and Bers [4], and recently by Husseini and Knopp [12].

An alternate method for describing the structure of these cohomology groups may be found in Ahlfors' paper [3]. The description of Ahlfors uses meromorphic Eichler integrals. In our approach only holomorphic Eichler integrals are considered. We must however, introduce the generalized Beltrami coefficients considered by Bers [6]. In section 8 we show how to

[1] Research partially supported by NSF grant GP - 12467.

recover Ahlfors' results [3] by our methods. This section contains a theorem (8.3) that may be considered as a Riemann-Roch theorem for (meromorphic) Eichler integrals.

2. Cohomology.

Let Γ be a (non-elementary) Kleinian group (that is, Γ is a group of fractional linear transformations of the extended complex plane, $\mathbb{C} \cup \{\infty\}$, whose limit set Λ is nowhere dense and consists of more than two points), and $q \geq 1$ an integer.

If D is an open subset of $\mathbb{C} \cup \{\infty\}$ and if $f: D \to \mathbb{C} \cup \{\infty\}$ is a holomorphic mapping, then for every function ϕ on $f(D)$ we define a function on D via

$$(2.1) \qquad (f^*_{r,s}\phi)(z) = \phi(f(z))\, f'(z)^r\, \overline{f'(z)}^s, \qquad z \in D .$$

We require only that r and s be half integers such that $r+s$ is an integer. It is clear that we have introduced a contravariant functor. We shall abbreviate $f^*_{r,0}$ by f^*_r.

Let Π_{2q-2} denote the vector space of complex polynomials in one variable of degree $\leq 2q-2$. The group Γ acts on the right on Π_{2q-2} via

$$(2.2) \qquad p\gamma = \gamma^*_{1-q}p, \quad p \in \Pi_{2q-2}, \qquad \gamma \in \Gamma .$$

A mapping $\chi: \Gamma \to \Pi_{2q-2}$ is called a *cocycle* if

$$(2.3) \qquad \chi_{\gamma_1 \circ \gamma_2} = \chi_{\gamma_1}\gamma_2 + \chi_{\gamma_2}, \qquad \gamma_1 \text{ and } \gamma_2 \in \Gamma ,$$

here χ_γ stands for $\chi(\gamma)$. If $p \in \Pi_{2q-2}$, its *coboundary* is the cocycle

$$(2.4) \qquad \gamma \longmapsto p\gamma - p, \qquad \gamma \in \Gamma .$$

The (first) *cohomology space* $H^1(\Gamma, \Pi_{2q-2})$ is the space of cocycles factored by the space of coboundaries.

LEMMA 2.1. *If* Γ *is generated by* N *elements, then*

$$\dim H^1(\Gamma, \mathbb{C}) \leq N \quad ,$$

and

$$\dim H^1(\Gamma, \Pi_{2q-2}) \leq (2q-1)(N-1) \quad \textit{for} \quad q \geq 2 \ .$$

Equalities hold whenever Γ *is a free group on* N *generators.*

Proof: Note that if $q = 1$, then $H^1(\Gamma, \Pi_{2q-2})$ consists of the vector space of homomorphisms from Γ into the additive complex numbers. A proof for $q \geq 2$ appears in Bers [6]. In this case, it suffices to show that the map from Π_{2q-2} to coboundaries is injective. Assume that there is a $p \in \Pi_{2q-2}$ such that

(2.5) $$p(\gamma z)\gamma'(z)^{1-q} = p(z), \qquad \gamma \in \Gamma, \qquad z \in \mathbb{C}.$$

Clearly, $\deg p > 0$. If p vanishes at z_0, then by (2.5) p also vanishes at $\gamma(z_0)$ for all $\gamma \in \Gamma$. Thus p has infinitely many zeros (because Γ is non-elementary), and hence $p = 0$.

3. Automorphic forms.

Let Δ be a Γ-invariant union of components of the region of discontinuity Ω of Γ, and λ the Poincaré metric on Δ. For convenience we assume that $\infty \in \Lambda$, the limit sets of Γ. A holomorphic function ϕ on Δ is called an *automorphic form of weight* $(-2q)$ on Δ if

(3.1) $$\gamma_q^* \phi = \phi \quad \text{all } \gamma \in \Gamma \ .$$

An automorphic form ϕ of weight $(-2q)$ on Δ is called *integrable* if

(3.2) $$\iint_{\Delta/\Gamma} \lambda(z)^{2-q} |\phi(z)\, dz \wedge d\bar{z}| < \infty \ .$$

We denote the Banach space of integrable automorphic forms on Δ by

$A_q(\Delta, \Gamma)$. The form ϕ is called *bounded* if

$$\sup\{\lambda(z)^{-q}|\phi(z)| ; \ z \in \Delta\} < \infty . \tag{3.3}$$

The Banach space of bounded automorphic forms on Δ is denoted by $B_q(\Delta,\Gamma)$.

For $\phi \in A_q(\Delta,\Gamma)$ and $\psi \in B_q(\Delta,\Gamma)$, we define the *Petersson scalar product* by

$$(\phi,\psi)_\Gamma = \iint_{\Delta/\Gamma} \lambda(z)^{2-2q}\phi(z)\overline{\psi(z)}\,dz \wedge d\bar{z} . \tag{3.4}$$

It is well known (see, for example, Ahlfors [1] for $q = 2$, Bers [4] for Γ a Fuchsian group) that for $q \geq 2$ the Petersson scalar product establishes an anti-linear topological isomorphism between $B_q(\Delta,\Gamma)$ and the dual space of $A_q(\Delta,\Gamma)$.

Remark. A duality theorem between $A_1(\Delta,\Gamma)$ and $B_1(\Delta,\Gamma)$ is not readily accessible. We shall interpret $A_1(\Delta,\Gamma)$ and $B_1(\Delta,\Gamma)$ as the Hilbert space of square integrable automorphic forms of weight (–2) with inner product defined by (3.4).

For fixed Δ, the above definitions and assertions remain valid for any subgroup of the group Γ. In particular, they hold for the trivial subgroup. We shall abbreviate $A_q(\Delta)$ for $A_q(\Delta,\{1\})$ etc... . Also, the Petersson inner product may clearly be extended to *measurable automorphic forms.*

For $\phi \in A_q(\Delta)$, define the *Poincaré series* of ϕ via

$$(\textcircled{H}_q\phi)(z) = \sum_{\gamma\in\Gamma} \phi(\gamma z)\gamma'(z)^q, \qquad z \in \Delta . \tag{3.5}$$

It is well known that

$$\textcircled{H}_q : A_q(\Delta) \to A_q(\Delta,\Gamma)$$

is a surjective, norm decreasing, linear mapping for $q \geq 2$.

Let $a_1, \ldots, a_{2q-2}$ $(q \geq 2)$ be fixed finite distinct points in Λ. For $z \in \Lambda - \{a_1, \ldots, a_{2q-2}\}$ we set

$$\phi^z(\zeta) = \frac{1}{2\pi i} \frac{(z-a_1)\cdots(z-a_{2q-2})}{(\zeta-z)(\zeta-a_1)\cdots(\zeta-a_{2q-2})}, \qquad \zeta \in \Delta . \tag{3.6}$$

It is easy to check that $\phi^z \in A_q(\Delta)$, $q \geq 2$.

PROBLEM 3.1. (a) Is $\{\phi^z;\ z \in \Lambda - \{a_1, \ldots, a_{2q-2}\}\}$ dense in $A_q(\Delta)$?
(b) Is $\{\text{Ⓗ}_q \phi^z;\ z \in \Lambda - \{a_1, \ldots, a_{2q-2}\}\}$ dense in $A_q(\Delta, \Gamma)$?
It is easy to check that for $\psi \in B_q(\Delta, \Gamma)$ we have

$$(\phi, \psi)_{\{1\}} = (\text{Ⓗ}_q \phi, \psi)_\Gamma, \quad \text{for} \quad \phi \in A_q(\Delta) . \tag{3.7}$$

Thus an affirmative solution to 3.1 (a) implies the same answer for 3.1 (b).

THEOREM 3.2. (Ahlfors [1], Bers [5]) *We have* $\{\phi^z;\ z \in \Lambda - \{a_1, a_2\}\}$ *is dense in* $A_2(\Delta)$.

Proof: (Bers [5]) Let ℓ be a continuous linear functional on $A_2(\Delta)$. Then

$$\ell(\phi) = \iint_\Delta \phi(z)\mu(z)\, dz \wedge d\bar{z} , \tag{3.8}$$

where μ is a bounded measurable function on Δ. (We may actually assume that $\lambda^2 \bar{\mu} \in B_2(\Delta)$. But this does not simplify the proof.) Set

$$h(z) = \frac{(z-a_1)(z-a_2)}{2\pi i} \iint_\Delta \frac{\mu(\zeta)\, d\zeta \wedge d\bar{\zeta}}{(\zeta-z)(\zeta-a_1)(\zeta-a_2)}, \qquad z \in C . \tag{3.9}$$

Then h is continuous everywhere, $h(a_1) = h(a_2) = 0$, h has locally square integrable generalized derivatives and

$$\frac{\partial h}{\partial \bar{z}} = \mu \quad \text{in } \Delta .$$

Furthermore,

$$|h(z)| = 0(|z|\log|z|), \quad z \to \infty , \tag{3.10}$$

and for every $R > 0$,

$$|h(z)-h(w)| \leq C(R)|z-w|\,|\log|z-w|| \quad \text{for} \quad |z|, |w| \leq R . \tag{3.11}$$

We must show that $h = 0$ on Λ implies that ℓ is the zero linear functional. Integration by parts would show this. However, the boundary of Δ can be quite bad and the functions considered (ϕ and μ) need not be continuous on the closure of Δ. The situation is remedied by use of a "mollifier" due to Ahlfors [1]. The details may be found in Bers [5].

Remark. Note that h is a potential (see 5.5) for the (generalized) Beltrami coefficient μ (the group here is the trivial group).

4. The structure of Ω/Γ.

Recall that Ω is the region of discontinuity of Γ and Δ is a Γ-invariant union of components of Ω. Let $\pi : \Delta \to \Delta/\Gamma$ be the natural projection map. The set Δ/Γ has a natural complex structure for which π is a holomorphic mapping. The mapping π is locally one-to-one except at the fixed points of the elliptic elements where it is n-to-one (n = order of the stability subgroup of fixed point of the elliptic element). Let $\Delta_1, \Delta_2, \ldots$ be a maximal inequivalent set of components of Δ. Let

$$\Gamma_j = \{\gamma \in \Gamma;\ \gamma(\Delta_j) = \Delta_j\}, \quad j = 1,2,\ldots . \tag{4.1}$$

Then it is clear that

$$\Delta/\Gamma = \cup_j (\Delta_j/\Gamma_j) \quad , \tag{4.2}$$

where each Δ_j/Γ_j is a (connected) Riemann surface.

We shall say that Δ/Γ is of *finite type* if

(i) Δ/Γ consists of finitely many (K) components,

(ii) the mapping $\pi : \Delta \to \Delta/\Gamma$ is ramified over finitely many points, and

(iii) for each j ($j = 1, 2, \ldots, K$), there exists a compact Riemann surface S_j $(= \overline{\Delta_j/\Gamma_j})$ of genus g_j such that $S_j - (\Delta_j/\Gamma_j)$ consists of a finite number of points.

We shall assign ramification number ∞ to each of the points $p \in S_j - (\Delta_j/\Gamma_j)$.

The following is proven in Ahlfors [1].

LEMMA 4.1. *Let* $\Delta_j/\Gamma_j = S - \{p\}$ *where* S *is a Riemann surface and* $p \in S$. *If there is a punctured neighborhood* M *of* p *such that* π *is unramified over* M, *then there exists a parabolic element* $\gamma \in \Gamma_j$ *with fixed point* $\zeta \in \Lambda$, *and there is a Möbius transformation* A *with the following properties:*

(i) $A(\infty) = \zeta$ *and* $A^{-1} \circ \gamma \circ A(z) = z + 1, \; z \in \mathbf{C}$,

(ii) $A^{-1}(\Delta_j)$ *contains a half plane,*

$$U_c = \{z \in \mathbf{C}\,;\ \operatorname{Im} z > c\}, \qquad \textit{for some } c > 0, \tag{4.3}$$

(iii) *two points* z_1 *and* z_2 *of* $A(U_c)$ *are equivalent under* Γ_j *if and only if* $z_2 = \gamma^n(z_1)$ *for some integer* n, *and*

(iv) *the image of* $A(U_c)$ *under* π *is a deleted neighborhood of* p *homeomorphic to a punctured disc.*

We shall call

$$A\{z \in \mathbf{C}\,;\ 0 \le \operatorname{Re} z < 1,\ \operatorname{Im} z > c\} \tag{4.4}$$

a *cusped region belonging to* p, and $A(U_c)$ a *half plane belonging to* p. The point p will be called a *puncture* on Δ_j/Γ_j. We shall also say that the parabolic element γ *corresponds* to the puncture $p \in \overline{\Delta_j/\Gamma_j}$.

The punctures play a significant role in the cohomology theory. They determine distinguished cohomology classes. A cohomology class $p \in H^1(\Gamma, \Pi_{2q-2})$ is called Δ-*parabolic* if for every parabolic transformation $B \in \Gamma$ that corresponds to a puncture on Δ/Γ, there is a $v \in \Pi_{2q-2}$ such that

$$p_B = vB - v, \tag{4.5}$$

for some (and hence every) cocycle that represents p. The space of Δ-parabolic cohomology classes is denoted by $PH^1_\Delta(\Gamma, \Pi_{2q-2})$.

Each open set $s \subset \Delta/\Gamma$ has a canonical area

(4.6) $$\text{Area } s = \iint_{\omega/\Gamma} \lambda(z)^2 \, \frac{dz \wedge d\bar{z}}{-2i} ,$$

where $\omega = \pi^{-1}(s)$. It is well known (Gauss-Bonnet formula) that

(4.7) $$\text{Area}(\Delta/\Gamma) < \infty \iff \Delta/\Gamma \text{ is of finite type},$$

and in this case

(4.8) $$\text{Area}(\Delta_j/\Gamma_j) = 2\pi\{2g_j - 2 + \sum_{p \in S_j} (1 - \frac{1}{\ell(p)})\}$$

where $\ell(p)$ is ramification number of p, and $1/\infty = 0$.

To obtain (4.8) we must use in (4.6) the Poincaré metric with constant negative curvature -1. For example, for Δ the upper half plane, $\lambda(z) = (\text{Im } z)^{-1}$, $z \in \Delta$.

Again, whenever Δ/Γ is of finite type and consists of K components, then

(4.9) $$A_q(\Delta,\Gamma) = B_q(\Delta,\Gamma), \quad q \geq 1 ,$$

(4.10) $$B_q(\Delta,\Gamma) = \bigoplus_{j=1}^{K} B_q(\Delta_j,\Gamma_j) , \quad q \geq 1 ,$$

and

(4.11) $$\dim B_q(\Delta_j,\Gamma_j) = \begin{cases} g_j & \text{if } q = 1, \\ (2q-1)(g_j - 1) + \sum_{p \in S_j} [q - \frac{q}{\ell(p)}], & \text{if } q \geq 2, \end{cases}$$

where $[x]$ is the largest integer $\leq x$, and $[q - \frac{q}{\infty}] = q - 1$. Equation (4.11) is a consequence of the Riemann-Roch theorem.

5. Generalized Beltrami Coefficients.

We shall consider measurable functions μ on Δ that satisfy (3.1) and (3.3), and denote the corresponding Banach space by $L_q^\infty(\Delta, \Gamma)$. We shall call such a μ a *measurable bounded automorphic form of weight* $(-2q)$ on Δ. Note that for such a μ, $\lambda^{2-2q}\bar{\mu}$ is a *generalized Beltrami coefficient* in the sense of Bers [6], that is

$$\gamma^*_{1-q,1}(\lambda^{2-2q}\bar{\mu}) = \lambda^{2-2q}\bar{\mu}\,, \tag{5.1}$$

and

$$|\lambda^{2-2q}\bar{\mu}| \leq \text{const.}\ \lambda^{2-q}\,. \tag{5.2}$$

We shall always consider μ to be defined on the entire extended plane and to vanish outside Δ (in particular, $\mu\,|\,\Lambda = 0$ all $\mu \in L_q^\infty(\Delta, \Gamma)$). A continuous function F on $\mathbf{C}$ will be called a *potential* for the generalized Beltrami coefficient $\lambda^{2-2q}\bar{\mu}$, if

$$F(z) = 0(|z|^{2q-2}), \quad z \to \infty, \text{ and} \tag{5.3}$$

$$\frac{\partial F}{\partial \bar{z}} = \lambda^{2-2q}\bar{\mu} \tag{5.4}$$

in the sense of distributions. If F is a potential for $\lambda^{2-2q}\bar{\mu}$ so is $F + p$ with $p \in \Pi_{2q-2}$. Conversely, if F_1 and F_2 are potentials for $\lambda^{2-2q}\bar{\mu}$, then $F_1 - F_2 \in \Pi_{2q-2}$.

LEMMA 5.1. *Let* $q \geq 2$, *and* $\mu \in L_q^\infty(\Delta, \Gamma)$. *If* $\{a_1, \ldots, a_{2q-2}\}$ *are distinct points in* Λ, *then*

$$F(z) = \frac{(z-a_1)\cdots(z-a_{2q-2})}{2\pi i} \iint_\Delta \frac{\lambda(\zeta)^{2-2q}\bar{\mu}(\zeta)\, d\zeta \wedge d\bar{\zeta}}{(\zeta-z)(\zeta-a_1)\cdots(\zeta-a_{2q-2})}, \quad z \in \mathbf{C}, \tag{5.5}$$

is a potential for the generalized Beltrami coefficient $\lambda^{2-2q}\bar{\mu}$ *that satisfies*

$$F(z) = 0(|z|^{2q-3}\log|z|), \quad z \to \infty\,. \tag{5.6}$$

Proof: Only the ideas behind the statement of the Lemma are non-trivial. These are due to Bers [6]. The verification is standard.

THEOREM 5.2. (a) (Ahlfors [1], Bers [6], Kra [14]) *There is a canonical anti-linear mapping*

$$\beta^*: L_q^\infty(\Delta,\Gamma) \to PH_\Delta^1(\Gamma,\Pi_{2q-2}), \quad q \geq 2.$$

(b) (Bers [6]) $\beta^*(L_q^\infty(\Delta,\Gamma)) = \beta^*(B_q(\Delta,\Gamma))$, $q \geq 2$.

(c) (Ahlfors [1]) $\beta^*|B_2(\Delta,\Gamma)$ *is injective.*

Proof: (a) Let $\mu \in L_q^\infty(\Delta,\Gamma)$ and let F be the potential for $\lambda^{2-2q}\bar{\mu}$ given by (5.5). For $\gamma \in \Gamma$, set

$$p_\gamma = \gamma_{1-q}^* F - F. \tag{5.7}$$

Equation (5.4) and the invariance properties of μ imply that p_γ is an entire function and (5.3) that $p_\gamma \in \Pi_{2q-2}$. Thus (5.7) defines a cocycle whose cohomology class $\beta^*\mu$ depends on μ and not F. To verify that $\beta^*\mu$ is a Δ-parabolic cohomology class, it suffices to consider a single puncture with corresponding parabolic element $A(z) = z+1$. It is now clear that p_A has degree $\leq 2q-3$. Thus there is an element $v \in \Pi_{2q-2}$ such that $p_A = vA - v$.

(b) This assertion will not be needed in the sequel. For a proof see Bers [6].

(c) Let $\phi \in B_2(\Delta,\Gamma)$. Let F be the potential for $\lambda^{-2}\bar{\phi}$ formed by (5.5). If $\beta^*(\phi) = 0$, then there is a $p \in \Pi_2$ such that

$$\gamma_{-1}^* F - F = \gamma_{-1}^* p - p, \quad \text{all } \gamma \in \Gamma. \tag{5.8}$$

We may, without loss of generality assume that $\{a_1, a_2, \infty\}$ are fixed points of loxodromic (including hyperbolic) elements of Γ. Since F vanishes at a_1 and a_2 so does p (the derivative of Möbius transformation at a fixed point of a loxodromic element is not equal to one (in absolute value)).

Since ∞ is a fixed point of a non-elliptic element of Γ, there is a $\gamma \epsilon \Gamma$ for which the left hand side of (5.8) is $0(|z| \log |z|)$, $z \to \infty$. We may assume $\gamma(z) = \lambda z$, $\lambda > 1$, $z \epsilon \mathbf{C}$. Thus the degree of p is ≤ 1. Since p vanishes at a_1 and a_2, $p = 0$. Thus

(5.9) $$\gamma^*_{-1} F = F, \text{ all } \gamma \epsilon \Gamma \ .$$

Hence, if $z \epsilon \Lambda - \{a_1, a_2, \infty\}$ is a fixed point of a loxodromic element, we have $F(z) = 0$. Thus $\phi = 0$ by Theorem 3.2 and the duality between $A_q(\Delta)$ and $B_q(\Delta)$.

DEFINITION. We shall call β^* the *Bers map* and $\beta^*(\phi)$ the *Bers cohomology class of* $\phi \epsilon B_q(\Delta, \Gamma)$.

COROLLARY. (Ahlfors [1]) *If* Γ *is a finitely generated Kleinian group with region of discontinuity* Ω, *then each component of* Ω/Γ *is a Riemann surface of finite type.*

Proof: Let S be a component of Ω/Γ and $\pi : \Omega \to \Omega/\Gamma$ the natural projection map. We have that $A_2(\pi^{-1}(S), \Gamma)$ is canonically isomorphic to a space of integrable (meromorphic) quadratic differentials on S. The surface S is of finite type if and only if $A_2(\pi^{-1}(S), \Gamma)$ is finite dimensional. See, for example, Nevanlinna [15]. By Theorem 5.2 and Lemma 2.1, $A_2(\pi^{-1}(S), \Gamma)$ is finite dimensional.

REMARK. We cannot at this point conclude that there are finitely many components in Ω/Γ, because a "thrice punctured sphere" carries no integrable quadratic differentials. This defficiency in Ahlfors' original argument [1] was subsequently removed by Greenberg [10], Bers [7], and Ahlfors [2]. The arguments of the next section will also allow us to conclude that only a finite number of thrice punctured spheres can appear in Ω/Γ. Our argument is in the spirit of Bers [7].

6. Structure of $H^1(\Gamma, \Pi_{2q-2})$ and $PH^1_\Delta(\Gamma, \Pi_{2q-2})$.

A holomorphic function F on Δ is called an *Eichler integral of order* $(1-q)$ on Δ if

$$p_\gamma = \gamma^*_{1-q} F - F \in \Pi_{2q-2}, \text{ all } \gamma \in \Gamma . \tag{6.1}$$

REMARK. (Ahlfors [3]) The crucial part of the above definition is that p_γ be the *same* polynomial on each component of Δ.

Let D^{2q-1} denote differentiation $2q-1$ times. As a consequence of the identity

$$D^{2q-1} \circ \gamma^*_{1-q} = \gamma^*_q \circ D^{2q-1}, \quad \gamma \in \Gamma, \tag{6.2}$$

discovered by G. Bol, we conclude that D^{2q-1} maps Eichler integrals of order $(1-q)$ into automorphic forms of weight $(-2q)$. We shall say that an Eichler integral F of order $1-q$ is *bounded* if

$$\phi = D^{2q-1} F \in B_q(\Delta, \Gamma) . \tag{6.3}$$

From now on we assume that Δ/Γ consists of an arbitrary union of surfaces of finite type. The projection of ϕ to Δ/Γ is a meromorphic q-differential on $\overline{\Delta/\Gamma}$ with order $\geq -(q-1)$ at the punctures of Δ/Γ. An Eichler integral F on Δ is called *quasi-bounded* if the projection of $D^{2q-1}F$ to Δ/Γ can be extended as a meromorphic q-differential to $\overline{\Delta/\Gamma}$ whose order at a puncture is $\geq -q$. For later use we describe explicitly the local behavior of an Eichler integral F in a cusped region belonging to a puncture. We assume that ∞ is the fixed point of the translation $A(z) = z+1$ corresponding to the puncture. The function F is holomorphic in a half plane that may be taken to be U_c (see (4.3)). Thus $\phi = D^{2q-1}F$ has a Fourier series expansion

$$\phi(z) = \sum_{n=-\infty}^{\infty} a_n e^{2\pi i n z}, \quad z \in U_c . \tag{6.4}$$

If F is quasi-bounded, then $a_n = 0$ for $n < 0$, and

$$(6.5) \qquad F(z) = \sum_{n=1}^{\infty} a_n (2\pi i n)^{1-2q} e^{2\pi i n z} + a_0 \frac{z^{2q-1}}{(2q-1)!} + v(z) \quad ,$$

where $v \in \Pi_{2q-2}$. If F is bounded, then $a_0 = 0$ in (6.5).

The space of bounded Eichler integrals modulo Π_{2q-2} will be denoted by $E^b_{1-q}(\Delta,\Gamma)$. Similarly, $E^c_{1-q}(\Delta,\Gamma)$ denotes the space of quasi-bounded Eichler integrals modulo Π_{2q-2}.

THEOREM 6.1. *Let Γ be a Kleinian group and Δ an invariant union of components of its region of discontinuity such that Δ/Γ is a union of surfaces of finite type. Then there exists a canonical linear mapping*

$$(6.6) \qquad \alpha : E^c_{1-q}(\Delta,\Gamma) \to H^1(\Gamma, \Pi_{2q-2}), \qquad q \geq 1 \ ,$$

such that

$$(6.7) \qquad \alpha(E^b_{1-q}(\Delta,\Gamma)) \subset PH^1_\Delta(\Gamma, \Pi_{2q-2}) \ .$$

Furthermore, α is injective whenever $q > 1$. For $q = 1$, the kernel of α consists of those holomorphic automorphic forms of weight 0 that reduce to constants on each component of Δ.

Proof: If F represents an element of $E^c_{1-q}(\Delta,\Gamma)$, then (6.1) defines a cocycle whose cohomology class depends only on the equivalence class of F in $E^c_{1-q}(\Delta,\Gamma)$. This defines the map α. If $\alpha(F) = 0$, then we may assume that p_γ defined by (6.1) is the zero polynomial for all $\gamma \in \Gamma$. Thus F projects to $(1-q)$-differential $\tilde{F}$ on Δ/Γ. We may without loss of generality assume that Δ/Γ is the single compact Riemann surface S of genus g. It is easy to show that if $p \in S$, then the order of $\tilde{F}$ of p is at least

$$(6.8) \qquad -[(1-q)(1-\frac{1}{\ell(p)})]$$

where $1/\infty = 0$. If $\tilde{F}$ is not identically zero, then its degree is equal to $(1-q)(2g-2)$. Thus,

$$- \sum_{p \in S} [(1-q)(1-\frac{1}{\ell(p)})] \leq (1-q)(2g-2) . \tag{6.9}$$

But it is known that the Poincaré area of S is positive (it is at least $\pi/21$), thus from (4.8) we obtain the contradiction

$$- \sum_{p \in S} [(1-q)(1-\frac{1}{\ell(p)})] \geq -(1-q) \sum_{p \in S} (1-\frac{1}{\ell(p)}) > (1-q)(2q-2) \tag{6.10}$$

whenever $q \geq 2$. The case $q = 1$ is obvious. Statement (6.7) is a simple consequence of (6.5) and the invariance of the parabolicity of cohomology classes under conjugation.

DEFINITION. We shall call α the *Eichler* or *period map,* and $\alpha(F)$ the *Eichler cohomology class of* $F \in E^c_{1-q}(\Delta, \Gamma)$.

THEOREM 6.2. *Let* Γ *be a Kleinian group and* Δ *an invariant union of components of its region of discontinuity such that* Δ/Γ *is of finite type. There exists a canonical anti-linear mapping*

$$\beta : H^1(\Gamma, \Pi_{2q-2}) \to B_q(\Delta, \Gamma), \quad q \geq 1, \tag{6.11}$$

such that

$$\beta \circ \beta^* = \text{id} \quad on \quad B_q(\Delta, \Gamma), \quad q \geq 2 . \tag{6.12}$$

In particular, β^* *is injective and* β *is surjective for* $q \geq 2$.

Proof: Let ω be a fundamental domain for Γ in Δ. We may assume that ω consists of finitely many components, that each of these components is a smooth Jordan polygon with sides pairwise identified by Γ, and that each component consists of a relatively compact subset of Δ and finitely many cusped regions belonging to the punctures on Δ/Γ.

Let η be a *partition of unity* for Γ on Δ (see [13]); that is, η is a smooth function on Δ that satisfies the following conditions:

(1) $0 \leq \eta(z) \leq 1$, all $z \in \Delta$,

(2) for each $z \in \Delta$, there is a neighborhood U of z and a finite subset $J \subset \Gamma$ such that $\eta | \gamma(U) = 0$ for $\gamma \in \Gamma - J$,

(3) $\Sigma_{\gamma \in \Gamma} \eta(\gamma z) = 1$, $z \in \Delta$, and

(4) if $V \subset \omega$ is a cusped region belonging to a puncture on Δ/Γ, and A is the corresponding transformation, then $\eta | \gamma(V) = 0$ all $\gamma \in \Gamma - \{1, A\}$.

The partition of unity is constructed by lifting one from Δ/Γ.

Let p be a cocycle representing a cohomology class of $H^1(\Gamma, \Pi_{2q-2})$. We construct a smooth function θ that satisfies:

(1) $p_\gamma = \gamma^*_{1-q}\theta - \theta$, $\gamma \in \Gamma$.

(2) If ∞ is a fixed point of a parabolic element of Γ that corresponds to a puncture on Δ/Γ and $\infty \in \bar{\omega}$ (the closure of ω), then

$$\theta(z) = 0(|z|^{2q-2+\varepsilon}), \quad z \to \infty, \quad z \in \omega. \tag{6.13}$$

For a corresponding finite fixed point $\zeta \in \bar{\omega}$, the condition is replaced by

$$\theta(z) = 0(|z-\zeta|^{-\varepsilon}), \quad z \to \zeta, \quad z \in \omega. \tag{6.13)'$$

In the above $\varepsilon = 1$. However if p represents a Δ-parabolic cohomology class, then we require that $\varepsilon = 0$.

(3) If $\mu(z) = \partial\theta/\partial\bar{z}$, $z \in \Delta$, then μ is a generalized Beltrami coefficient.

REMARK. Conditions (6.13) and (6.13)′ are invariant under conjugation. That is, if we have constructed θ that satisfies the above conditions for the cocycle p for the group Γ, then for any Möbius transformation B, $B^*_{1-q}\theta$ satisfies these properties for the cocycle p^B for the group $B^{-1}\Gamma B$, where

$$p^B_{B^{-1}\gamma B} = p_\gamma B = B^*_{1-q} p_\gamma, \quad \gamma \in \Gamma. \tag{6.14}$$

If Δ/Γ has no punctures, then

$$\theta(z) = -\sum_{\gamma \in \Gamma} \eta(\gamma z)\, p_\gamma(z), \quad z \in \Delta. \tag{6.15}$$

If Δ/Γ has precisely one puncture, it suffices to assume that the corresponding parabolic element is $A(z) = z+1$, $z \in \mathbf{C}$. Choose a polynomial v of degree $\leq 2q-1$ such that

$$p_A(z) = v(z+1) - v(z), \quad z \in \mathbf{C}.$$

Note that $v \in \Pi_{2q-2}$ if p represents a class in $PH^1_\Delta(\Gamma, \Pi_{2q-2})$. Set

$$(6.16) \qquad \hat{\theta}(z) = -\sum_{\gamma \in \Gamma} \eta(\gamma z)[p_\gamma(z) - v(\gamma z)\gamma'(z)^{1-q} + v(z)], \quad z \in \Delta,$$

and

$$(6.17) \qquad \theta = \hat{\theta} + v .$$

Note that the function θ that we have constructed is actually holomorphic in a half plane belonging to the puncture.

If Δ/Γ has more than one puncture, one uses induction (see [14]). We now define

$$(6.18) \qquad \ell(\phi) = \iint_{\Delta/\Gamma} \phi(z)\mu(z)\, dz \wedge d\bar{z}, \quad \phi \in A_q(\Delta, \Gamma) .$$

If $q \geq 2$, then it is clear that

$$(6.19) \quad |\ell(\phi)| \leq \left(\iint_{\Delta/\Gamma} \lambda(z)^{2-q}|\phi(z)\, dz \wedge d\bar{z}|\right)(\sup\{\lambda(z)^{q-2}|\mu(z)|;\ z \in \Delta\}) .$$

Thus ℓ is bounded linear functional on $A_q(\Delta, \Gamma)$ which may be identified with an element $\beta p \in B_q(\Delta, \Gamma)$ via the Petersson scalar product. If Δ/Γ has no punctures, then integration by part shows that βp is independent of the choices made and depends only on the cohomology class of the cocycle p. If punctures are present, we must use a device due to Bers [6]. The details may be found in [14].

It is clear that for $q \geq 2$,

$$(6.20) \qquad \beta \circ \beta^*(\phi) = \phi, \text{ all } \phi \in B_q(\Delta, \Gamma) .$$

Thus,

(1) β^* is injective (Theorem 3 of [6]),

(2) β is surjective (even when restricted to $PH^1_\Delta(\Gamma, \Pi_{2q-2})$), and

(3) β is canonical.

If $q = 1$, then (6.18) induces a bounded linear functional on the Hilbert space $B_1(\Delta, \Gamma)$. However, in this case we cannot conclude surjectivity of the β map.

COROLLARY 1. *For* $q \geq 2$, β^* *is injective whenever* Δ/Γ *is a union of surfaces of finite type.*

Proof: For finite unions, we have already established the result by (6.12). Let $\Delta = \cup_j \Delta_j$, where Δ_j is an invariant union of components of Δ such that Δ_j/Γ represents a single Riemann surface. Then $\beta^*| B_q(\Delta_j, \Gamma)$ is injective. Thus $\{\textcircled{H}_q \phi^z;\ z \in \Lambda - \{a_1, \ldots, a_{2q-2}\}\}$ is dense in $A_q(\Delta_j, \Gamma)$ for each j. Thus these same functions are also dense in $A_q(\Delta, \Gamma)$. Thus $\beta^*| B_q(\Delta, \Gamma)$ is injective. (We have and shall continue to use the notation and conventions introduced in sections 3 and 4.)

COROLLARY 2. *If* Γ *is generated by* N *elements, then* Δ/Γ *is of finite type and*

$$\sum_{j=1}^{K} \{(2q-1)(g_j - 1) + \sum_{p \in S_j} [q - \frac{q}{\ell(p)}]\} \leq (2q-1)(N-1), \qquad q \geq 2.$$

Proof: The first assertion holds because $\dim B_q(\Delta_j, \Gamma) \geq 1$ for $q \geq 44$.

REMARK. The above contains Ahlfors' finiteness theorem [1].

Following Bers [6] we divide by q and let $q \to \infty$. We obtain Bers' area theorem [6] in

COROLLARY 3. *If* Γ *is generated by* N *elements, then*

$$\text{Area}\,(\Delta/\Gamma) \leq 4\pi(N-1) .$$

CONJECTURE. (Maskit) If Γ is generated by N elements, then $K \leq 2(N-1)$, where K is the number of components of Ω/Γ.

(a) Since Area $(\Delta_j/\Gamma_j) \geq \pi/21$, $K \leq 84(N-1)$ (Bers [6]).

(b) A better estimate is obtained as follows. In [2] Ahlfors observes that $\dim B_4(\Delta_j, \Gamma_j) + \dim B_6(\Delta_j, \Gamma_j) \geq 1$, thus $K \leq 18(N-1)$.

(c) If no elliptic elements are present, then Area $(\Delta_j/\Gamma_j) \geq 2\pi$. Thus the conjecture has been verified in this case (Maskit).

(d) If Γ is purely loxodromic, then $K \leq N/2$. This result has been obtained by Marden using topological arguments. For a simple and elegant proof see Marden's note in these Proceedings. Note that this is a stronger result that is obtainable by our methods. If Γ is purely loxodromic, then Area $(\Delta_j/\Gamma_j) \geq 4\pi$. Thus we obtain (trivially), that $K \leq N-1$ in this case.

THEOREM 6.3. *Let Γ be a Kleinian group and Δ an invariant union of components of its region of discontinuity. If Δ/Γ is of finite type, then for $q \geq 2$ the following is a commutative diagram with exact rows:*

$$\begin{array}{ccccccccc} 0 & \longrightarrow & E^b_{1-q}(\Delta,\Gamma) & \xrightarrow{\alpha} & PH^1_\Delta(\Gamma,\Pi_{2q-2}) & \xrightarrow{\beta} & B_q(\Delta,\Gamma) & \longrightarrow & 0 \\ & & \downarrow & & \downarrow & & \downarrow \text{id} & & \\ 0 & \longrightarrow & E^c_{1-q}(\Delta,\Gamma) & \xrightarrow{\alpha} & H^1(\Gamma,\Pi_{2q-2}) & \xrightarrow{\beta} & B_q(\Delta,\Gamma) & \longrightarrow & 0 . \end{array}$$

Proof: The injectivity of the α maps (Theorem 6.1) as well as the surjectivity of the β maps (Theorem 6.2) have already been verified. The arguments in the proof of Theorem 6.2 show that

$$\text{Image } \alpha \subset \text{Kernel } \beta .$$

To verify the reverse inclusion we follow the notation in the proof of Theorem 6.2 and arguments of Bers [6]. Assume that $\beta(p) = 0$ for some cocycle p; that is, the linear functional defined by (6.18) is the zero linear functional. Let F be the potential for the generalized Beltrami coefficient μ defined by (5.5).

For $z \in \Lambda - \{a_1, \ldots, a_{2q-2}\}$, $Ⓗ_q \phi^z \in A_q(\Delta, \Gamma)$, and hence

$$(6.21) \quad F(z) = (\phi^z, \lambda^{2q-2}\bar{\mu})_{\{1\}} = (Ⓗ_q \phi^z, \lambda^{2q-2}\bar{\mu})_\Gamma = \ell(\phi^z) = 0 .$$

Thus $F|\Lambda = 0$, and hence $\gamma^*_{1-q}F - F = 0$ on $\mathbf{C}$ for all $\gamma \in \Gamma$. Now $\partial(\theta - F)/\partial\bar{z} = 0$, and $\theta - F$ also induces the cocycle p via (6.1). Clearly, $\theta - F$ is an Eichler integral. The construction of θ in the proof of Theorem 6.2 shows that $\theta - F$ is quasi-bounded, and bounded whenever p represents a Δ-parabolic cohomology class.

COROLLARY 1. *Under the hypothesis of the theorem, every cohomology class* $p \in H^1(\Gamma, \Pi_{2q-2})$ *can be written uniquely as*

$$p = p_B + p_E ,$$

where p_B *is the Bers cohomology class of a unique* $\phi \in B_q(\Delta, \Gamma)$, *and* p_E *is the Eichler cohomology class of a unique* $f \in E^c_{1-q}(\Delta, \Gamma)$. *Furthermore,* $p \in PH^1_\Delta(\Gamma, \Pi_{2q-2})$ *if and only if* $f \in E^b_{1-q}(\Delta, \Gamma)$.

COROLLARY 2. *Let* Γ *be a finitely generated Kleinian group with an invariant domain* Δ, *then*

$$\dim B_q(\Omega, \Gamma) \le \dim PH^1_\Delta(\Gamma, \Pi_{2q-2}) \le 2 \dim B_q(\Delta, \Gamma) ,$$

and

$$\dim B_q(\Omega, \Gamma) \le \dim H^1(\Gamma, \Pi_{2q-2}) \le 2 \dim B_q(\Delta, \Gamma) + n ,$$

where n *is the number of punctures on* Δ/Γ. *Furthermore, if* Δ *is connected and simply connected, then*

$$\dim PH^1_\Delta(\Gamma, \Pi_{2q-2}) = 2 \dim B_q(\Delta, \Gamma) ,$$

and

$$\dim H^1(\Gamma, \Pi_{2q-2}) = 2 \dim B_q(\Delta, \Gamma) + n .$$

Proof: The map $D^{2q-1}: E^b_{1-q}(\Delta,\Gamma) \to B_q(\Delta,\Gamma)$ is injective. It is surjective when Δ is also simply connected. The dimension of the image of $E^c_{1-q}(\Delta,\Gamma)$ under this map exceeds the dimension of $B_q(\Delta,\Gamma)$ by at most n (exactly n when Δ is simply connected).

Dividing the first inequality of the corollary by q and letting $q \to \infty$ we obtain Bers' second area inequality [6].

COROLLARY 3. *Let Γ be a finitely generated Kleinian group with an invariant domain Δ, then*

$$\text{Area}\,(\Omega/\Gamma) \leq 2\,\text{Area}\,(\Delta/\Gamma) \ .$$

It is of some interest to see when the Bers map is surjective. A dimension count yields

COROLLARY 4. *If Γ is a finitely generated Schottky group, then*

$$H^1(\Gamma,\Pi_{2q-2}) = \beta^*(B_q(\Omega,\Gamma)) \ .$$

REMARKS. (1) For Δ connected and simply connected, the dimensions of $H^1(\Gamma,\Pi_{2q-2})$ and $PH^1_\Delta(\Gamma,\Pi_{2q-2})$ can also be computed by purely algebraic techniques. See Eichler [9], Weil [16], Ahlfors [3], and Curran [8].

(2) If Γ is a finitely generated degenerate Kleinian group (connected, simply connected region of discontinuity) then β^* is not surjective. In fact, in this case,

$$\dim PH^1_\Omega(\Gamma,\Pi_{2q-2}) = 2 \dim B_q(\Omega,\Gamma) \ .$$

An Eichler integral $F \in E^b_{1-q}(\Delta,\Gamma)$ is called *trivial* if $D^{2q-1}F = 0$. The space of trivial Eichler integrals (modulo polynomials) is denoted by $E^0_{1-q}(\Delta,\Gamma)$.

THEOREM 6.4. *Let Γ be a finitely generated Kleinian group with an invariant domain Δ. If $\Omega-\Delta$ is non-empty, then for $q \geq 2$ every cohomology class $p \in PH^1_{\Omega-\Delta}(\Gamma,\Pi_{2q-2})$ can be written as the sum of a Bers cohomology* *class of some* $\phi \in B_q(\Omega, \Gamma)$ *and an Eichler cohomology class of a trivial Eichler integral* $F \in E^0_{1-q}(\Omega - \Delta, \Gamma)$.

Proof: We begin with $\Omega = \Omega_1 \cup \Omega_2$ where each Ω_j is an open invariant subset of Ω with $\Omega_1 \cap \Omega_2$ empty. We shall define a bounded anti-linear operator (Bers [6])

(6.22) $$L_{\Omega_1}: B_q(\Omega_1, \Gamma) \to B_q(\Omega_2, \Gamma)$$

as follows. Let $\phi \in B_q(\Omega_1, \Gamma)$, and let F be a potneital for $\lambda^{2-2q}\bar{\phi}$. Then F is holomorphic on Ω_2. We set $\psi = L_{\Omega_1}\phi = D^{2q-1}F$. It is easy to show that

(6.23) $$\psi(z) = \frac{(2q-1)!}{2\pi i} \iint_{\Omega_1} \frac{\lambda(\zeta)^{2-2q}\overline{\phi(\zeta)}}{(\zeta - z)^{2q}}\, d\zeta \wedge d\bar{\zeta}, \qquad z \in \Omega_2 \ .$$

From this integral representation of ψ it follows that this operator is "self-adjoint" in the sense that

(6.24) $$(\phi, L_{\Omega_1}\psi)_\Gamma = -\overline{(L_{\Omega_2}\phi, \psi)_\Gamma}, \quad \phi \in A_q(\Omega_2, \Gamma), \quad \psi \in B_q(\Omega_1, \Gamma) \ .$$

(Note that in the first inner produce integration is over Ω_2/Γ, and in the second over Ω_1/Γ.) Observe, also that (by abuse of notation)

(6.25) $$L = D^{2q-1} \circ \beta^* \ .$$

It follows from (6.25) that L_{Ω_1} is injective whenever Ω_2 is connected. From the finite dimensionality of the B_q spaces and (6.24) we also conclude that L_{Ω_1} is surjective whenever Ω_1 is connected.

Now let $p \in PH^1_{\Omega - \Delta}(\Gamma, \Pi_{2q-2})$. By Corollary 1 to Theorem 6.3

(6.26) $$p = \beta^*(\phi_0) + \alpha(F_0) \ ,$$

where $\phi_0 \in B_q(\Omega - \Delta, \Gamma)$ and $F_0 \in E^b_{1-q}(\Omega - \Delta, \Gamma)$. Now

$$\psi = D^{2q-1}F_0 \in B_q(\Omega - \Delta, \Gamma) \ .$$

Let $\psi = L_\Delta \phi$ with $\phi \in B_q(\Delta, \Gamma)$ and let F be a potential for $\lambda^{2-2q}\bar{\phi}$. Then

$$p = \beta^*(\phi_0 + \phi) + \alpha(F_0 - F) \; . \tag{6.27}$$

Since $D^{2q-1}F = \psi$, $F_0 - F \in E^0_{1-q}(\Omega - \Delta, \Gamma)$.

COROLLARY 1. *Let* Γ *be a finitely generated Kleinian group with two invariant components,* Δ *and* $\Omega - \Delta$. *Then for* $q \geq 2$,

$$PH^1_\Omega(\Gamma, \Pi_{2q-2}) = PH^1_\Delta(\Gamma, \Pi_{2q-2}) = \beta^*(B_q(\Omega, \Gamma)) \; .$$

COROLLARY 2. *Let* Γ *be a Kleinian group, then* $E^b_{1-q}(\Omega, \Gamma) = \{0\}$, *whenever* $q \geq 2$, *and*

(a) Γ *is finitely generated with two invariant components,*

(b) Γ *is a finitely generated Schottky group, or*

(c) Γ *is any (non-elementary) Fuchsian group.*

REMARK. Statement (c) has not been proven here. See [13] for a similar argument.

7. The case $q = 1$.

We shall obtain

THEOREM 7.1. *Let* Γ *be a finitely generated Kleinian group with a simply connected invariant component* Δ. *The following is a commutative diagram with exact rows:*

$$\begin{array}{ccccccccc} 0 & \longrightarrow & E^b_0(\Delta, \Gamma) & \xrightarrow{\alpha} & PH^1_\Delta(\Gamma, C) & \xrightarrow{\beta} & B_1(\Delta, \Gamma) & \longrightarrow & 0 \\ & & \downarrow & & \downarrow & & \downarrow \mathrm{id} & & \\ 0 & \longrightarrow & E^c_0(\Delta, \Gamma) & \xrightarrow{\alpha} & H^1(\Gamma, C) & \xrightarrow{\beta} & B_1(\Delta, \Gamma) & \longrightarrow & 0 \end{array}$$

In the above, we interpret $B_1(\Delta, \Gamma)$ as the Hilbert space of square integrable automorphic forms of weight (-2).

We recall that every finitely generated Kleinian group with a simply connected invariant component has a signature consisting of integers (or ∞)

$$\{g, n; \nu_1, \nu_2, \ldots, \nu_n\} ,$$

where $0 \leq g < \infty$, $0 \leq n < \infty$, $2 \leq \nu_1 \leq \nu_2 \cdots \leq \nu_n \leq \infty$. Let k be the largest integer $\leq n$ such that $\nu_k \neq \infty$ ($k = 0$ if $n = 0$), then $\overline{\Delta/\Gamma}$ is a compact surface of genus g, the complement of Δ/Γ in $\overline{\Delta/\Gamma}$ consists of $n-k$ points, and the natural projection map $\pi : \Delta \to \Delta/\Gamma$ is ramified over precisely k points with ramification numbers $\nu_1, \ldots, \nu_k$. Furthermore, Γ can be represented by generators $\{a_i, b_i, c_j$ ($i = 1, \ldots, g$ and $j = 1, \ldots, n$)$\}$ subject to the relations

$$\text{(7.1)} \qquad a_1 b_1 a_1^{-1} b_1^{-1} \cdots a_g b_g a_g^{-1} b_g^{-1} c_1 \cdots c_n = 1 ,$$

$$\text{(7.2)} \qquad c_j^{\nu_j} = 1, \qquad j = 1, \ldots, k .$$

Furthermore, the elements c_j, $j = k+1, \ldots, n$ are non-conjugate parabolic elements corresponding to the pucntures on Δ/Γ.

LEMMA 7.2. $\dim PH^1_\Delta(\Gamma, C) = 2g$, *and*

$$\dim H^1(\Gamma, C) = \begin{cases} 2g & \textit{if } n-k = 0, \\ 2g + (n-k-1) & \textit{if } n-k \geq 1 . \end{cases}$$

We leave the proof of this lemma to the reader.

Proof of Theorem 7.1. The arguments of section 6 show the existence and injectivity of the α maps. The construction of the β maps also applies, if we represent a cohomology class by a function θ such that

$$\mu = \partial\theta/\partial\bar{z} \in L^2(\Delta/\Gamma) ,$$

and view the Petersson scalar product as the inner product in the Hilbert space $B_1(\Delta, \Gamma)$. It is quite obvious that

$$\text{Image } \alpha \subset \text{Kernel } \beta \ .$$

We show that β is surjective. Let $\phi \in B_1(\Delta, \Gamma)$. Let F be a holomorphic function such that $F' = \phi$. Then $\bar{\partial} \bar{F} = \bar{\phi}$. Clearly $\bar{F}$ induces an element of $PH^1_\Delta(\Gamma, C)$ via

$$p_\gamma = \gamma_0^* \bar{F} - \bar{F}, \quad \gamma \in \Gamma \ . \tag{7.3}$$

We conclude from the surjectivity of the β maps and the Riemann-Roch theorem ($\dim B_1(\Delta, \Gamma) = g$) that the dimension of the kernel of the first β map is g, and of the second β map

$$\begin{array}{ll} g, & \text{if } n-k = 0 \\ g+(n-k-1), & \text{if } n-k \geq 1 \ . \end{array} \tag{7.4}$$

From the injectivity of the α maps and the Riemann-Roch theorem we conclude that

$$\dim \text{Image } \alpha = \dim \text{Kernel } \beta \ .$$

Thus

$$\text{Image } \alpha = \text{Kernel } \beta \ .$$

REMARKS. (1) The dimension of $PH^1_\Delta(\Gamma, C)$ has been computed by Eichler [9]. The decomposition of the groups $PH^1_\Delta(\Gamma, C)$ and $H^1(\Gamma, C)$ appears to be new.

(2) We conjecture that a similar theorem holds for arbitrary finitely generated (non-elementary) Kleinian groups.

(3) It is well known that whenever Γ acts freely on Δ (that is, Γ does not contain any elements of finite order), then

$$H^1(\Gamma, C) \cong H^1(\Delta/\Gamma, C) \ . \tag{7.5}$$

If Γ is an arbitrary Kleinian group, then Γ can be extended to a group of conformal self-mappings of U, the upper half space in R^3. If Γ acts freely on U, then we have

$$H^1(\Gamma, C) \cong H^1(U/\Gamma, C) . \tag{7.6}$$

Thus a description of $H^1(\Gamma, C)$ for Kleinian groups Γ would yield the first Betti number for a wide class of 3 dimensional manifolds with boundary. A simple dimension count shows that (7.5) is also valid when Γ contains elements with fixed points in Δ. Even without dimension counts, it is quite easy to show that (7.5) and (7.6) hold for groups Γ that have elements with fixed points in Δ or U.

(4) As a simple consequence of our theorem we obtain the well known fact that the only holomorphic abelian integral of the first kind with zero periods is the zero integral. See [13] for a generalization of this fact to the case $q \geq 2$.

8. On a Theorem of Ahlfors.

Throughout this section we assume that Δ/Γ is of finite type.

A meromorphic function F on Δ is called a *meromorphic Eichler integral of order* $1-q$ on Δ if (6.1) holds and if the projection of $\phi = D^{2q-1}F$ to Δ/Γ extends as a meromorphic q-differential to $\overline{\Delta/\Gamma}$. A meromorphic Eichler integral F has a Fourier series expansion similar to (6.5) in each cusped region in Δ. However, the index n appearing in the sum of (6.5) may start at any negative integer. We shall say that F is a *parabolic* integral if the Fourier series for ϕ, in each cusped region contained in Δ, has zero constant term. We shall denote the space of meromorphic Eichler integrals on Δ modulo Π_{2q-2} by $M_{1-q}(\Delta, \Gamma)$ and the space of parabolic meromorphic Eichler integrals modulo Π_{2q-2} by $PM_{1-q}(\Delta, \Gamma)$. It is clear that the period map α extends to these spaces of Eichler integrals.

Let ω_0 be a *fundamental set* for Γ in Δ; that is, ω_0 contains precisely one point from each orbit. By a *divisor* d we mean a formal sum

(8.1) $$d = \sum_{p \in \omega_0} n(p)p ,$$

where $n(p) \in \mathbf{Z}$, the integers, and $n(p) = 0$ for all but finitely many points p.

CONVENTION. For the sake of simplicity we shall consider only divisors d for which $n(p) = 0$ for all elliptic fixed points $p \in \omega_0$.

From now on we shall consider only (meromorphic) automorphic forms ϕ on Δ that project to meromorphic differentials on $\overline{\Delta/\Gamma}$. It is clear how to assign a divisor (ϕ) to such a form. Denote by $B_q(d)$ the space of meromorphic automorphic forms of weight $(-2q)$ that are regular at the punctures (if $q > 0$, this means that the Fourier series for ϕ at ∞ contains only terms with positive indices, and if $q \leq 0$, only non-negative indices appear) and whose divisors are multiples of d.

It is not possible to define a corresponding space of Eichler integrals for an arbitrary divisor d. However, for $d \leq 0$ (that is $n(p) \leq 0$ all $p \in \omega_0$), we can clearly introduce spaces of meromorphic Eichler integrals whose polar divisors are multiples of d. (Note that the zero set of an Eichler integral is not invariant under Γ.) We obtain this way $M_{1-q}(d)$, and the corresponding space of parabolic integrals, $PM_{1-q}(d)$. We again require the elements of $M_{1-q}(d)$ and $PM_{1-q}(d)$ to be regular over the punctures (that is, for $f \in PM_{1-q}(d)$, $D^{2q-1}f$ is regular over the punctures.) Here $q \geq 2$. Note that $M_{1-q}(0) = E^c_{1-q}(\Delta, \Gamma)$ and $PM_{1-q}(0) = E^b_{1-q}(\Delta, \Gamma)$.

Let $q \geq 2$. Let $0 < k = \dim B_q(\Delta, \Gamma)$. (Recall that $B_q(\Delta, \Gamma) = B_q(0)$.) Choose k divisors

(8.2) $$0 \leq d_1 < d_2 < \cdots < d_k ,$$

$$d_i = \sum_{j=1}^{\ell_i} n_{ij} p_j , \quad i = 1, 2, \ldots, k, \text{ with } n_{ij} \geq 0 ,$$

satisfying the following conditions:

(8.3) $$d_{i+1} = d_i + n_i q_i , \quad n_i > 0, \quad q_i \in \omega_0, \quad i = 1, \ldots, k-1 ,$$

(8.4) there exists a basis $\{\phi_1, \ldots, \phi_k\}$ for $B_q(0)$ such that

order $(\phi_i, p_j) = n_{ij}$, $i = 1, 2, \ldots, k$, and $(\phi_i) \not\geq d_i + q_i$.

Assume ∞ is neither a limit point nor an elliptic fixed point for Γ (this involves no loss of generality). Consider

$$h_1(z, \zeta) = \Sigma_{\gamma \in \Gamma} (z - \gamma\zeta)^{-1} \gamma'(\zeta)^q . \tag{8.5}$$

Ahlfors [3] has shown that for fixed ζ, $h_1(z, \zeta)$ is an Eichler integral with a simple pole at $z = \zeta$. Furthermore this integral is regular over the punctures. To obtain an Eichler integral with a pole of order $\nu \geq 1$ at $z = \zeta$ one simply sets

$$h_\nu(z, \zeta) = \partial^{\nu-1} h_1 / \partial \zeta^{\nu-1} . \tag{8.6}$$

If $d > 0$ is any positive divisor,

$$d = \sum_{i=1}^{\ell} n_i p_i , \qquad n_i > 0 , \tag{8.7}$$

then

$$\sum_{i=1}^{\ell} h_{n_i}(z, p_i) \tag{8.8}$$

is a meromorphic Eichler integral whose polar divisor is $-d$. Consider the space of Eichler integrals generated by the quasi-bounded Eichler integrals and the meromorphic Eichler integrals

$$F_1, F_2, \ldots, F_k , \tag{8.9}$$

whose polar divisors are

$$-(d_1 + q_1), -(d_2 + q_2), \ldots, -(d_{k-1} + q_{k-1}), -(d_k + q_k) , \tag{8.10}$$

where $q_k \in \omega_0$ is arbitrary, such that $(\phi_k) \not\geq d_k + q_k$. It is clear that this space has dimension

$$\dim E^c_{1-q}(\Delta, \Gamma) + \dim B_q(\Delta, \Gamma) = \dim H^1(\Gamma, \Pi_{2q-2}) . \tag{8.11}$$

Furthermore, the period map α restricted to this space is injective. For if F belongs to this space, then

$$(8.12) \qquad F = F_0 + \sum_{i=1}^{k} c_i F_i \ , \quad \text{where } F_0 \in E^c_{1-q}(\Delta, \Gamma) \ .$$

Let m be the largest integer $\leq k$ such that $c_m \neq 0$ ($m = 0$, if $c_i = 0$, $i = 1, \ldots, k$). If $\alpha(F) = 0$, then we may assume that F induces the zero cocycle; that is, F is an automorphic form of weight $-2(1-q)$. If $m = 0$, then we have already seen that $\alpha(F) = 0$ implies $F = 0$. Assume that $F \neq 0$ and $m > 0$. If $\alpha(F) = 0$, then $F\phi_m$ is an automorphic form of weight -2, whose projection to Δ/Γ has a simple pole at $\pi(q_m)$ and no other poles. With this contradiction we have obtained

THEOREM 8.1. *We have*

$$\alpha(M_{1-q}(\Delta, \Gamma)) = H^1(\Gamma, \Pi_{2q-2}) \ .$$

Furthermore, for $F \in M_{1-q}(\Delta, \Gamma)$, $\alpha(F) \in PH^1_\Delta(\Gamma, \Pi_{2q-2})$ *if and only if* $F \in PM_{1-q}(\Delta, \Gamma)$.

We have

$$(8.13) \qquad M_{1-q}(\Delta, \Gamma) \xrightarrow{\ \alpha\ } H^1(\Gamma, \Pi_{2q-2}) \xrightarrow{\ \beta\ } A_q(\Delta, \Gamma)^* \ ,$$

where we view β is the linear map from cohomology classes to linear functionals on the space of integrable automorphic forms.

THEOREM 8.2. *Let* $F \in M_{1-q}(\Delta, \Gamma)$, $q \geq 2$, *and let*

$$F^* = (\beta \circ \alpha)(F) \ .$$

If $\phi \in A_q(\Delta, \Gamma)$, *then*

$$F^*(\phi) = 2\pi i \sum_{z \in \omega_0} - \operatorname{Res}(F\phi, z) \ .$$

In the above the residues of $F\phi$ are summed over a fundamental set ω_0

for Γ in Δ. At an elliptic fixed or order $n \geq 2$, the residue must, of course, be divided by n. There is a contribution to the sum of residues from the parabolic fixed points corresponding to the punctures on Δ/Γ. If ∞ is a fixed point corresponding to a puncture and $A(z) = z+1$ is the generator of the stability subgroup of ∞, then in some half plane, U_c, $F\phi$ has a Fourier series expansion

$$(8.14) \qquad (F\phi)(z) = \sum_{n=k}^{\infty} a_n e^{2\pi i n z} + v(z) \sum_{n=1}^{\infty} e^{2\pi i n z}, \quad z \in U_c ,$$

where $k \in \mathbf{Z}$ and v is a polynomial of degree $\leq 2q-1$. We must set

$$(8.15) \qquad \text{Res}\,(F\phi, \infty) = a_0/2\pi i \ .$$

Proof: Choose a fundamental domain ω for Γ in Δ as in the proof of Theorem 6.2. We assume that F is regular on $\partial\omega$ as well as over the punctures. Choose a smooth function χ on Δ such that

$$\gamma_0^* \chi = \chi, \quad \text{all } \gamma \in \Gamma$$

$$0 \leq \chi \leq 1,$$

$\chi = 0$ in a neighborhood of the singularities of $F\phi$, and

$\chi = 1$ in a neighborhood of $\partial\omega$.

Let θ be any function that represents the cocycle $\alpha(F)$ as in the proof of Theorem 6.2. It is clear, that

$$(8.16) \qquad \chi F + (1-\chi)\theta = \theta_1$$

also represents this cocycle. Now

$$(8.17) \qquad F^*(\phi) = \iint_\omega \mu(z)\phi(z)\, dz \wedge d\bar{z} ,$$

where $\mu(z) = \partial\theta_1/\partial\bar{z}$, $z \in \Delta$. By a standard application of Stokes' theorem we obtain

$$\text{(8.18)} \qquad F^*(\phi) = \lim_n \left(-\int_{\partial\omega_n} \theta_1(z)\phi(z)\,dz \right) ,$$

where $\{\omega_n\}$ is a suitably chosen sequence (see [14]) of subsets of ω with $\omega_1 \subset \omega_2 \subset \ldots$, and $\cup_n \omega_n = \omega$. Since $\theta_1 = F$ on $\partial\omega_n$ for n sufficiently large n, we are done by the residue theorem.

Next we assume that F is an automorphic form of weight $-2(1-q)$. Then $F\phi$ is an automorphic form of weight -2. Its projection to $\overline{\Delta/\Gamma}$ is an abelian differential. Hence the sum of its residues is zero.

For the general case, choose an Eichler integral F_0 whose singularities are in the interior of ω such that $\alpha(F) = \alpha(F_0)$. We may then assume that $F - F_0$ is an automorphic form of weight $-2(1-q)$.

THEOREM 8.3. *For* $q \geq 2$, *the following is a commutative diagram with exact rows for every divisor* $d \leq 0$:

$$\begin{array}{ccccccccc}
0 \longrightarrow & B_{1-q}(d) & \xrightarrow{\ i\ } & PM_{1-q}(d) & \xrightarrow{\ \alpha\ } & PH^1_\Delta(\Gamma,\Pi_{2q-2}) & \xrightarrow{\ \beta\ } & B_q(-d)^* & \longrightarrow 0 \\
 & \downarrow \text{id} & & \cap & & \cap & & \downarrow \text{id} & \\
0 \longrightarrow & B_{1-q}(d) & \xrightarrow{\ i\ } & M_{1-q}(d) & \xrightarrow{\ \alpha\ } & H^1(\Gamma,\Pi_{2q-2}) & \xrightarrow{\ \beta\ } & B_q(-d)^* & \longrightarrow 0
\end{array} ,$$

where i *is the inclusion map,* $B_q(-d)^*$ *is the space of linear functionals on* $B_q(-d)$, *and* β *is the restriction of the map of* (8.13).

Proof: It is clear that i is injective (there are no automorphic polynomials). It is also obvious that Image i = Kernel α, Image $\alpha \subset$ Kernel β, and β is surjective. It remains to verify that Kernel $\beta \subset$ Image α. There is nothing to verify if $\dim B_q(-d) = 0$, as a result of the arguments in the proof of Theorem 8.1. If $\dim B_q(-d) > 0$, then we may choose the k divisors of (8.2) that satisfy (8.3), (8.4) and for some j, $0 \leq j < k$,

$$d_j < -d \leq d_{j+1}, \text{ where } d_0 = 0 .$$

We have seen that $p \in H^1(\Gamma,\Pi_{2q-2})$, then $p = \alpha(F)$ where F is given by (8.12). It is now apparent that Kernel β = Image α.

9. Generalizations.

Let χ be a character on the group Γ; that is, χ is a homomorphism of Γ into the (multiplicative) group of complex numbers of modulus one. Using χ, we can introduce an action of Γ on Π_{2q-2} by

$$p\gamma = \chi(\gamma)(\gamma^*_{1-q}p), \quad p \in \Pi_{2q-2}, \quad \gamma \in \Gamma . \tag{9.1}$$

We have developed the theory for the trivial character χ. It is quite clear that everything goes through for an arbitrary character. Furthermore, instead of using an integer $q \geq 2$, we may use a half-integer $q \geq 2$.

APPENDIX

Remarks on Automorphic Forms for Kleinian Groups

Duality and completeness theorems for automorphic forms for Kleinian groups have been used recently by many authors (see for example, Bers [6], [19], and Kra [13]). They have played a significant role in the cohomology theory outlined above. The purpose of this appendix is to outline a relatively painless way to arrive at these results. Although these theorems are well known, the simplifications in the proofs are probably new to many readers. A second purpose for this note is to state some open and (perhaps) interesting problems. In the last paragraph we shall give some references for the theorems in this appendix. The notation in the appendix is slightly different from the rest of the paper. It may be read independently of the rest of the paper.

Throughout this section D represents an open (not necessarily connected) set in the extended complex plane, $\mathbb{C} \cup \{\infty\}$, whose boundary consists of more than two points. Let p and q be half-integers such that $p + q$ is an integer. Let $f\colon D \to \mathbb{C} \cup \{\infty\}$ be a conformal map. For every function ϕ on $f(D)$ we may introduce a function $f^*_{p,q}\phi$ on D via

$$(f^*_{p,q}\phi)(z) = \phi(f(z))f'(z)^p\,\overline{f'(z)^q}, \quad z \in D.$$

It is clear that we have introduced a contravariant functor. For brevity we shall write f_p^* for $f_{p,0}^*$. Let λ_D (when there is no room for confusion we shall drop the subscript) be the Poincaré metric for D, and G be a discrete (hence discontinuous) group of conformal self-mappings on D. (In the usual situation, G is a Kleinian group, and D its region of discontinuity.) We fix an integer $q \geq 2$.

A *measurable automorphic form* of weight $(-2q)$ is an equivalence class of measurable functions μ that satisfy $A_q^*\mu = \mu$, all $A \in G$. We require that all definitions be invariant under mappings f_q^* with f conformal. For any real number $p \geq 1$, the measurable automorphic forms with

$$\|\mu\|_{q,p,G}^p = \iint_{D/G} \lambda(z)^{2-qp} |\mu(z)|^p |dz \wedge d\bar{z}| < \infty$$

form a Banach space $L_q^p(D, G)$ of *p-integrable* forms. The forms with

$$\|\mu\|_{q,\infty} = \sup\{\lambda(z)^{-q}|\mu(z)|, \ z \in D\} < \infty$$

form a Banach space $L_q^\infty(D, G)$ of *bounded* forms. For $\mu \in L_q^p(D, G)$ and $\nu \in L_q^{p'}(D, G)$ with $1/p + 1/p' = 1$ $(1/\infty = 0)$, the *Petersson* scalar product is defined by

$$(\mu, \nu)_{p,G} = \iint_{D/G} \lambda(z)^{2-2q} \mu(z) \overline{\nu(z)} \, dz \wedge d\bar{z} \ .$$

The holomorphic functions in $L_q^p(D, G)$ are denoted by $A_q^p(D, G)$. When $G = \{id\}$, it is dropped from the notation. We shall outline the proofs of the following four theorems.

THEOREM 1. *There exists a function* K_D *on* $D \times D$ *with the following properties:*

(1.1) *for fixed* $\zeta \in D$, $K_D(z, \zeta) = K_D^\zeta(z)$ *belongs to* $A_q^p(D)$ *for every* $1 \leq p \leq \infty$,

(1.2) $K_D(\zeta, z) = -\overline{K_D(z, \zeta)}$ *for* $\{z, \zeta\} \subset D$,

(1.3) *for every (one-to-one) conformal map* $f: D \to \mathbb{C} \cup \{\infty\}$ *and*

$$\{z, \zeta\} \subset D, K_{f(D)}(f(z), f(\zeta)) f'(z)^q \overline{f'(\zeta)^q} = K_D(z, \zeta),$$

(1.4) $$\iint_D \lambda(\zeta)^{2-q} |K(z, \zeta) d\zeta \wedge d\bar{\zeta}| \leq c_q \lambda(z)^q, \quad z \in D$$

where $c_q = (2q-1)/(q-1))$, *and*

(1.5) *for* $z \in D$ *and* $\phi \in A_q^p(D, G)$,

$$\phi(z) = \iint_D \lambda(\zeta)^{2-2q} K(z, \zeta) \, \phi(\zeta) \, d\zeta \wedge d\bar{\zeta} \, .$$

For a measurable function μ on D, set

$$(\beta_q \mu)(z) = \iint_D \lambda(\zeta)^{2-2q} K(z, \zeta) \mu(\zeta) \, d\zeta \wedge d\bar{\zeta} \, , \qquad z \in D,$$

whenever the integral converges absolutely.

THEOREM 2. $\beta_q: L_q^p(D, G) \to A_q^p(D, G)$ *is a bounded (surjective) projection of norm* $\leq c_q$.

For ϕ holomorphic on D, define the *Poincaré series* of ϕ by

$$(\text{Ⓗ}_q \phi)(z) = \Sigma_{A \in G} \, \phi(Az) A'(z)^q, \qquad z \in D,$$

whenever the sum converges normally.

THEOREM 3. (Bers [4]) $\text{Ⓗ}_q: A_q^1(D) \to A_q^1(D, G)$ *is a bounded surjective linear mapping of norm* ≤ 1. *Furthermore, for all* $p, 1 \leq p \leq \infty$, *and all* $\psi \in A_q^p(D, G)$, *there is a* $\phi \in A_q^p(D)$ *such that* $\psi = \text{Ⓗ}_q \phi$ *and* $\|\phi\|_{q,p} \leq c_q \|\psi\|_{q,p,G}$.

REMARK. The existence of the β_q map will show that every automorphic form is a Poincaré series. For $p = 1$, surjectivity of the Ⓗ$_q$ maps follows also from the following two facts. The adjoint of the Ⓗ$_q$ map is the inclusion map. For $p = \infty$, the inclusion map is injective with closed range.

THEOREM 4. (Bers [4]) *The Petersson scalar product establishes an anti-linear isomorphism between* $A_q^{p'}(D, G)$ *and the dual space to* $A_q^p(D, G)$, *where* $1 \leq p < \infty$ *and* $1/p + 1/p' = 1$.

Outline of proof. (1) If f is a (one-to-one) conformal map, then

$$f_q^*: A_q^p(f(D), fGf^{-1}) \to A_q^p(D, G)$$

is a surjective isometry. Furthermore for $\phi \in A_q^p(f(D), fGf^{-1})$ and $\psi \in A_q^{p'}(f(D), fGf^{-1})$,

$$(f_q^*\phi, f_q^*\psi)_{q,G} = (\phi, \psi)_{q, fGf^{-1}} .$$

(2) Assume that D is conformally equivalent to the unit disk Δ. Recall that $\lambda_\Delta(z) = (1-|z|^2)^{-1}$, $z \in \Delta$, and that $\lambda(z)|dz|$ is a conformal invariant. The *Bergmen kernel* function k_D may be defined by setting $k_\Delta(z, \zeta) = [\pi(1-z\bar{\zeta})^2]^{-1}$, $\{z, \zeta\} \subset \Delta$, and requiring that $k_D(z, \zeta)\, dz \wedge d\bar{\zeta}$ be a conformal invariant.

LEMMA 1. *For* $z \in D$, $\iint_D \lambda(\zeta)^{2-q} |k_D(z, \zeta)^q d\zeta \wedge d\bar{\zeta}| = (2\pi^{1-q}/q-1)\lambda(z)^q$.

Using conformal invariance, the proof of the Lemma is reduced to the case $D = \Delta$ and $z = 0$. This calculation is completely trivial.

Set for $\{z, \zeta\} \subset D$

$$K_D(z, \zeta) = (1/2)(2q-1)\pi^{q-1}\, i\, k_D(z, \zeta)^q .$$

It is obvious that K satisfies (1.1), (1.2), (1.3). Also (1.4) holds (with equality) as a result of Lemma 1.

(3) Using Fubini, Cauchy-Schwarz, and the invariance properties of K and λ, it is easily shown that β_q is a bounded operator in $L_q^p(D, G)$ of norm $\leq c_q$. Using Fubini, once again, we conclude that $(\beta_q\mu)(z)$ converges absolutely for almost all $z \in D$ whenever $\mu \in L_q^p(D, G)$.

(4) Assume $D = \Delta$. For fixed $\zeta \in \Delta$, $K_\Delta(z, \zeta)$ is bounded and bounded away from zero. Thus $(\beta_q\mu)(z)$ converge absolutely for all $z \in \Delta$. Since

we may differentiate $\beta_q \mu$ under the integral sign (and obtain a convergent integral) $\beta_q\mu \in A_q^p(D,G)$. Thus for $\phi \in A_q^p(D,G)$ we may compute $\beta_q\phi$ formally. To evaluate $(\beta_q\phi)(z)$ it suffices to assume $z = 0$ (because of the presence of many isometries). Starting from the mean value property it is easy to show that $(\beta_q\phi)(0) = \phi(0)$. The proof of Theorems 1 and 2 is complete for connected and simply connected D.

(5) The operator β is "self-adjoint" in the sense that $(\beta_q\mu,\nu)_{q,G} = (\mu,\beta_q\nu)_{q,G}$, for $\mu \in L_q^p(D,G)$ and $\nu \in L_q^{p'}(D,G)$.

(6) Using (5), Hahn-Banach, and F. Riesz, it is quite clear how to prove Theorem 4 from Theorems 1 and 2. We also have two corrollaries to Theorem 4.

COROLLARY 1. *Let* $\mu \in L_q^p(D,G)$. *Then* $\beta_q\mu = 0$ *if and only if* $(\mu,\psi)_{q,G} = 0$, *all* $\psi \in A_q^{p'}(D,G)$.

COROLLARY 2. Let $\nu \in L_q^{p'}(D,G)$. *Then* $\nu \in A_q^{p'}(D,G)$ *if and only if* $(\mu,\nu)_{q,G} = (\beta_q\mu,\nu)_{q,G}$, *all* $\mu \in L_q^p(D,G)$.

(7) Now we prove Theorem 3. The existence of the $Ⓗ_q$ map for $p = 1$ is quite standard by now. Bers' [4] proof cannot be improved, and applies to this general setting. To show surjectivity, let χ be the characteristic function of a fundamental domain for G in D. Then for $\mu \in L_q^p(D,G)$ we have, $\beta_q\mu = (Ⓗ_q \circ \beta_q)(\chi\mu)$.

COROLLARY. *There exists a bounded linear projection of* $A_q^\infty(D)$ *onto* $A_q^\infty(D,G)$.

It remains to verify Theorems 1 and 2 for arbitrary D.

(8) We return to the case D is conformally equivalent to Δ. Set for $\{z,\zeta\}$,

$$F_D(z,\zeta) = \sum_{A\in G} K_D(Az,\zeta)\,A'(z)^q \ .$$

THEOREM 1′. *We have*

$$F(\zeta,z) = -\overline{F(z,\zeta)}\ , \qquad (1.1)'$$

(1.2)′ $$F(Az,\zeta)A'(z)^q = F(z,\zeta), \quad A \in G,$$

(1.3)′ $F(Az,A\zeta)\,A'(z)^q\overline{A'(\zeta)^q} = F(z,\zeta)$, $A \in N(G)$, *the normalizer of* G *in the group of all conformal self-maps of* D,

(1.4)′ *if* $F^{\zeta}(z) = F(z,\zeta)$, *then* $F^{\zeta} \in A_q^p(D,G)$ *for every* $1 \le p \le \infty$, *and*

(1.5)′ $$\phi(z) = (\phi, -F^z)_{q,G}, \ \phi \in A_q^p(D,G) .$$

Only the fact that $F^{\zeta} \in A_q^{\infty}(D,G)$ needs verification. This is a consequence of

LEMMA 2. (Godement [23]) *Let* G *be a discrete group of conformal selfmappings of* Δ. *Then for any* $z \in \Delta$,

$$\Sigma_{A\in G}(1-|Az|^2)^q \le C(q,G) .$$

(9) We now assume that D is connected. Let $\pi : \Delta \to D$ be a universal covering map for D, and let Γ be the covering group of π. Let

$$H = \{h;\ h \text{ is a Möbius transformation and } \pi \circ h = g \circ \pi \text{ for some } g \in G\} .$$

Define K_D on $D \times D$ by

$$K_D(\pi(z),\pi(\zeta))\,\pi'(z)^q\,\overline{\pi'(\zeta)^q} = F_{\Delta}(z,\zeta) ,$$

where F is formed with respect to the covering group Γ. It is easy to check that K is well defined and satisfies the properties of Theorem 1.

(10) For general D, write $D = \cup_i D_i$ with D_i components of D. We define

$$K_D(z,\zeta) = \begin{cases} K_{D_i}(z,\zeta) & \text{if } \{z,\zeta\} \subset D_i \text{ and} \\ 0 & \text{if } z \in D_i \text{ and } \zeta \in D_j \text{ with } i \neq j . \end{cases}$$

This completes the proofs of Theorems 1 and 2, and hence Theorems 3 and 4 as well.

(11) Let us return to the case of connected D. The operator

$$\pi_q^*: L_q^p(D,G) \to L_q^p(\Delta,H)$$

is a surjective isometry. Furthermore $\beta_q \circ \pi_q^* = \pi_q^* \circ \beta_q$. This observation gives an alternate way of lifting the results from Δ to arbitrary connected D. Standard tricks may now be used for the disconnected case.

PROBLEMS. (1) Is $A_q^p(D,G) \subset A_q^\infty(D,G)$? For D connected and simply connected and G the trivial group, the answer is (trivially) 'yes.' Drasin and Earle [20] have given an affirmative answer for G a finitely generated Fuchsian group and D the unit circle. The general case is open.

(2) Clearly $A_q^2(D,G)$ is a Hilbert space. Thus the function K_D is the familiar Bergman kernel formed with respect to this Hilbert space. Is there another characterization of the function K_D?

BIBLIOGRAPHICAL DATA. The completeness of Poincaré series (for certain Fuchsian groups) was first proven by Poincaré. The scalar product, as well as some completeness proofs are due to Petersson [25]. The development schetched in this paper originated with two papers by Bers, [18] and [4]. It was subsequently simplified by Earle [21]. Some of the ideas of this note were discovered independently by Earle (see [22]). The problem of finding reproducing formulas for holomorphic functions is also treated by Innis [24]. Ahlfors [17] gave an elegant and independent proof of Godement's inequality [23].

REFERENCES

[1] L. V. Ahlfors, "Finitely generated Kleinian groups," *Amer. J. Math.*, 86 (1964), 413-429 and 87 (1965), 759.

[2] L. V. Ahlfors, "Eichler integrals and Bers' area theorem," *Mich. Math. J.*, 15 (1968), 257-263.

[3] L. V. Ahlfors, "The structure of a finitely generated Kleinian group," *Acta Math.*, 122 (1969), 1-17.

[4] L. Bers, "Automorphic forms and Poincaré series for infinitely generated Fuchsian groups," *Amer. J. Math.*, 87 (1965), 169-214.

[5] L. Bers, "An approximation theorem," *J. Analyse Math.*, 14 (1965), 1-4.

[6] L. Bers, "Inequalities for finitely generated Kleinian groups," *J. Analyse Math.*, 18 (1967), 23-41.

[7] L. Bers, "On Ahlfors' finiteness theorem," *Amer. J. Math.*, 89 (1967), 1078-1082.

[8] P. Curran, "Cohomology of F-groups," (to appear).

[9] M. Eichler, "Eine Verallgemeinerung der Abelschen Integrale," *Math. Z.*, 67 (1957), 267-298.

[10] L. Greenberg, "On a theorem of Ahlfors and conjugate subgroups of Kleinian groups," *Amer. J. Math.*, 89 (1967), 56-68.

[11] R. C. Gunning, "The Eichler cohomology groups and automorphic forms," *Trans. Amer. Math. Soc.*, 100 (1961), 44-62.

[12] S. Y. Husseini and M. I. Knopp, "Eichler cohomology and automorphic forms," (to appear).

[13] I. Kra, "On cohomology of Kleinian groups," *Ann. of Math.*, 89 (1969), 533-556.

[14] I. Kra, "On cohomology of Kleinian groups: II," *Ann. of Math.*, 90 (1969), 576-590.

[15] R. Nevanlinna, *Uniformisierung*, Springer, Berlin, 1953.

[16] A. Weil, "Remarks on cohomology of groups," *Ann. of Math.*, 80 (1964), 149-157.

ADDITIONAL REFERENCES FOR APPENDIX

[17] L. V. Ahlfors, "Eine Bemerkung über Fucchssche Gruppen," *Math. Z.*, 84 (1964), 244-245.

[18] L. Bers, "Completeness theorems for Poincaré series in one variable," *Proc. Internat. Symp. on Linear Spaces*, Jerusalem, 1960, 88-100.

[19] L. Bers, "Extremal quasiconformal mappings," (to appear).

[20] D. Drasin and C. J. Earle, "On the boundedness of automorphic forms," *Proc. Amer. Math. Soc.*, 19 (1968), 1039-1042.

[21] C. J. Earle, "A reproducing formula for integrable automorphic forms," *Amer. J. Math.*, 88 (1966), 867-870.

[22] C. J. Earle, "Some remarks on Poincaré series," *Compos. Math.*, 21 (1969), 167-176.

[23] R. Godement, "Séries de Poincaré et Spitzenformen," *Sém. Henri Cartan, 10e anée* (1957/1958), exposé 10.

[24] G. S. Innis Jr., "Some reproducing kernels for the unit disc," *Pac. J. Math.*, 14 (1964), 177-187.

[25] H. Petersson, "Theorie der automorphen Formen beliebigen reeler Dimension und ihre Darstellung eine neue Art Poincaréscher Reihen," *Math. Ann.*, 103 (1930), 369-436.

State University of New York at Stony Brook
Stony Brook, New York

ON THE DEGENERATION OF RIEMANN SURFACES

by Aaron Lebowitz

1. Introduction

This paper deals with compactifying the embedding of the Torelli space in the Siegel upper half-plane [see 5]. The question that arises is, what happens if one tends to the "boundary" of Torelli space? Or, to put it another way, how do the period matrices behave?

In this paper we find results confirming one's expectations in certain sufficiently typical types of approach to the boundary, namely, in the cases where a handle is dropped by being pinched off and where the surface splits into two of lower genera by pinching a waist. The results are enunciated in theorems one and two below. Although the theorems seem special we note that by a Siegel modular transformation we have actually a large class of cases, an illustration of which is found in the last section.

The machinery for the most general case has not yet been developed. One could anticipate that the results would somehow be a mixture of the two types studied here.

2. Basic Definitions and Formulas

Let S be a Riemann surface of genus g. We will denote by (Γ, Δ), with $\Gamma = (\gamma_1, \gamma_2, \dots, \gamma_g)$ and $\Delta = (\delta_1, \delta_2, \dots, \delta_g)$, a fixed one-dimensional homology basis on S. The cycles, γ_i and δ_i, satisfy

$$\begin{aligned} KI(\gamma_i, \gamma_j) &= 0 \\ KI(\delta_i, \delta_j) &= 0 \qquad\qquad i, j = 1, 2, \dots, g\ . \\ KI(\gamma_i, \delta_j) &= \delta_{ij} \end{aligned}$$

KI denotes the skew-symmetric, bilinear, integral valued intersection number.

Note that the genus of any surface may be computed from the number of independent cycles on that surface which satisfy the intersection property $KI(\gamma_i, \delta_j) = \delta_{ij}$. If we have 2g cycles with this property the genus of the surface is g.

Consider a basis, $d\zeta_1, d\zeta_2, \ldots, d\zeta_g$, for the abelian differentials of the first kind on S, such that

$$\int_{\gamma_j} d\zeta_i = \delta_{ij} \qquad \gamma_j \in \Gamma \ .$$

The periods, π_{ij}, of the differentials with respect to (Γ, Δ) are defined to be

$$\pi_{ij} = \int_{\delta_j} d\zeta_i \qquad \delta_j \in \Delta,\ i, j = 1, \ldots, g \ . \tag{2.1}$$

These periods may be arranged in the form of a $g \times g$ matrix. This matrix is called the Riemann period matrix for the surface with respect to (Γ, Δ) and is denoted by (π_{ij}).

The period matrix is symmetric and has positive definite imaginary part.

There are other matrices which have these two properties but, nevertheless, are not period matrices for any Riemann surface. The totality of all such matrices form the Siegel upper half plane, denoted by $\mathfrak{S}_g$.

We recall the definitions of two groups. The homogeneous Siegel modular group, $S_p(g, Z)$, of degree g, is the set of $2g \times 2g$ matrices M, with integral entries, satisfying

$$M\begin{pmatrix} O_g & I_g \\ -I_g & O_g \end{pmatrix} M^T = \begin{pmatrix} D & C \\ B & A \end{pmatrix}\begin{pmatrix} O_g & I_g \\ -I_g & O_g \end{pmatrix}\begin{pmatrix} D & C \\ B & A \end{pmatrix}^T = \begin{pmatrix} O_g & I_g \\ -I_g & O_g \end{pmatrix}$$

where T denotes transpose, O_g and I_g are respectively the $g \times g$ zero and unit matrices, and A, B, C, D are $g \times g$ submatrices in the indicated positions. The inhomogeneous Siegel modular group, $\mathfrak{M}^g$, is isomorphic to $S_p(g, Z)/\{I_{2g}, -I_{2g}\}$, where I_{2g} is the $2g \times 2g$ unit matrix, and acts on $\mathfrak{S}_g$ by

$$(\pi'_{ij}) = (A\pi + B)(C\pi + D)^{-1} \equiv M \circ (\pi_{ij}) \tag{2.2}$$

where (π_{ij}) and $(\pi'_{ij}) \in \mathfrak{S}_g$ and the operations are matrix operations. The homogeneous and inhomogeneous groups both have the multiplication structures defined by the conventional left composition (matrix multiplication for the homogeneous group).

In equation (2.2), the period matrices are calculated with respect to canonical homology bases which are related by the following:

$$(\Gamma', \Delta') = M(\Gamma, \Delta) \tag{2.3}$$

where M operates on (Γ, Δ) on the left as a column vector. (See [5].)

The following notation is adapted from Patt [2]. Capital letters, X, Y, P, Q, will denote points on S and the corresponding small letters, x, y, p, q, local parameters about the points, X, Y, P, Q, such that $x(X) = 0$, etc. and $x(Y)$ will denote x evaluated at the point Y, where Y is in the parameter disk of the local parameter, x. It will not generally be necessary to indicate what local parameters are being used. The parameters will be indicated, where necessary, for differentiation and integration. Functions and integrals, on the surface, will have points as variables unless a parameter is required.

Let $\partial\pi$ be the boundary of the simple connected surface, π, which is obtained by the canonical dissection of S determined by the fixed canonical basis (Γ, Δ). Then

$$\partial\pi = \sum_{i=1}^{g} \gamma_i^+ + \delta_i^- - \gamma_i^- - \delta_i^+ \quad .$$

The superscripts + and − are determined so that if a right-angled coordinate system is set up with the intersection of γ_i and δ_i as the origin, and the positive x-axis is taken to be the + direction along γ_i, then the positive y-axis is the + direction for δ_i, γ_i^+ will represent the side of γ_i toward the upper half plane, etc.

Let α and β be any abelian integrals on S, α and $d\beta$ having at most poles in the interior of π and being analytic on $\partial\pi$. Then

$$\int_{\partial\pi} \alpha\, d\beta = \sum_{i=1}^{g} \left[\int_{\gamma_i^+} \alpha\, d\beta - \int_{\gamma_i^-} \alpha\, d\beta + \int_{\delta_i^-} \alpha\, d\beta - \int_{\delta_i^+} \alpha\, d\beta \right]$$

(2.4)
$$= \sum_{i=1}^{g} \left[\int_{\gamma_i} (\alpha^+ - \alpha^-)\, d\beta + \int_{\delta_i} (\alpha^- - \alpha^+)\, d\beta \right]$$

$$= \sum_{i=1}^{g} \left[\int_{\gamma_i} d\alpha \int_{\delta_i} d\beta - \int_{\delta_i} d\alpha \int_{\gamma_i} d\beta \right]$$

where α^+ is the value of α on γ^+, etc.

Let η_{XY} denote an integral of the third kind on S with residue -1 at X and $+1$ at Y, and with zero γ periods, $\int_{\gamma_i} d\eta_{XY} = 0$, $i = 1, \ldots, g$. Such an integral is called a normal integral of the third kind.

In equation (2.4), let $\alpha = \zeta_i$ and $\beta = \eta_{XY}$. Then, applying the residue theorem to the left hand side of (2.4), we obtain

(2.5)
$$\zeta_i(Y) - \zeta_i(X) = \frac{1}{2\pi i} \int_{\delta_i} d\eta_{XY} = \frac{1}{2\pi i} \int_{\delta_i} \frac{\partial \eta_{XY}(z)}{\partial z}\, dz \quad .$$

If we differentiate (2.5) with respect to any parameter, y, about Y, we obtain

$$\frac{\partial \zeta_i(y)}{\partial y} = \frac{1}{2\pi i} \int_{\delta_i} \frac{\partial^2}{\partial y\, \partial z} \eta_{Xy}(z)\, dz \quad .$$

We substitute this result in the integral of equation (2.1) defining π_{ij}, and we obtain

(2.6)
$$\pi_{ij} = \int_{\delta_j} d\zeta_i = \int_{\delta_j} \frac{\partial \zeta_i(y)}{\partial y}\, dy = \frac{1}{2\pi i} \int_{\delta_j} \int_{\delta_i} \frac{\partial^2}{\partial y\, \partial z} \eta_{Xy}(z)\, dz\, dy \quad .$$

The period matrix is symmetric, that is

(2.7)
$$\pi_{ij} = \pi_{ji} \quad .$$

We also obtain the "law of the interchange of argument and parameter," formula (2.8), below. Apply equation (2.4), letting $\alpha = \eta_{XY}$ and $\beta = \eta_{ZW}$. We make η_{XY} single valued in π by slitting π in an arc, λ, from X to Y, with λ completely in the interior of π. Then, applying the residue theorem Then, applying the residue theorem to $\int \alpha \, d\beta$ to the domain obtained after slitting along λ, we get

$$\eta_{XY}(W) - \eta_{XY}(Z) = \frac{1}{2\pi i} \int_{\partial\pi + \lambda} \eta_{XY} \, d\eta_{ZW} = \frac{1}{2\pi i} \int_{\partial\pi} \eta_{XY} \, d\eta_{ZW} + \frac{1}{2\pi i} \int_{\lambda} \eta_{XY} \, d\eta_{ZW} \ ,$$

note that the λ under the integral really stands for both sides of the slit λ.

The integral over $\partial\pi$ is zero by equation (2.4) since η_{XY} and η_{ZW} are normal integrals of the third kind on S.

$$\frac{1}{2\pi i} \int_{\lambda} \eta_{XY} \, d\eta_{ZW} = \eta_{ZW}(Y) - \eta_{ZW}(X),$$

since η_{XY} has a jump of $2\pi i$ from one side of λ to the other. We thus obtain

$$\eta_{XY}(W) - \eta_{XY}(Z) = \eta_{ZW}(Y) - \eta_{ZW}(X) \ , \tag{2.8}$$

independent of the branch of η_{XY} chosen.

We will be dealing with more than one surface in this paper. If a surface is designated by S with subscripts or superscripts, we will denote its periods, integrals, boundary of its canonical dissection, etc. with the corresponding subscript or superscript.

3. Degeneration by handle removal

We summarize now the procedure used in this section.

Start with a specific surface S of genus g and construct a specific surface S* of genus $g^* = g + 1$. Obtain formulas relating differentials on the two surfaces S and S*. Using these formulas we calculate the periods π^*_{ij} on S* in terms of the π_{ij} on S wherever possible. We get estimates for all of the π^*_{ij}, i, j = 1, ..., g* = g + 1.

We now give the prescription for changing the surface S of genus g to obtain the surface S* of genus g + 1.

Let $\eta_{P_0Q_0}$ be a normal integral of third kind on S where P_0 and Q_0 are not on any of the fixed cycles of the canonical homology basis, (Γ, Δ), for S. For convenience, we will index the g cycles in Γ and Δ by i = 2, 3, ..., g + 1.

Consider the local parameters

$$t_1 \equiv e^{+\eta_{P_0Q_0}} \quad \text{on S about } P_0 \tag{3.1}$$

and

$$t_2 \equiv e^{-\eta_{P_0Q_0}} \quad \text{on S about } Q_0 \quad . \tag{3.2}$$

Let D_{P_0} and D_{Q_0} be parameter disks about P_0 and Q_0, respectively. These disks are chosen so as to have null intersection with the basis cycles, (Γ, Δ). In the parameter disk D_{P_0}, the set of points P on S, near P_0, such that $\operatorname{Re} \eta_{P_0Q_0}(P) = \log \epsilon$, forms a circle of radius ϵ, $|t_1| = \epsilon$. Denote this circle by C_{P_0}. Similarly, the set of points Q on S, near Q_0, such that $\operatorname{Re} \eta_{P_0Q_0}(Q) = \log 1/\epsilon$, forms a circle of radius ϵ in the parameter disk D_{Q_0}, $|t_2| = \epsilon$. Denote this circle by C_{Q_0}. Take fixed points P' and Q' in D_{P_0} and D_{Q_0}, respectively, on C_{P_0} and C_{Q_0}, respectively, with

$$\eta_{P_0Q_0}(P') = \log \epsilon + i\alpha'$$

and

$$\eta_{P_0Q_0}(Q') = \log \frac{1}{\epsilon} + i\beta' \ .$$

Then

$$\eta_{P_0Q_0}(P') = \eta_{P_0Q_0}(Q') + 2 \log \epsilon + i(\alpha' - \beta') \ . \tag{3.3}$$

We identify the points P' and Q' and, furhter, we identify the points P and Q on C_{P_0} and C_{Q_0}, respectively, if they satisfy the following equation:

$$\eta_{P_0Q_0}(P) = \eta_{P_0Q_0}(Q) + 2\log\epsilon + i(\alpha' - \beta') \ . \tag{3.4}$$

Prescription A: Let a Jordan arc, γ_1, be drawn on S from P' to Q' which does not intersect any of the cycles in (Γ,Δ). Delete the interiors of C_{P_0} and C_{Q_0} and identify the boundaries of C_{P_0} and C_{Q_0} as described above. We call the identified curve, δ_1. We see that we have formed from S a new topological surface $S^{\#}$, which will serve as a base for a new Riemann surface S^*.

LEMMA 1. *The genus of* $S^{\#}$ *is* $g+1$.

We denote the basis by $(\Gamma^{\#},\Delta^{\#})$ where

$$\Gamma^{\#} = (\gamma_1^{\#},\gamma_2^{\#},\dots,\gamma_{g+1}^{\#})$$

and

$$\Delta^{\#} = (\delta_1^{\#},\delta_2^{\#},\dots,\delta_{g+1}^{\#}) \ .$$

$$\gamma_1^{\#} = \gamma_1 \quad , \quad \delta_1^{\#} = \delta_1 \quad \text{and} \quad \gamma_i^{\#} = \gamma_i$$

$$\delta_i^{\#} = \delta_i \quad , \qquad i = 2,\dots,g+1 \ ,$$

the γ_i and δ_i being the fixed canonical basis for S.

Denote the domain which S and $S^{\#}$ have in common by S_0. Then $\partial S_0 = C_{P_0} + C_{Q_0}$. We orient the curves C_{P_0} and C_{Q_0} so that when we traverse these curves in the positive sense S_0 is on the left. S_0 itself is all of S with the interiors of C_{P_0} and C_{Q_0} excised.

LEMMA 2. $S^{\#}$ *may be endowed with an analytic structure to make it a Riemann surface,* S^*.

Proof: In the interior of S_0 we may use the same local parameters on $S^{\#}$ as were used on S.

We need only find local parameters about points on the curve δ_1. A neighborhood of δ_1 is made up of two "annuli," A_{P_0} and A_{Q_0}, on S. The "annulus" A_{P_0} is mapped by t_1, equation (3.1), to an annulus with concentric circles for boundaries. Similarly, A_{Q_0} is mapped by t_2, equation (3.2), to an annulus with concentric circles as boundaries. The inner boundaries are C_{P_0} and C_{Q_0}, respectively.

Now C_{P_0} and C_{Q_0} have been identified by equation (3.4) and thus if we invert A_{Q_0} with respect to the circle C_{Q_0}, we can weld the two annuli A_{P_0} and A_{Q_0} to form a single annulus, A, with the identified curve δ_1 in the interior of A. The inversion process is performed by means of the function $Z = \epsilon^2/t_2$, which is analytic in A_{Q_0}. Now we choose local parameters with centers on δ_1 in the annulus A.

We note that these parameters will overlap only half neighborhoods of parameters from S. These half neighborhoods will be related analytically with the parameters about points on δ_1 using either t_1 or t_2.

We have thus constructed a new Riemann surface, S*, from S of genus $g^* = g + 1$ where we use as a specified fixed set of cycles $(\Gamma^*, \Delta^*) = (\Gamma^{\#}, \Delta^{\#})$.

Definition 1. We define the "pinching off a handle" to be the degeneration of S*, caused by allowing ϵ to tend to zero, where S* is constructed by means of the prescription A, above, with the limit surface to which S* degenerates being S.

We obtain now a variational formula which enables us to compare integrals of the third kind on S and S*. Formulas of this type can be found in Rauch [3, 4], Patt [2] and Schiffer and Spencer [6].

LEMMA 3. *Let* S *and* S* *be two Riemann surfaces which have the domain* S_0 *in common. Assume that we have a set of cycles in* S_0 *which is* *canonical with respect to* S *and can be extended to a canonical basis for* S^*. *Then for any four points* X, Y, Z *and* W *in* S_0,

$$\Delta\eta_{XY}(Z) - \Delta\eta_{XY}(W) = \frac{1}{2\pi i}\int_{\partial S_0} \eta_{XY}\, d\eta^*_{ZW} , \tag{3.5}$$

and η_{XY} *and* η^*_{ZW} *are normal integrals of the third kind on* S *and* S^*, *respectively, and* $\Delta\eta_{XY} = \eta^*_{XY} - \eta_{XY}$.

Proof: Let λ be a slit in S_0, from X to Y, which does not intersect any of the elements in (Γ, Δ). On the region obtained by a canonical dissection of S_0 along the canonical basis for S and slitting S_0 along λ we find that a determination of η_{XY} is a single valued analytic function. Similarly, a determination of η^*_{XY} will be a single valued analytic function.

We integrate $\eta_{XY}\, d\eta^*_{ZW}$ around the boundary of this region, that is, around $\partial\pi + \partial S_0 + \lambda$. Apply Cauchy's residue theorem and we obtain

$$\eta_{XY}(W) - \eta_{XY}(Z) = \frac{1}{2\pi i}\int_{\partial\pi + \partial S_0 + \lambda} \eta_{XY}\, d\eta^*_{ZW} . \tag{3.6}$$

λ stands for both sides of λ.

The integral on the right hand side of (3.6) may be evaluated as follows:

$$\frac{1}{2\pi i}\int_{\partial\pi} \eta_{XY}\, d\eta^*_{ZW} + \frac{1}{2\pi i}\int_{\lambda} \eta_{XY}\, d\eta^*_{ZW} + \frac{1}{2\pi i}\int_{\partial S_0} \eta_{XY}\, d\eta^*_{ZW} . \tag{3.7}$$

The integral over $\partial\pi$ is zero when we apply equation (2.4) since both η_{XY} and η^*_{ZW} are normal integrals of the third kind.

η_{XY} has a jump of $2\pi i$ across the curve λ and thus we obtain

$$\frac{1}{2\pi i}\int_{\lambda} \eta_{XY}\, d\eta^*_{ZW} = \eta^*_{ZW}(Y) - \eta^*_{ZW}(X) . \tag{3.8}$$

By the law of the interchange of argument and parameter equation (2.8), we obtain

$$\eta^*_{ZW}(Y) - \eta^*_{ZW}(X) = \eta^*_{XY}(W) - \eta^*_{XY}(Z) . \tag{3.9}$$

Combining the results of (3.6), (3.7), (3.8), and (3.9) we obtain

$$(3.5) \qquad \Delta \eta_{XY}(Z) - \Delta \eta_{XY}(W) = \frac{1}{2\pi i} \int_{\partial S_0} \eta_{XY}\, d\eta^*_{ZW} \; .$$

We are now ready to study the periods on S^* with respect to (Γ^*, Δ^*), defined above.

Consider the local parameters t_1 and t_2 defined by equations (3.1) and (3.2). On the circles, C_{P_0} and C_{Q_0}, the parameters t_1 and t_2 are related by equation (3.4) and we obtain

$$(3.10) \qquad t_1 = \frac{\epsilon^2 \phi}{t_2} \quad \text{with} \quad |\phi| = 1 \; .$$

Using our definition for the periods in section 2, equation (2.1), we have

$$\pi^*_{ij} = \int_{\delta^*_j} d\zeta^*_i \qquad i, j = 1, \ldots, g^* = g + 1 \; .$$

Let us consider the following:

$$(3.11) \qquad \frac{1}{2\pi i} \int_{\delta^*_1} d\eta^*_{ZW} = \frac{1}{2\pi i} \int_{C_{P_0}} d\eta^*_{ZW} = \zeta^*_1(W) - \zeta^*_1(Z) \; .$$

The integral on the left can be looked on as a function of W. Then when W crosses the cycle, δ^*_1, the integral jumps by one, and is regular everywhere else in S^*. Now consider the function $\eta_{P_0Q_0}$, it is regular everywhere on S^* and it has a jump of $2 \log \epsilon + i(\alpha' - \beta')$ across the cycle, δ^*_1. Thus, the integral considered as a function of W, and

$$\frac{\eta_{P_0Q_0}}{2 \log \epsilon + i(\alpha' - \beta')}$$

will differ on S^* by at most a constant. We obtain the following:

$$(3.12) \qquad \zeta^*_1(W) - \zeta^*_1(Z) = \frac{1}{2\pi i} \int_{\delta^*_1} d\eta^*_{ZW} = \frac{\eta_{P_0Q_0}(W)}{2 \log \epsilon + i(\alpha' - \beta')} + C \; .$$

Differentiate equation (3.12) with respect to a local parameter w about W and we obtain

$$d\zeta_1^*(w) = \frac{d\eta_{P_0Q_0}(w)}{2\log\epsilon + i(\alpha'-\beta')} . \tag{3.13}$$

We use this equation to find π_{1j}^*. We obtain

$$\pi_{1j}^* = \int_{\delta_j^*} d\zeta_1^* = \int_{\delta_j^*} \frac{d\eta_{P_0Q_0}(w)}{2\log\epsilon + i(\alpha'-\beta')}\, dw, \quad j = 1, \ldots, g^* = g+1 . \tag{3.14}$$

For $j = 2, \ldots, g+1$, $\delta_j^* = \delta_j$ and so we have

$$\pi_{1j}^* = \frac{1}{2\log\epsilon + i(\alpha'-\beta')} [\zeta_j(Q_0) - \zeta_j(P_0)] . \tag{3.15}$$

The expression in brackets is independent of ϵ and so we find that

$$\pi_{1j}^* = 0\left(\frac{1}{\log\epsilon}\right) \qquad \text{for } j = 2, \ldots, g+1 .$$

Since we are now integrating a function on S we may replace δ_1^* by either C_{P_0} or C_{Q_0} so long as we maintain the proper orientation. C_{P_0} and C_{Q_0} each surround a pole of $d\eta_{P_0Q_0}$ and so we find

$$\pi_{11}^* = \int_{\delta_1^*} \frac{d\eta_{P_0Q_0}(w)}{2\log\epsilon + i(\alpha'-\beta')}\, dw = \frac{2\pi i}{2\log\epsilon + i(\alpha'-\beta')} . \tag{3.16}$$

We obtain

$$\pi_{11}^* = 0\left(\frac{1}{\log\epsilon}\right) .$$

We state the preceding results in the form of the following lemma.

LEMMA 4. *Let* S^* *be constructed from* S *as described above. Then when we pinch a handle on* S^*, *definition 1, above, we have*

$$\pi^*_{1j} = O(\frac{1}{\log \epsilon}) \qquad j = 1,\dots,g^* = g+1 \ .$$

By equation (2.7), $\pi^*_{1j} = \pi^*_{j1}$ and so we also have that

$$\pi^*_{j1} = O(\frac{1}{\log \epsilon}) \qquad j = 1,\dots,g^* = g+1 \ .$$

We wish to examine the remaining periods in (π^*_{j1}), that is, π^*_{ij} for $i, j = 2,\dots,g+1$.

In order to accomplish this end we consider the integral in equation (3.5), above. η_{XY} is defined on S and so we expand $\eta_{XY}(t)$ in the local parameter t_1 about P_0. We obtain

$$(3.17) \qquad \eta_{XY}(t) = \eta_{XY}(P_0) + \eta'_{XY}(P_0)\, t_1 + \frac{1}{2!}\eta''_{XY}(P_0)\, t_1^2 + \cdots .$$

The integral over ∂S_0 is the same as the integral over C_{P_0} and C_{Q_0} with proper orientation. We will consider first the integral over C_{P_0}. We substitute for η_{XY} using equation (3.17) and we obtain

$$(3.18) \qquad \frac{1}{2\pi i}\int_{C_{P_0}} \eta_{XY}\, d\eta^*_{ZW} = \frac{1}{2\pi i}\left[\eta_{XY}(P_0)\int_{C_{P_0}} d\eta^*_{ZW} + \eta'_{XY}(P_0)\int_{C_{P_0}} t_1 d\eta^*_{ZW} + \frac{\eta''_{XY}(P_0)}{2!}\int_{C_{P_0}} t_1^2\, d\eta^*_{ZW} + \cdots\right] .$$

By equations (3.12) and (3.16), above, the first integral on the right is $O(1/\log \epsilon)$. For the second, and all succeeding integrals, we replace t_1 by t_2 using equation (3.10), above, and we replace integration over C_{P_0} by the appropriate integral over C_{Q_0}. We obtain

$$(3.19) \qquad \eta'_{XY}(P_0)\,\epsilon^2\phi \int_{C_{Q_0}} \frac{1}{t_2}\, d\eta^*_{ZW} + \text{terms with } \epsilon^4 \text{ at least.}$$

We may replace integration over C_{Q_0} by some curve, C, which is homologous to C_{Q_0} so long as Z and W are not in the region bounded by C_{Q_0} and C.

Then, by an argument in Schiffer and Spencer [6, Chapter 7], (3.19) is $0(\epsilon^2)$.

We get similar results when integrating over C_{Q_0} and so we obtain

$$\Delta\eta_{XY}(Z) - \Delta\eta_{XY}(W) = 0(\frac{1}{\log\epsilon})$$

for Z and W in a closed subdomain of S_0.

We differentiate with respect to any local parameter z about Z and we obtain

$$d\eta^*_{XY}(z) - d\eta_{XY}(z) = 0(\frac{1}{\log\epsilon}) \quad , \tag{3.20}$$

this result being obtained by differentiating under the integral sign in (3.18).

In the equation defining π^*_{ij} we replace $d\eta^*$ by $d\eta$ and we find that the error is just $0(1/\log\epsilon)$, and we obtain

$$\begin{aligned}\pi^*_{ij} &= \int_{\delta^*_j}\int_{\delta^*_i} \frac{\partial^2\eta^*_{Xy}(z)}{\partial y\,\partial z}\,dz\,dy \\ &= \int_{\delta^*_j}\int_{\delta^*_i} \frac{\partial^2\eta_{Xy}(z)}{\partial y\,\partial z}\,dz\,dy + 0(\frac{1}{\log\epsilon}) \quad .\end{aligned} \tag{3.21}$$

For $i, j = 2,\ldots,g^* = g+1$, we replace δ^*_i by δ_i and δ^*_j by δ_j, and so we have

$$\pi^*_{ij} = \pi_{ij} + 0(\frac{1}{\log\epsilon}) \qquad i, j = 2,\ldots,g^* = g+1 \ . \tag{3.22}$$

We combine the result of equation (3.22) with Lemma 4, above, as the following theorem.

THEOREM 1. *Let* S^* *be constructed from* S *as described above. Then when we pinch a handle on* S^*, *definition 1, above, we have*

$$\pi^*_{1j} = \pi^*_{j1} = 0(\frac{1}{\log\epsilon}) \ , \qquad j = 1,\ldots,g+1, \text{ and} \tag{3.23}$$

$$\pi_{ij}^{*} = \pi_{ij} + O\left(\frac{1}{\log \epsilon}\right), \quad i, j = 2, \ldots, g+1 . \tag{3.22}$$

4. Degeneration by Splitting into Surfaces of Lower Genera

We turn now to study the degeneration of a surface which occurs when a curve which is homologous to zero on the surface is shrunk to a point. Our procedure here is similar to that of Section 3. Start with two fixed surfaces and construct from them a specific surface, S^*. The shrinking will be analogous to the process described in Section 3.

If the two surfaces, S_1 and S_2, which are fused to form S^* are of genera, g_1 and g_2, respectively, S^* will be of genus $g^* = g_1 + g_2$.

We study the period matrix of S^* with respect to a fixed canonical one-dimensional homology basis, (Γ^*, Δ^*), in two blocks, a left hand block, $(g_1 + g_2) \times g_1$ and a right hand block $(g_1 + g_2) \times g_2$. We obtain the asymptotic behavior of the period matrix for these two blocks separately.

We now give the prescription by which we obtain S^* from the two given surfaces S_1 and S_2 of genera g_1 and g_2, respectively.

Let (Γ_1, Δ_1) and (Γ_2, Δ_2) be fixed canonical one-dimensional homology bases for S_1 and S_2, respectively. We take two normal integrals of the third kind, $\eta_{P_1Q_1,1}$ on S_1 and $\eta_{P_2Q_2,2}$ on S_2, respectively. The points $P_1, Q_1 \in S_1$ not on any of the fixed cycles in (Γ_1, Δ_1) and, similarly, P_2, $Q_2 \in S_2$ not on any fixed cycle in (Γ_2, Δ_2).

Consider the local parameters

$$t_1 \equiv e^{+\eta_{P_1Q_1,1}} \quad \text{on } S_1 \text{ about } P_1 , \tag{4.1}$$

and

$$t_2 \equiv e^{-\eta_{P_2Q_2,2}} \quad \text{on } S_2 \text{ about } Q_2 . \tag{4.2}$$

We can take parameter disks about P_1 and about Q_2 and get circles C_{P_1} and C_{Q_2} of radius ϵ about P_1 and Q_2 respectively, and identify the boundaries of C_{P_1} and C_{Q_2} in a manner similar to the procedure used in Section 3. We could then obtain:

Prescription B: Delete the interiors of C_{P_1} and C_{Q_2} and identify the boundaries of C_{P_1} and C_{Q_2} as described above. We call the identified curve, C. We see that we have formed from S_1 and S_2 a new topological surface $S^{\#}$, which will serve as a base for a new Riemann surface, S^*.

LEMMA 5. *The genus of* $S^{\#}$ *is* $g_1 + g_2$.

As a matter of convenience we will denote the elements in (Γ_2, Δ_2) by

$$\Gamma_2 = (\gamma_{g_1+1}, \ldots, \gamma_{g_1+g_2}) \text{ and } \Delta_2 = (\delta_{g_1+1}, \ldots, \delta_{g_1+g_1}) .$$

Thus

$$\Gamma^{\#} = (\gamma_1^{\#}, \ldots, \gamma_{g_1}^{\#}, \gamma_{g_1+1}^{\#}, \ldots, \gamma_{g_1+g_2}^{\#})$$

and

$$\Delta^{\#} = (\delta_1^{\#}, \ldots, \delta_{g_1}^{\#}, \delta_{g_1+1}^{\#}, \ldots, \delta_{g_1+g_2}^{\#}) .$$

$$\gamma_i \in \Gamma_1, \ \delta_i \in \Delta_1, \ i = 1, \ldots, g_1$$

$$\gamma_i^{\#} = \gamma_i ; \ \delta_i^{\#} = \delta_i$$

$$\gamma_i \in \Gamma_2, \ \delta_i \in \Delta_2, \ i = 1, \ldots, g_1 + g_2 .$$

Denote $S_1 \cap S^{\#}$ by $S_{1,0}$ and $S_2 \cap S^{\#}$ by $S_{2,0}$ and call the union of $S_{1,0}$ and $S_{2,0}$, S_0. S_0 will have two have two boundary components C_{P_1} C_{Q_2} which we have identified and called C, on $S^{\#}$.

LEMMA 6. $S^{\#}$ *may be endowed with an analytic structure to make it an abstract Riemann surface,* S^*.

Proof: The proof is exactly the same as the proof of Lemma 2, Section 3.

We have thus constructed a new Riemann surface, S^*, from S_1 and S_2 of genus $g^* = g_1 + g_2$, where we use as a specified fixed set of cycles $(\Gamma^*, \Delta^*) = (\Gamma^{\#}, \Delta^{\#})$.

Definition 2: We define the "splitting into surfaced of lower genera" to be the degeneration of S^*, caused by letting ϵ go to zero, where S^* is constructed by means of prescription B, above, with the limit surfaces being S_1 and S_2.

We derive the basic lemma which emables us to compare special functions on S_1 and S_2 with an integral of the third kind of S^*.

Define

$$\tilde{\eta}_{XY} = \begin{cases} \eta_{XY,1} & \text{on } S_1 \text{ where X and Y are in the interior of } S_{1,0} \\ 0 & \text{on } S_2 \end{cases} \tag{4.3}$$

Recall that $\eta_{XY,1}$ is a normal integral of the third kind on S_1. $\tilde{\eta}_{XY}$ is not an integral of the third kind on S^*, but so long as we keep away from the curve C, we may differentiate $\tilde{\eta}_{XY}$.

LEMMA 7: *Let* S_1 *and* S_2 *and* S^* *be Riemann surfaces. Assume that* $S_{1,0} = S_1 \cap S^*$ *contains a specific set of cycles which are canonical with respect to* S_1 *and that* $S_{2,0} = S_2 \cap S^*$ *contains a specific set of cycles which are canonical with respect to* S_2. *Assume that these sets of cycles can be extended to form a basis for the cycles on* S^*. *Denote* $S_{1,0} \cup S_{2,0}$ *by* S_0. *Then for points* X, Y *in* $S_{1,0}$ *and* Z, W *both in the interior of* $S_{1,0}$ *or both in the interior of* $S_{2,0}$ *we have*

$$\Delta\eta_{XY}(Z) - \Delta\eta_{XY}(W) = \frac{1}{2\pi i}\int_{\partial S_0} \tilde{\eta}_{XY}\, d\eta^*_{ZW}\ , \tag{4.4}$$

where η^* *is a normal integral of the third kind on* S^* *and* $\tilde{\eta}$ *is defined by Equation (4.3), above, and* $\Delta\eta_{XY} = \eta^*_{XY} - \eta_{XY}$.

Proof: Let λ be a slit in $S_{1,0}$ from X to Y which does not intersect any of the cycles of (Γ_1, Δ_1). Then on the region obtained by a canonical dissection of S_0 along the canonical bases for S_1 and S_2 and slitting along λ we find that a determination of $\tilde{\eta}_{XY}$ is a single valued analytic function except for the curve C.

We integrate $\tilde{\eta}_{XY}\, d\eta^*_{ZW}$ around $\partial\pi_1 + \partial\pi_2 + \partial S_0 + \lambda$. Using Cauchy's residue theorem we obtain

$$(4.5) \qquad \tilde{\eta}_{XY}(W) - \tilde{\eta}_{XY}(Z) = \frac{1}{2\pi i}\int_{\partial\pi_1+\partial\pi_2+\partial S_0+\lambda} \tilde{\eta}_{XY}\, d\eta^*_{ZW} \, .$$

The integral on the right may be evaluated as the sum of four integrals as follows:

$$(4.6) \qquad \frac{1}{2\pi i}\int_{\partial\pi_1} \tilde{\eta}_{XY}\, d\eta^*_{ZW} + \frac{1}{2\pi i}\int_{\partial\pi_2} \tilde{\eta}_{XY}\, d\eta^*_{ZW}$$

$$+ \frac{1}{2\pi i}\int_{\partial S_0} \tilde{\eta}_{XY}\, d\eta^*_{ZW} + \frac{1}{2\pi i}\int_{\lambda} \tilde{\eta}_{XY}\, d\eta^*_{ZW} \, .$$

Note that the λ under the integral really stands for both sides of the slit λ.

For the first integral replace $\tilde{\eta}_{XY}$ by $\eta_{XY,1}$ and we find by equation (2.4), that the integral is zero. For the integral over $\partial\pi_2$ replace $\tilde{\eta}_{XY}$ by zero and thus the second integral is zero.

λ is in $S_{1,0}$ and so we may replace $\tilde{\eta}_{XY}$ by $\eta_{XY,1}$ and we obtain

$$(4.7) \qquad \frac{1}{2\pi i}\int_{\lambda} \tilde{\eta}_{XY}\, d\eta^*_{ZW} = \frac{1}{2\pi i}\int_{\lambda} \eta_{XY,1}\, d\eta^*_{ZW} \, .$$

$\eta_{XY,1}$ has a jump of $2\pi i$ across λ and so

$$(4.8) \qquad \frac{1}{2\pi i}\int_{\lambda} \eta_{XY,1}\, d\eta^*_{ZW} = \eta^*_{XY}(W) - \eta^*_{XY}(Z)$$

Using equation (2.8), we find

$$(4.9) \qquad \eta^*_{ZW}(Y) - \eta^*_{ZW}(X) = \eta^*_{XY}\,(W) - \eta^*_{XY}\,(Z) \, .$$

Combining equations (4.5), (4.6), (4.7), (4.8) and (4.9) we obtain

$$(4.4) \qquad \Delta\eta_{XY}(Z) - \Delta\eta_{XY}(W) = \frac{1}{2\pi i}\int_{\partial S_0} \tilde{\eta}_{XY}\, d\eta^*_{ZW} \, .$$

We are now ready to study the periods on S^* with respect to (Γ^*,Δ^*). We will estimate the integral in equation (4.4), above.

The local parameters t_1 and t_2, are related on the circles C_{P_1} and C_{Q_2} by an equation similar to (3.3) and we obtain

$$t_1 = \frac{\epsilon^2\phi}{t_2} \quad \text{with } |\phi| = 1 \; . \tag{4.10}$$

Now

(4.11)
$$\frac{1}{2\pi i}\int_{\partial S_0} \tilde{\eta}_{XY}\, d\eta^*_{XW} = \frac{1}{2\pi i}\int_{C_{P_1}} \tilde{\eta}_{XY}\, d\eta^*_{ZW} + \frac{1}{2\pi i}\int_{C_{Q_2}} \tilde{\eta}_{XY} d\eta^*_{ZW}$$

since $\partial S_0 = C_{P_1} + C_{Q_2}$, oriented properly.

The second integral on the right in equation (4.11) is zero since $\tilde{\eta}_{XY} = 0$ on S_2.

For the first integral we replace $\tilde{\eta}_{XY}$ by $\eta_{XY,1}$ and we expand $\eta_{XY,1}$ in terms of the local parameter t_1 which is valid on C_{P_1}. We obtain

$$\frac{1}{2\pi i}\Bigg[\eta_{XY,1}(P_1)\int_{C_{P_1}} d\eta^*_{ZW} + \eta'_{XY,1}(P_1)\int_{C_{P_1}} t_1\, d\eta^*_{ZW} + \frac{\eta''_{XY,1}(P_1)}{2!}\int_{C_{P_1}} t_1{}^2\, d\eta^*_{ZW} + \cdots\Bigg] . \tag{4.12}$$

The first integral in the brackets is zero as the sum of the residues of $d\eta^*_{ZW}$ in the region on S^* bounded by C_{P_1} is zero. For the other integrals we replace t_1 by t_2 using equation (4.10), above, and we replace integration over C_{P_1} by the appropriate integral over C_{Q_2}. We obtain

$$\frac{1}{2\pi i}\left[\eta'_{XY,1}(P_1)\,\epsilon^2\phi\int_{C_{Q_2}} \frac{1}{t_2}\, d\eta^*_{ZW} + \text{terms with } \epsilon^4 \text{ at least}\right] . \tag{4.13}$$

We may replace integration over C_{Q_2} by integration over some other curve homologous to C_{Q_2} so long as Z and W are not in the region bounded by the two curves. We find that all of equation (4.13) is $o(\epsilon)$, (see [6]).

Thus we find that

$$\Delta\eta_{XY}(Z) - \Delta\eta_{XY}(W) = o(\epsilon) \quad . \tag{4.14}$$

Differentiate with respect to any local parameter z about Z and we obtain

$$d\eta^*_{XY}(z) - d\tilde{\eta}_{XY}(z) = o(\epsilon) \quad . \tag{4.15}$$

Thus using our definition for $\tilde{\eta}_{XY}$, equation (4.3), we have

$$\begin{aligned} &d\eta^*_{XY}(z) - d\eta_{XY,1}(z) = o(\epsilon) && z \in S_{1,0} \\ &d\eta^*_{XY}(z) = o(\epsilon) && z \in S_{2,0} \end{aligned} \tag{4.16}$$

We can now replace $d\eta^*$ in the equation defining π^*_{ij} with respect to our fixed cycles (Γ^*,Δ^*) creating an error at most $o(\epsilon)$. We obtain the following:

$$\pi^*_{ij} = \int_{\delta^*_j}\int_{\delta^*_i} \frac{\partial^2\eta^*_{Xy}(z)}{\partial y\,\partial z}\,dz\,dy = \begin{cases} \displaystyle\int_{\delta^*_j}\int_{\delta^*_i} \frac{\partial^2\eta_{Xy,1}}{\partial y\,\partial z}(z)\,dz\,dy + o(\epsilon) \\ o(\epsilon) \end{cases} \tag{4.17}$$

Since Y must be in $S_{1,0}$, j can only run from 1 to g_1. Since Z can be in either $S_{1,0}$ or $S_{2,0}$, i can run from $1,\ldots, g_1, g_1+1, \ldots, g_1+g_2$. Thus we obtain the following.

LEMMA 8: *Let S^* be constructed from S_1 and S_2 as described above. Then when we split the surface S^* into two surfaces of lower genera, g_1 and g_2, definition 2, above, we have*

a) $\pi^*_{ij} = \pi_{ij,1} + o(\epsilon) \qquad i, j = 1, \ldots, g_1$,

b) $\pi^*_{ij} = o(\epsilon) \qquad i = g_1 + 1, \ldots, g_1 + g_2 \,;\; j = 1, \ldots, g_1$.

Proof: a) In the right hand side of equation (4.17), above, replace δ^*_j by δ_j and use the definition of $\pi_{ij,1}$ for $i, j = 1, \ldots, g_1$. For b), use the second half of equation (4.17).

We may repeat Lemma 7, above, replacing $\tilde{\eta}_{XY}$ by $\hat{\eta}_{XY}$ defined as follows:

$$\hat{\eta}_{XY} = \begin{cases} 0 \text{ on } S_1 \\ \eta_{XY,2} \text{ on } S_2 \text{ where X and Y are in the interior of } S_{2,0} \,. \end{cases}$$

We could then prove the following lemma.

LEMMA 9: *Let* S* *be constructed from* S_1 *and* S_2 *as described above. Then we split the surface* S* *into two surfaces of lower genera,* g_1 *and* g_2, *definition* 2, *above, we have*

a) $\pi^*_{ij} = o(\epsilon) \qquad i = 1, \ldots, g_1 \,;\; j = g_1 + 1, \ldots, g_1 + g_2$.

b) $\pi^*_{ij} = \pi_{ij,2} + o(\epsilon) \qquad i, j = g_1 + 1, \ldots, g_1 + g_2$.

We combine the two lemmas, 8 and 9, in the form of the following theorem.

THEOREM 2: *Let* S* *be constructed from* S_1 and S_2 *of genera* g_1 *and* g_2, *respectively. Then the periods of* S* *with respect to the canonical basis* (Γ^*, Δ^*) *satisfy the following, when we split into two surfaces of lower genera, definition* 2, *above:*

(4.18) $\pi^*_{ij} = \pi_{ij,1} + o(\epsilon) \qquad i, j = 1, \ldots, g_1$,

(4.19) $\pi^*_{ij} = \pi_{ij,2} + o(\epsilon) \qquad i, j = g_1 + 1, \ldots, g_1 + g_2$;

(4.20) $\pi^*_{ij} = \pi^*_{ji} = o(\epsilon) \qquad i = 1, \ldots, g_1 \quad j = g_1 + 1, \ldots, g_1 + g_2$.

5. Application of a Siegel Modular Transformation

In section 2, a process was given for obtaining the period matrix of a given Riemann surface relative to a new set of basis cycles, when we already know what the period matrix is with respect to a given fixed set of cycles. We give here an example where we make the period π_{11}^*, of Section 3, become a new period which goes to infinity like a logarithm.

Let a new basis (Γ', Δ') be related to (Γ^*, Δ^*) as follows:

$$(\Gamma', \Delta') = M(\Gamma^*, \Delta^*) \quad , \tag{2.3}$$

where

$$M = \begin{pmatrix} D & C \\ B & A \end{pmatrix} .$$

The matrices A and D are the same, with elements as follows:

$$a_{11} = 0, \quad a_{ij} = 0 \quad i \neq j, \quad a_{ii} = 1, \quad i = 2, \dots, g+1,$$

B, has $a_{11} = -1$ and all other $a_{ij} = 0$. C, has $a_{11} = 1$ and all other $a_{ij} = 0$.

We find that

$$\gamma_1' = \delta_1^*, \ \delta_1' = -\gamma_1^* \quad \text{and} \quad \gamma_i' = \gamma_i^* \quad \delta_i' = \delta_i^* \quad \text{for } i = 2, \dots, g+1.$$

By means of a simple calculation with Equation (2.2) one then obtains:

$$\pi_{11}' = -\frac{1}{\pi_{11}^*}, \ \pi_{1j}' = \frac{\pi_{1j}^*}{\pi_{11}^*} = \pi_{j1}', \qquad j = 2, \dots, g+1 .$$

$$\pi_{ij}' = \pi_{ij}^* - \frac{\pi_{1j}^* \cdot \pi_{i1}^*}{\pi_{11}^*} = \pi_{ji}' \qquad i, j = 2, \dots, g+1$$

π_{11}' goes to infinity like a logarithm. The π_{1j}' are bounded and the π_{ij}' differ from π_{ij}^* by $0(\frac{1}{\log \epsilon})$ for $i, j = 2, \dots, g+1$.

BIBLIOGRAPHY

[1] L. V. Ahlfors and L. Sario, *Riemann Surfaces*, Princeton University Press, 1960.

[2] C. Patt, Variations of Teichmüller and Torelli Surfaces, *J. d'Analyse Math.* 11 (1963), 221-247.

[3] H. E. Rauch, Weierstrass points, branch points, and the moduli of Riemann surfaces, *Comm. Pure Appl. Math.*, 12 (1959), 543-560.

[4] ______, Variational methods in the problem of the moduli of Riemann surfaces, *Contributions to Finction Theory*, Tata Institute of Fundamental Research, Bombay, 1960, 17-40.

[5] ______, A transcendental view of the space of algebraic Riemann surfaces, *Bull. Amer. Math. Soc.*, 71 (1965), 1-39.

[6] M. Schiffer and D. C. Spencer, *Functionals of Finite Riemann Surfaces*, Princeton University Press, 1954.

Lehman College of
City University of New York

SINGULAR RIEMANN MATRICES

Joseph Lewittes*

Introduction

In the first section of this paper we give a sufficient condition for a Riemann matrix to be singular; that is, to admit more than one principal matrix. Our main result is that the normalized period matrix of a compact Riemann surface of genus greater than one is a singular Riemann matrix if the surface admits an automorphism of order two, other than the hyperelliptic involution. Then a connection is pointed out between these singular period matrices and a new class of theta functions on the surface. In the second section we illustrate some of the theory by a concrete numerical example. *Note:* We use Conforto's book [1] as a standard reference for the terminology and basic facts about Riemann matrices.

I.

Let g be a positive integer, Ω a $g \times 2g$ Riemann matrix. Recall that a principal matrix for Ω is a $2g \times 2g$ rational, skew-symmetric, non-singular matrix P such that

(1) $$\Omega P {}^t\Omega = 0$$

(2) $$i\Omega P {}^t\overline{\Omega} > 0$$

Here ${}^t\Omega$ denotes the transpose of Ω, $i = \sqrt{-1}$, $\overline{\Omega}$ is the complex conjugate, and $H > 0$ for a hermitian matrix means H is positive definite. If P is a principal matrix, P^{-1} is called a characteristic matrix for Ω. In general, Ω has only one principal matrix—up to rational multiples, see [1, p. 129].

* Research supported in part by N.S.F. Grant GP 9466.

If Ω has more than one principal matrix it is called singular.

Denote by $\mathcal{P} = \mathcal{P}(\Omega)$ the set of all principal matrices for Ω, $\Sigma = \Sigma(\Omega)$ the set of all $2g \times 2g$ rational skew symmetric matrices Q satisfying $\Omega Q\, {}^t\bar{\Omega} = 0$ and by $\mathfrak{N} = \mathfrak{N}(\Omega)$ the set of all $2g \times 2g$ rational matrices N satisfying $\Omega N\, {}^t\Omega = 0$. It follows directly that $\mathcal{P} \subset \Sigma \subset \mathfrak{N}$ and $\Sigma, \mathfrak{N}$ are finite dimensional vector spaces over the field of rational numbers. Whenever we use the concepts of dimension, linear dependence, etc. in Σ or $\mathfrak{N}$ it is to be understood that it is with respect to the field of rational numbers.

THEOREM 1: (a) $\mathcal{P}$ *is a 'convex cone' in* Σ; *i.e., if* $P_1, P_2 \in \mathcal{P}$ *and* t_1, t_2 *are non-negative rational numbers, then also* $t_1P_1 + t_2P_2 \in \mathcal{P}$.

(b) *Every* $Q \in \Sigma$ *is the difference of two matrices in* $\mathcal{P}$. *In fact, given any* $Q \in \Sigma$, $P_1 \in \mathcal{P}$ *there are* $P \in \mathcal{P}$ *and rational* $t_1 > 0$ *such that* $Q = P - t_1P_1$.

(c) $\mathcal{P}$ *contains a basis for* Σ.

Proof: (a) is clear from the definitions.

(b) Let $K = i\Omega Q\, {}^t\bar{\Omega}$ and $H_1 = i\Omega P_1\, {}^t\bar{\Omega}$. Then K, H_1 are hermitian matrices and the corresponding hermitian forms $K(z) = {}^tzK\bar{z}$, $H_1(z) = {}^tzH_1\bar{z}$ are real valued for every column vector $z \in C^g$. Also $K(z), H_1(z)$ are bounded from below on the unit sphere $\|z\| = 1$ in C^g, say $K(z) \geq k$, $H_1(z) \geq h_1$, for all z of length one, and $h_1 > 0$ since $H_1 > 0$. Thus for a suitable positive rational t_1 we have $k + t_1h_1 > 0$ so that $K + t_1H_1 > 0$ and so $Q + t_1P_1 = P \in \mathcal{P}$ and $Q = P - t_1P_1$, as asserted.

(c) Let $d =$ dimension of Σ. Since Ω is a Riemann matrix there is at least one principal matrix, so it suffices to show that if $r \geq 1$ linearly independent principal matrices, say $P_1, \ldots, P_r$, have been found, and $r < d$, then we can find a $P \in \mathcal{P}$ which is not a linear combination of $P_1, \ldots, P_r$. Since $r < d$, there is a $Q \in \Sigma$ not dependent on $P_1, \ldots, P_r$. By (b) above, $Q = P - t_1P_1$ for $P \in \mathcal{P}$, t_1 rational. Now P is not a linear combination of $P_1, \ldots, P_r$ for Q is not. This completes the proof of the theorem.

The space $\mathfrak{N}$ is related to the notion of 'complex multiplication.'

Namely, Ω is said to admit a complex multiplication if there are K, a $g \times g$ complex matrix, and R, a $2g \times 2g$ rational matrix, such that $K\Omega = \Omega R$. Of course every Ω has trivial multiplications where K, R are rational scalar matrices. The set of all these multipliers R forms an algebra over the rationals, denote it, say, by $\mathfrak{R}$. Then it is known that if $P \in \mathcal{P}$ is fixed, $R \to RP$ is a vector space isomorphism of $\mathfrak{R}$ onto $\mathfrak{N}$. That $RP \in \mathfrak{N}$ is clear, for $\Omega RP\,{}^t\Omega = K\Omega P\,{}^t\Omega = 0$, while the converse, that every $N \in \mathfrak{N}$ is of the form RP is also straightforward once Ω has been reduced to a normal form. See [1, p. 132] for details.

Consider now the Riemann matrix $\Omega = (E, Z)$ where E is the $g \times g$ identity matrix and Z a $g \times g$ complex symmetric matrix with $\operatorname{Im} Z > 0$: thus $Z \in \mathcal{H}^g$ the generalized Siegel upper-half plane of degree g. Such an Ω has $J = \begin{pmatrix} 0 & E \\ -E & 0 \end{pmatrix}$ as a principal matrix. Let $\Gamma = \Gamma^g$ be the Siegel modular group of degree g, namely the group of $2g \times 2g$ integral matrices M such that ${}^tMJM = J$. Each $M \in \Gamma$ defines an automorphism of $\mathcal{H}^g$ as follows: if $M = \begin{pmatrix} A & B \\ C & D \end{pmatrix}$ and $Z \in \mathcal{H}^g$, then $M(Z) = (AZ + B)(CZ + D)^{-1}$.

Assuming still $\Omega = (E,Z)$, let $\mathfrak{R}$ be the algebra of multipliers of Ω and consider $\begin{pmatrix} A & B \\ C & D \end{pmatrix} = M \in \mathfrak{R} \cap \Gamma$. Then there is a complex matrix K such that $K\Omega = \Omega M$, which gives $K = A + ZC$, $KZ = B + ZD$. Eliminating K, and recalling ${}^tZ = Z$, we obtain, $M^*(Z) = Z$ where $M^* = \begin{pmatrix} {}^tD & {}^tB \\ {}^tC & {}^tA \end{pmatrix}$. If we let $\mathfrak{S} = \begin{pmatrix} -E & 0 \\ 0 & E \end{pmatrix}$, so $\mathfrak{S} = \mathfrak{S}^{-1}$, then $\mathfrak{S} \notin \Gamma$, but $\mathfrak{S}\Gamma\mathfrak{S}^{-1} = \Gamma$ and recalling that $M^{-1} = \begin{pmatrix} {}^tD & -{}^tB \\ -{}^tC & {}^tA \end{pmatrix}$ one sees that $M^* = \mathfrak{S}M^{-1}\mathfrak{S}^{-1}$ so that $M \to M^*$ is an anti-automorphism of Γ or order 2, $(M^*)^* = M$. Thus if $M \in \mathfrak{R} \cap \Gamma$, then $M^*(Z) = Z$ or $M^* \in \Gamma_Z$, the subgroup of Γ fixing Z, and clearly the argument is reversible so that we have the following theorem.

THEOREM 2. *Keeping the above notation, if* $\Omega = (E, Z)$ *then* $M \to M^*$ *is an anti-isomorphism of* $\mathfrak{R}(\Omega)$ *onto* Γ_Z.

THEOREM 3. *If* Γ_Z *contains an element of order* 2 *other than* $-E$, *then the Riemann matrix* $\Omega = (E, Z)$ *is singular.*

Proof: If $M^* \epsilon \Gamma_Z$ has order 2 and $\neq -E$ then $M \epsilon \mathfrak{R} \cap \Gamma$ has order 2 and $\neq -E$. Since $R \to RJ$ is an isomorphism of $\mathfrak{R}$ onto $\mathfrak{N}$, the matrix $Q = MJ \epsilon \mathfrak{N}$. As $M \epsilon \Gamma$, $MJ\,{}^tM = J$ or $MJ = J\,{}^tM^{-1}$ and by our hypothesis that M has order 2, ${}^tM^{-1} = {}^tM$ so $MJ = J\,{}^tM$. Thus ${}^tQ = {}^tJ\,{}^tM = -J\,{}^tM = -MJ = -Q$. Thus Q is skew symmetric, so $Q \epsilon \Sigma$. Also Q is not dependent on J, for if so $MJ = tJ$ for some rational t which implies $t = \pm 1$, $M = \pm E$ which has been excluded. Thus Σ has dimension greater than one and by Theorem 1, (c), $\mathfrak{P}$ contains at least two linearly independent principal matrices, so that Ω is singular, as asserted.

THEOREM 4. *Let* S *be a compact Riemann surface of genus* $g > 1$ *and suppose that* S *admits an automorphism* h *of order* 2 *(an involution) which, in case* S *is hyperelliptic, is not the standard hyperelliptic involution. Then any normalized period matrix* $\Omega = (E, Z)$ *of* S *is singular.*

Proof: Choose a canonical homology basis and the corresponding dual basis for the abelian differentials of first kind giving the normalized period matrix $\Omega = (E, Z)$. Let h_* denote the automorphism induced by h on the first homology group and h^* the natural induced action (pull-back) on the differentials. These are related by the formula $\int_\gamma h^*\phi = \int_{h_*\gamma} \phi$ for any differential ϕ and cycle γ. Thus if K, M are the matrices of h^*, h_*, respectively, with respect to the chosen bases, we obtain the complex multiplication $K\Omega = \Omega M$. To be perfectly explicit we remark that K is obtained so that $h^*\Phi = K\Phi$ where Φ is the g-column vector whose entries form the basis of the differentials, while $h_*\Delta = \Delta M$ where Δ is the 2g-row vector whose entries form the canonical homology basis. Now $M \epsilon \mathfrak{R}$ and also $M \epsilon \Gamma$ for h_* takes a canonical homology basis into a canonical homology basis. Thus $M \epsilon \mathfrak{R} \cap \Gamma$ and $M^2 = E$ since h^2 is the identity. The theorem now follows from Theorem 2 and 3 once we show $M \neq \pm E$. If $M = E$, then $K = E$ but—and here we actually need $g > 1$—it is known [2, p. 737] that for $g > 1$ this cannot occur for an automorphism different from the identity. If $M = -E$ then $K = -E$ so that there are no differentials invariant under

h^* from which it follows [2, p. 741] that the quotient surface $S/\{id, h\}$ has genus zero, so that S is a 2-sheeted cover of the sphere with h the sheet interchange—so that S is hyperelliptic and h the standard involution, which has been ruled out. Thus $M \neq \pm E$, and the proof is complete.

COROLLARY. *Let S be a non-hyperelliptic compact Riemann surface of genus > 1 and suppose that the full group of automorphisms of S has even order. Then any period matrix of S is singular.*

Proof: Since the group of automorphisms has even order it contains an involution.

Concerning surfaces S with singular period matrix we would like to sketch briefly a connection with the theory of theta functions. Again let $\Omega = (E, Z)$ be a normalized period matrix for S with respect to a canonical homology basis and suppose Ω singular. Pick a $P \in \mathcal{P}$ which is not a rational multiple of J and consider the characteristic matrix $C = P^{-1}$. We assume C is primitive, that is, C has integer entries whose greatest common divisor is one, for this can always be attained by multiplying P by a positive rational number. By the standard reduction theory, [1], there are a unimodular U and complex K such that $K\Omega U = (B, W)$ and ${}^tUCU = \begin{pmatrix} 0 & -B \\ B & 0 \end{pmatrix}$, where B is a diagonal matrix with diagonal entries $b_1, \ldots, b_g$, positive integers, $b_1 = 1$ and b_1 divides b_2 divides $b_3 \cdots$ and W is symmetric with positive definite imaginary part. We then construct the det B-dimension space of first order theta functions belonging to the Riemann matrix (B, W). If f(z) is any one of these thetas, then $F(z) = f(Kz)$ will be a multiplicative function—but in general not a theta—with respect to the lattice $\mathcal{L}$ in C^g generated by the columns of Ω. Using an embedding of S into $C^g/\mathcal{L}$ which is the Jacobian variety of S, we obtain a corresponding multiplicative function F on S. The number of zeros n(F) of such an F can be computed, see [3, p. 41], as follows. Write $C = \begin{pmatrix} C_1 & C_2 \\ C_3 & C_4 \end{pmatrix}$, then $n(F) = \text{trace } C_3$. In particular n(F) does not depend of F but only on (B, W), hence ultimately only on P. Also one can obtain abelian sum relations for the divisor of zeros of F. It seems that this area warrants closer investigation. In particular it would be interesting to obtain links between the arithmetic nature of Ω and the function-theoretic properties of S.

II.

As an illustration of some of the above theory we take as S that remarkable surface of genus 3, sometimes called Klein's surface after Felix Klein who studied it in connection with the modular group. S is the unique surface of genus 3 admitting a group of 168 automorphisms, the largest possible number. Recently Professor H. E. Rauch and the author [4] have computed a normalized period matrix for S; it is $\Omega = (E, Z)$ where

$$Z = \begin{pmatrix} -\frac{1}{8} + \frac{3\sqrt{7}}{8}i, & -\frac{1}{4} - \frac{\sqrt{7}}{4}i, & -\frac{3}{8} + \frac{\sqrt{7}}{8}i \\ -\frac{1}{4} - \frac{\sqrt{7}}{4}i, & \frac{1}{2} + \frac{\sqrt{7}}{2}i, & -\frac{1}{4} - \frac{\sqrt{7}}{4}i \\ -\frac{3}{8} + \frac{\sqrt{7}}{8}i, & -\frac{1}{4} - \frac{\sqrt{7}}{4}i, & \frac{7}{8} + \frac{3\sqrt{7}}{8}i \end{pmatrix}$$

Write $Z = X + i\sqrt{7}Y$ so that X, Y are rational symmetric matrices. The condition that $N = \begin{pmatrix} N_1 & N_2 \\ N_3 & N_4 \end{pmatrix} \in \mathfrak{N}(\Omega)$ is $\Omega N {}^t\Omega = 0$, which upon multiplying out and separating into real and imaginary parts, gives:

$$N_1 + XN_3 + N_2X + XN_4X - 7YN_4Y = 0 \tag{3}$$

$$\sqrt{7}(YN_3 + N_2Y + XN_4Y + YN_4X) = 0 \ . \tag{4}$$

The $\sqrt{7}$ may be cancelled from (4) and these matrix equations become a linear homogeneous system of 18 equations, with rational coefficients, in the 36 unknown rational entries in N. This system can be solved explicitly although we have not carried out the calculations. In any case we have that $\dim \mathfrak{N}(\Omega) \geq 18$ so that the dimension of the algebra of complex multiplications of Ω is ≥ 18. If one wants to find an $N \in \Sigma(\Omega)$, we must restrict N to be a skew symmetric. One sees that this cuts down the above system to 6 equations in 15 unknowns, so that $\dim \Sigma(\Omega) \geq 9$. In particular, Ω admits at least 9 linearly independent principal matrices. Again we have not done all the calculations but by specializing have found at least some principal matrices other than J. In fact, let

$$K = \begin{pmatrix} 0 & 1 & 0 \\ 0 & -2 & 0 \\ 0 & 1 & 0 \end{pmatrix}$$

then a simple calculation shows that KZ is symmetric, $KZ = Z^tK$ hence $\begin{pmatrix} 0 & K \\ -{}^tK & 0 \end{pmatrix} \in \Sigma$. Note also that since KZ symmetric, $KX = X^tK$ and $KY = Y^tK$. It follows that for any rational numbers a, b the matrix $P = P(a, b) = aJ + b\begin{pmatrix} 0 & K \\ -{}^tK & 0 \end{pmatrix} \in \Sigma$. Write $P = \begin{pmatrix} 0 & L \\ -{}^tL & 0 \end{pmatrix}$ where L =

$$L = \begin{pmatrix} a & b & 0 \\ 0 & a-2b & 0 \\ 0 & b & a \end{pmatrix} .$$

Then $P \in \mathcal{P}$ if and only if the symmetric matrix LY is positive definite. But

$$LY = \frac{1}{8} \begin{pmatrix} 3a-2b, & -2a+4b, & a-2b \\ -2a+4b, & 4a-8b, & -2a+4b \\ a-2b, & -2a+4b, & 3a-2b \end{pmatrix}$$

Adding ½ the second row to the first and third rows, then adding ½ the second column to the first and third columns, one obtains a diagonal matrix with entries $a/4$, $a/2-b$, $a/4$. Thus $LY > 0$ provided $a > 0$, $a > 2b$, and the matrix $P(a, b) \in \mathcal{P}$ provided $a > 0$, $a > 2b$. For $a = 1$, $b = 0$, $P(a,b) = J$ and for $a = 2b+1$, $b > 0$, say, $P(a,b)$ is not a multiple of J and is a second principal matrix for Ω.

Finally, since

$$L^{-1} = \begin{pmatrix} \frac{1}{a} & \frac{b}{a(2b-a)}, & 0 \\ 0 & \frac{1}{a-2b}, & 0 \\ 0 & \frac{b}{a(2b-a)}, & \frac{1}{a} \end{pmatrix}$$

we see that if $a > 0$, $a > 2b$, a, b integers which are relatively prime, then the primitive characteristic matrix C corresponding to $P^{-1}_{(a,b)}$ is given by $C = \begin{pmatrix} 0 & -{}^tD \\ D & 0 \end{pmatrix}$ where

$$D = \begin{pmatrix} a-2b, & -b, & 0 \\ 0, & a, & 0 \\ 0, & -b, & a-2b \end{pmatrix}$$

In particular, by the remarks at the end of Section I, there is a multiplicative function on S with $3a-4b$ zeros, whenever a, b satisfy the above conditions. Specializing further, taking $b \geq 0$, $a = 2b+1$ we obtain a function with $2b+3$ zeros for each value of the integer b.

BIBLIOGRAPHY

[1] Conforto, F., *Abelsche Funktionen und Algebraische Geometrie*, Springer-Verlag, 1956.

[2] Lewittes, J., 'Automorphisms of Compact Riemann Surfaces,' *American Journal of Mathematics*, Vol. 85, Oct. 1963, pp. 734-752.

[3] ______, 'Riemann Surfaces and the Theta Function,' *Acta Mathematica*, Vol. 111, 1964, pp. 37-61.

[4] Rauch, H. E. and Lewittes, J., 'The Riemann Surface of Klein with 168 Automorphisms,' *Proceedings of the Symposium on Problems in Analysis*, in honor of Salomon Bochner, held at Princeton University, April 1-3, 1969.

Lehman College and
The City University of New York, Graduate Center

AN INEQUALITY FOR KLEINIAN GROUPS[1]

Albert Marden

The purpose of this note is to point out that an elementary 3-dimensional topological argument yields the following result.

THEOREM. Suppose G is a Kleinian group with N generators, Ω is its set of ordinary points, and $\Omega/G = \cup S_i$ is the decomposition of Ω/G into its components. If g_i is the genus of S_i then

$$\Sigma g_i \leq N .$$

COROLLARY. If G is purely loxodromic then Ω/G has at most N/2 components.

This corollary, which is an immediate consequence of Ahlfors' finiteness theorem [1] and of the preceding theorem, is apparently not a consequence of the deep analytic methods employed by Ahlfors, Bers, and Kra to study Kleinian groups (cf. Kra's contribution to these Proceedings).

Our theorem follows from the following classical lemma.

LEMMA. Let $\mathfrak{M}$ be an orientable 3-manifold with boundary $\partial\mathfrak{M}$ and suppose Z_1, Z_2 are 1-cycles on $\partial\mathfrak{M}$ such that the intersection number $Z_1 \times Z_2 \neq 0$. Then at most one of Z_1, Z_2 is homologous to zero in $\mathfrak{M}$.

Proof (Papakyriakopoulos [2]). Suppose $Z_2 = \partial C$ for a 2-chain C. The lemma follows from doubling $\mathfrak{M}$ and considering the intersection number of Z_1 and the 2-cycle $C - C^*$ where C^* is the double of C.

---

1 Supported in part by the National Science Foundation.

Proof of the theorem. G has a unique extension from the Riemann sphere $\partial\mathcal{B}$ to a group of Möbius transformations acting on the open unit 3-ball $\mathcal{B}$. If G has no elliptic elements then $\mathcal{M} = \mathcal{B} \cup \Omega/G$ is a 3-manifold with boundary $\partial\mathcal{M} = \cup S_i$. If G contains elliptic elements, for for example $\mathcal{M} = (\mathcal{B} - \zeta) \cup \Omega/G$ where ζ is the isolated set of points in $\mathcal{B}$ where two or more axes of rotation of these elements intersect. In either case $\operatorname{rank} H_1(\mathcal{M}) \leq N$ where $H_1(\mathcal{M})$ is the first homology group over the integers of $\mathcal{M}$ (it is really only $H_1(\mathcal{M})$ that plays a role in the theorem).

Select a homology basis $\{A_j^{(i)}, B_j^{(i)}\}$, modulo the dividing cycles, on each component S_i of $\partial\mathcal{M}$ satisfying $A_j^{(i)} \times B_k^{(i)} = \delta_{jk}$, and $A_j^{(i)} \times A_k^{(i)} = B_j^{(i)} \times B_k^{(i)} = 0$ for all j, k. Applying the lemma inductively and relabeling if necessary, we can assume the cycles $\{A_j^{(i)}\}$ are independent in $H_1(\mathcal{M})$ for all i and j. But $H_1(\mathcal{M})$ has at most N generators.

A deeper study of Kleinian groups from the point of view of 3-dimensional topology is in preparation. In general the problem is to investigate the manner in which $\mathcal{M}$ can be non-compact. Although the lemma above continues to play a very critical role, it must be supplemented by other techniques.

REFERENCES

[1] L. V. Ahlfors, Finitely generated Kleinian groups, *Amer. J. Math.*, 86 (1964), 413-429 and 87 (1965), 759.

[2] C. D. Papakyriakopoulos, On the ends of the fundamental groups of 3-manifolds with boundary, *Comment. Math. Helv.* 32 (1957), 85-92.

University of Minnesota
Institute for Advanced Study

ON KLEIN'S COMBINATION THEOREM III

Bernard Maskit[1]

The basic idea of Klein's combination theorem is as follows. One is given two Kleinian groups G_1 and G_2 satisfying certain algebraic and geometric conditions. One concludes that G, the group generated by G_1 and G_2 is again Kleinian, one can find a fundamental set for G, and the algebraic structure of G is determined.

In Klein's original combination theorem [4], G is the free product of G_1 and G_2. In the first paper of this series [7], G is the free product of G_1 and G_2 with an amalgamated cyclic subgroup. In the second paper [8], G_2 is cyclic, and the generator of G_2 conjugates a pair of cyclic subgroups of G_1. In this paper, we redo both of these theorems, but without the restrictions that the subgroups be cyclic. The major new tool is Koebe's theorem [5, 6], which also enables one to prove that if the limit sets of G_1 and G_2 both have measure zero, then the limit set of G has measure zero. Using the fact that Fuchsian, quasi-Fuchsian, and elementary groups all have limit sets of measure zero, these two theorems give a class of groups for which the limit sets have measure zero. A precise definition of this class will not be given here; it is known that there is no inclusion relation between this class and the class of groups for which Ahlfors [1] has proven that the limit sets have measure zero.

This paper is essentially self-contained. Some of the arguments also appear in [7] and [8]; they are included here for the sake of completeness. The point of view here is also somewhat more algebraic.

---

[1] Research supported by NSF contract number GP-9142.

§1. BASICS.

We will denote the Riemann sphere, or extended complex plane, by $\hat{C}$

A group G, of fractional linear transformations acting on $\hat{C}$, is said to be *discontinuous at* $z_0 \in \hat{C}$, if there is a neighborhood U of z_0 so that $g(U) \cap U = \emptyset$, for all $g \in G$, $g \neq 1$.

If G, as above, is discontinuous at at least one point, then G is called a *Kleinian group*. The set of points at which G is discontinuous is denoted by R(G). $\hat{C} - R(G)$, the complement of R(G), is denoted by L(G).

In what follows, we will be concerned with the measure of L(G). For any statement involving this measure, we assume that G has been normalized so that $\infty \in R(G)$; then the measure referred to is the ordinary 2-dimensional Lebesgue measure.

Remark 1: L(G), as defined here, consists of both limit points and isolated fixed points of elliptic transformations. From the group-theoretic point of view, it is convenient to deal only with R(G), where G acts freely. Note that the limit set of G has measure zero if and only if L(G) has measure zero.

Let G be a Kleinian group. A set D is called a *fundamental set* (FS) *for* G if the following hold.

a) $D \subset R(G)$.

b) If $x \in D$, $y \in D$, $g \in G$, $g(x) = y$, then $g = 1$, and $x = y$.

c) If $x \in R(G)$, then there is a $g \in G$, and there is a $y \in D$, so that $g(x) = y$.

A *conglomerate* $(G, D, H, \Delta, \gamma,)$ is a collection of objects as follows. G is a Kleinian group, D is a FS for G, H is a subgroup of G, Δ is a FS for H, and γ is a simple closed curve satisfying the following properties.

d) $D \subset \Delta$.

e) γ is invariant under H.

f) There is a neighborhood V of γ, so that $\Delta \cap V \subset D$.

A *big conglomerate* $(G, D, H, \Delta, \gamma, B)$ is a conglomerate $(G, D, H, \Delta, \gamma)$, where B is one of the open topological discs bounded by γ, B is invariant under H, and the following additional properties are satisfied.

g) $\Delta \cap (B \cup \gamma) = D \cap (B \cup \gamma)$.

h) $D \cap (B \cup \gamma) \neq D$.

i) For every $g \in G$, $g \notin H$, $g(\gamma) \cap \gamma = \emptyset$.

§2. STATEMENTS OF THEOREMS

COMBINATION THEOREM I: *Let* $(G_1, D_1, H, \Delta, \gamma, B_1)$ *and* $(G_2, D_2, H, \Delta, \gamma, B_2)$ *be big conglomerates where* $B_1 \cap B_2 = \emptyset$. *Let* G *be the group generated by* G_1 *and* G_2 *and let* $D = D_1 \cap D_2$. *Then*

1) G *is Kleinian;*

2) G *is the free product of* G_1 *and* G_2 *with amalgamated subgroup* H*;*

3) D *is a* FS *for* G*;*

4) $(G, D, H, \Delta, \gamma,)$ *is a conglomerate;*

5) *if* $g \in G$ *is elliptic or parabolic, then* g *is conjugate in* G *to an element of either* G_1 *or* G_2*; and*

6) *if* $L(G_1)$ *and* $L(G_2)$ *both have measure zero, then* $L(G)$ *has measure zero.*

Remark 2: Conclusion 5 was also observed by V. Chuckrow [2].

COMBINATION THEOREM II: *Let* $(G_1, D_1, H_1, \Delta_1, \gamma_1, B_1)$ *and* $(G_1, D_1, H_2, \Delta_2, \gamma_2, B_2)$ *be big conglomerates, and let* G_2 *be a cyclic Kleinian group generated by the transformation* f. *Assume that*

j) $D_1 - \{(D_1 \cap B_1) \cup (D_1 \cap B_2.\}$ *has a non-empty interior;*

k) *For every* $g \in G_1$, $g(B_1 \cup \gamma_1) \cap (B_2 \cup \gamma_2) = \emptyset$;

l) $f(\gamma_1) = \gamma_2$;

m) $f \circ H_1 \circ f^{-1} = H_2$; *and*

n) $f(B_1) \cap B_2 = \emptyset$.

Let $D_2 = \hat{C} - \{B_1 \cup B_2 \cup \gamma_2\}$. *Let* G *be the group generated by* G_1 *and* G_2, *and set* $D = D_1 \cap D_2$. *Then*

1) G *is Kleinian;*
2) *every relation in* G *is a consequence of the relations of* G_1, *and the relations of hypothesis* (m);
3) D *is a* FS *for* G;
4) *there is a* FS D' *for* G, *so that* $(G, D', H_1, \Delta_1, \gamma_1)$ *is a conglomerate;*
5) *if* $g \in G$ *is elliptic or parabolic, then* g *is conjugate in* G *to an element of* G_1; *and*
6) *if* $L(G_1)$ *has measure zero, then* $L(G)$ *has measure zero.*

Remark III: Hypotheses (1) and (n) imply that D_2 is a FS for G_2.

§3. KOEBE'S THEOREM.

For this section, we assume that $(G, D, H, \Delta, \gamma)$ is a conglomerate where G is normalized so that $\infty \in R(G)$, and so that γ, and all of its translates under G, are bounded.

We recall some notions about isometric circles; these are described more fully in Ford [3].

Every $g \in G$ is a transformation of the form $z \to (az + b)/(cz + d)$, where $a, b, c, d \in C$, and $ad - bc = 1$. The isometric circle $I(g)$ of the transformation g is the circle $\{z \mid |cz + d| = 1\}$. The isometric circle $I'(g)$ of g^{-1} is the circle $\{z \mid |cz - a| = 1\}$. The centers of $I(g)$ and $I'(g)$ are denoted by $p(g)$ and $p'(g)$, respectively. Then $p(g) = -d/c$, and $p'(g) = a/c$. The radius of $I(g)$ is denoted by $r(g)$; then $r(g) = |c|^{-1}$.

The transformation g can be decomposed into a product of three motions, $g = t_3 \circ t_2 \circ t_1$, where t_1 is inversion in $I(g)$, t_2 is reflection about the perpendicular bisector of $p(g)$ and $p'(g)$, and t_3 is a Euclidean rotation with center at $p'(g)$. t_2 and t_3 are Euclidean motions, and so if we are given a set S and want to measure the size of $g(S)$, it suffices to measure the size of $t_1(S)$.

It is well known (see Ford [3] pg. 104) that if we enumerate the elements of G as g_i, then

$$\sum_i r^4(g_i) < \infty . \tag{1}$$

If S is a set in the plane, then the Euclidean diameter of S will be denoted by dia (S).

KOEBE'S THEOREM: *Let* $(G, D, H, \Delta, \gamma,)$ *be a conglomerate where* G *is normalized so that* $\infty \in R(G)$, *and so that* γ *and all its translates under* G *are bounded. Let* $G = \Sigma_i g_i H$ *be a coset decomposition. Then*

$$\sum_i \text{dia}^2 g_i(\gamma) < \infty . \tag{2}$$

Remark 4: The basic idea of the proof given below is due to Koebe [5, 6]. It is difficult to precisely state Koebe's original theorem; he used it in [5] to prove that the limit set of a Schottky-type group has measure zero.

Proof of Koebe's theorem: It is clear that (2) is independent of which coset representatives we choose. We observe that $p(g) = g^{-1}(\infty)$, and so if $h \in H$, then $p(g \circ h) = (g \circ h)^{-1}(\infty) = h^{-1}(p(g))$; i.e., the set of points $\{p(g \circ h)\}$, as h varies through H, is invariant under H. Also, since $\infty \in R(G) \subset R(H)$, $p(g) \in R(H)$ for every $g \in G$. Therefore, for each i, there is a unique $h_i \in H$, so that $p(g_i \circ h_i) \in \Delta$. We assume that in fact, $p(g_i) \in \Delta$.

By property (f), there is a neighborhood V of γ, so that $\Delta \cap V \subset D$. Then there is at most one i, so that $p(g_i) \in V$. If we let δ_i be the distance from $p(g_i)$ to γ, then we have shown that there is a $\delta_0 > 0$, so that $\delta_i > \delta_0$ for almost all i.

For fixed i, let C_i be the circle, of radius δ_i, centered at $p(g_i)$. Denoting inversion in $I(g_i)$ by t_i, one sees at once that

$$\text{dia } t_i(\gamma) \leq \text{dia } t_i(C_i) = r^2(g_i)/\delta_i \quad .$$

Hence, using (1),

$$\sum_i \mathrm{dia}^2\, g_i(\gamma) = \sum_i \mathrm{dia}^2\, t_i(\gamma) \le 2\delta_0^{-2} \sum_i r^4(g_i) < \infty .$$

§4. PROOF OF COMBINATION THEOREM I.

Throughout this section, we assume that we are given two big conglomerates $(G_1, D_1, H, \Delta, \gamma, B_1)$ and $(G_2, D_2, H, \Delta, \gamma, B_2)$, and that the hypotheses of combination Theorem I hold. Recall that G is the group generated by G_1 and G_2, and that $D = D_1 \cap D_2$.

Since H is a subgroup of both G_1 and G_2, every element of $G-H$ (i.e., elements of G which are not in H) can be written in the *normal form*

3) $$g = g_n \circ \cdots \circ g_1 \;,$$

where either $g_{2i} \in G_1 - H$, $g_{2i+1} \in G_2 - H$, or $g_{2i} \in G_2 - H$, $g_{2i+1} \in G_1 - H$.

Property (g) asserts that $\Delta \cap B_1 = D_1 \cap B_1$. Another way of stating this is to say that two points of B_1 are equivalent under G_1 if and only if they are equivalent under H. We conclude that if $z \in B_1$, and $g_i \in G_1 - H$, then $g_i(z) \in B_2$. We also conclude that if $z \in B_1$, and $g_i \in G_1 - H$, then $g_i(z) \notin D$. We similarly observe that if $z \in B_2$ and $g_i \in G_2 - H$, then $g_i(z) \in B_1$, and $g_i(z) \notin D$.

Now let z be some point of D, and let $g_n \circ \cdots \circ g_1$ be some word in normal form; we assume for convenience that $g_{2i} \in G_2 - H$, $g_{2i+1} \in G_1 - H$. Then $g_1(z) \in B_2$, $g_1(z) \notin D$; $g_2 \circ g_1(z) \in B_1$, $g_2 \circ g_1(z) \notin D$; and so on. We state this as

PROPOSITION 1: If $z \in D$, and $g_n \circ \cdots \circ g_1$, $n > 0$, is a word in normal form, then $g(z) \notin D$.

Property (h) asserts that $D_1 \cap B_2 \neq \emptyset$. Of course $D_1 \cap B_2 \subset \Delta \cap B_2 = D_2 \cap B_2$, and so $D \cap B_2 \neq \emptyset$. Similarly, we obtain $D \cap B_1 \neq \emptyset$. Since D is not empty, Proposition 1 says that no non-trivial word in normal form can be the identity; i.e., G is the free product of G_1 and G_2 with amalgamated subgroup H. Conclusion 2 has been established.

Since $D \subset \Delta$, it is obvious that $h(D) \cap D = \emptyset$ for all non-trivial $h \in H$.

We restate this remark together with Proposition 1 as,

PROPOSITION 2: If $z \in D$, $g \in G$, $g \neq 1$, then $g(z) \notin D$.

It is well-known and easy to verify that the union of the translates under G_1 of $D_1 \cap B_2$ is open in $\hat{C}$. Since this set is non-empty, and since $D_1 \cap B_2 \subset D$, we conclude that for every $z_0 \in D \cap B_2$, there is a neighborhood U, so that every $z \in U$ is a translate under G_1, of a point of D. By Proposition 2, no two points of U are equivalent under G; i.e., if $g \in G$, $g \neq 1$, then $g(U) \cap U = \emptyset$. This proves conclusion 1.

Our next task is to show that $D \subset R(G)$. We have already observed that $D \cap B_2 = D_1 \cap B_2$, and $D \cap B_1 = D_2 \cap B_1$. We have also observed that for every $z \in D_1 \cap B_2$, there is a neighborhood U so that $g(U) \cap U = \emptyset$ for all $g \in G$, $g \neq 1$. Analogous reasoning shows that $D_2 \cap B_1 \subset R(G)$. We next consider $D \cap \gamma$. By property (f), there are neighborhoods V_1 and V_2 of γ, so that $\Delta \cap V_1 \subset D_1$, and $\Delta \cap V_2 \subset D_2$. Set $V = V_1 \cap V_2$, and observe that $\Delta \cap V \subset D$. We now trivially have $D \cap \gamma = \Delta \cap \gamma \subset \Delta \cap V$. Let z_0 be a point of $D \cap \gamma$, then there is a neighborhood U of z_0, so that $U \subset V$. For each $z \in U$, there is a $w \in \Delta$, and an $h \in H$, so that $h(z) = w$. We have to show that if we make U sufficiently small, then each $z \in U$ is equivalent under H to a point of $\Delta \cap V$. If not, we would have a sequence $z_n \to z_0$, and a sequence $\{h_n\}$ of elements of H, so that $h_n(z_n) \notin V$ for every n. We choose a subsequence so that $h_n(z_n) \to w_0$. Then by Lemma 2 in [7], w_0 is a limit point of H, and hence lies on γ. This is a contradiction, and so we conclude that there is a neighborhood U of z_0, so that every $z \in U$ is equivalent under H to a point of $\Delta \cap V \subset D$; i.e., $z_0 \in R(G)$.

We summarize the above arguments as

PROPOSITION 3: $D \subset R(G)$.

The above arguments actually show a little bit more. We have shown that there is a neighborhood V of γ so that $\Delta \cap V \subset D \subset R(G)$. We don't as yet know that D is a FS for G, but by Propositions 2 and 3 we know that there is a FS $D' \supset D$. We then have

PROPOSITION 4: Let $D' \supset D$ be a FS for G. Then $(G, D', H, \Delta, \gamma)$ is a conglomerate.

In order to use Koebe's theorem, we assume from here on that G has been normalized so that $\infty \in D$, and so that γ and all its translates under G are bounded.

We have already observed that every $g \in G - H$ can be written in the normal form (3). This representation is not unique, but as is well known for amalgamated free products, the number of factors n in the right hand side of (3) is determined by the element g. We call n the *length* of g, and denote the length by $|g|$. If $g \in H$, we set $|g| = 0$. Furthermore, writing g in the normal form (3), we say that g is *positive* if $g_1 \in G_1 - H$, and we say that g is *negative* if $g_1 \in G_2 - H$.

We decompose G into cosets modulo H, and observe that for a fixed coset gH, every element of gH has the same length, and either every element is positive or every element is negative. We now choose some set of coset representatives and write

$$G = H + \sum_{n,m} a_{nm}H + \sum_{n,m} b_{nm}H ,$$

where $|a_{nm}| = |b_{nm}| = n$, $a_{nm} > 0$, and $b_{nm} < 0$.

For fixed $n > 0$, we set

$$T_n = \bigcup_m \{z | z \in a_{nm}(B_1)\} \cup \bigcup_m \{z | z \in b_{nm}(B_2)\} .$$

For fixed n, let S_n be the complement of T_n. Set $S = \bigcup_n S_n$, and let T be the complement of S. The main results of this section, which we prove below, are as follows. First, the points of S are all known; i.e., if $z \in S$, then there is a $g \in G$, so that $g(z) \in D \cup L(G_1) \cup L(G_2)$. The points of T are more difficult to describe, but we will show that $T \subset L(G)$, and that T has measure zero.

PROPOSITION 5: For $n > 1$, $T_n \subset T_{n-1}$.

Proof: Let z be some point of T_n, where $n > 1$. Then either there is an $a_{nm} > 0$ so that $z \in a_{nm}(B_1)$, or there is a $b_{nm} < 0$ so that $z \in b_{nm}(B_2)$, and we assume for simplicity that the former case holds. We write $a_{nm} = g_n \circ \cdots \circ g_1$ in normal form, and since $a_{nm} > 0$, we know that $g_1 \in G_1 - H$. It follows at once that $g_1(B_1) \subset B_2$, and hence $a_{nm}(B_1) \subset g_n \circ \cdots \circ g_2(B_2)$: The proof is completed by observing that $g_n \circ \cdots \circ g_2$ has length $n-1$ and is negative.

We are now in a position to investigate S. We first observe that if $z \in T_1 \cap B_2$ then there is a $g \in G_1 - H$, so that $g(z) \in B_1$. Hence if $z \in S_1 \cap B_2$, either $z \in L(G_1)$, or there is a $g \in G_1$, so that $g(z) \in D_1 \cap B_2 \subset D$. We similarly observe that if $z \in S_1 \cap B_1$, either $z \in L(G_2)$,, or there is a $g \in G_2$ so that $g(z) \in D$. Finally, if $z \in S_1 \cap \gamma$, then either $z \in L(H) \subset L(G_1) \cap L(G_2)$, or there is an $h \in H$, so that $h(z) \in \Delta \cap \gamma = D \cap \gamma$. We have proved

PROPOSITION 6. If $z \in S_1$, there is a $g \in G$, so that

$$g(z) \in D \cup L(G_1) \cup L(G_2) \quad .$$

PROPOSITION 7. If $z \in S$, there is a $g \in G$, so that

$$g(z) \in D \cup L(G_1) \cup L(G_2) \quad .$$

Proof: Since $S_i \subset S_{i+1}$, for every i, there is an n, so that $z \in S_n$, $z \notin S_{n-1}$. Proposition 6 takes care of the case $n = 1$, and so we assume $n > 1$. Then $z \in T_{n-1}$, and so either there is a $w \in B_1$, and an $a_{n-1,m}$ with $z = a_{n-1,m}(w)$, or there is a $w \in B_2$, and a $b_{n-1,m}$, with $z = b_{n-1,m}(w)$. For simplicity, we assume that the former case holds; the proof for the latter case is analogous. Now w is either a point of T_1 of of S_1. If w were in T_1, then there would be a $g \in G_2 - H$, and a point $x \in B_2$, so that $w = g(x)$. Then we would have $z = a_{n-1,m} \circ g(x)$, where $|a_{n-1,m} \circ g| = n$, $a_{n-1,m} \circ g < 0$, contradicting the fact that $z \notin T_n$.

We conclude that for every $z \in S_n$, $n > 1$, there is a $g \in G$, and a $w \in S_1$, so that $g(z) = w$. The result now follows at once from Proposition 6.

We remark incidentally that Proposition 7 characterizes S.

We now turn our attention to T. We first look at T_n, for fixed n. We observe at once that $a_{nm}(B_1) \subset B_2$ if and only if n is odd, and $b_{nm}(B_2) \subset B_1$ if and only if n is odd. We conclude that for fixed n, $a_{nm}(B_1) \cap b_{nk}(B_2) = \emptyset$. Now consider two distinct coset representatives a_{nm} and a_{nk}, of the same length. We write $a_{nm}^{-1} \circ a_{nk} = g_s \circ \cdots \circ g_1$ in normal form, and observe that $s > 0$, $g_1 \in G_1 - H$, $g_s \in G_1 - H$. We conclude that $a_{nm}^{-1} \circ a_{nk}(B_1) \subset B_2$ and so $a_{nk}(B_1) \cap a_{nm}(B_1) = \emptyset$. In exactly the same manner, we show that if b_{nm} and b_{nk} represent distinct cosets, then $b_{nm}(B_2) \cap b_{nk}(B_2) = \emptyset$.

The above remarks, together with the trivial Proposition 9, below, yield

PROPOSITION 8. Each connected component T_{nm} of T_n is either a set of the form $\{z \mid z \in a_{nm}(B_1)\}$ or a set of the form $\{z \mid z \in b_{nm}(B_2)\}$.

PROPOSITION 9. If $g \in G - H$, then $g(\gamma) \cap \gamma = \emptyset$.

Proof: Write g in the normal form (3), and observe that $n > 0$. Assume, for simplicity, that $g > 0$. Then by property (i), $g_1(\gamma) \subset B_2$. Then $g_2 \circ g_1(\gamma) \subset B_1$ and so on.

We remark that $D \subset S_1$, and so each T_n is bounded.

We will no longer need to differentiate between positive and negative cosets. We now rewrite the coset decomposition as

$$G = \sum_{n,m} c_{nm} H$$

where $|c_{nm}| = n$, and T_{nm} is bounded by $c_{nm}(\gamma)$.

Now let z_0 be some point of T, then $z_0 \in T_n$ for every n, and so for each n, there is an $m(n)$, with $z_0 \in T_{n,m(n)}$.

By Proposition 4, Koebe's theorem is applicable, and so

4) $$\sum_{n,m} \operatorname{dia}^2 c_{nm}(\gamma) = \sum_{n,m} \operatorname{dia}^2 T_{nm} = \sum_n \sum_m \operatorname{dia}^2(T_{nm}) < \infty .$$

We conclude that

5) $$\lim_{n\to\infty} \sum_m \operatorname{dia}^2(T_{nm}) = 0 .$$

As a special case of (5), we observe that

6) $$\lim_{n\to\infty} \operatorname{dia}(T_{n,m(n)}) = 0 .$$

Equation (6) asserts that for every w on γ, $\lim_{n\to\infty} c_{n,m(n)}(w) = z_0$; i.e., since the $c_{n,m(n)}$ are all distinct, z_0 is a limit point of G. We restate this as

PROPOSITION 10. $T \subset L(G)$.

Since S and T are complementary, Propositions 7 and 10 together with Propositions 2 and 3 assert that D is a FS for G; i.e., we have proven conclusion 3.

The above remark, together with Proposition 4 yields conclusion 4.

We return to equation (5) and observe that since $T \subset T_n$, for all n,

$$\operatorname{meas}(T) \leq \operatorname{meas}(T_n) = \sum_m \operatorname{meas}(T_{nm}) \leq \pi/4\Sigma_m \operatorname{dia}^2(T_{nm}) \to 0 .$$

We restate this as

Proposition 11: T has measure zero.

To prove conclusion 6, we observe that $L(G) = (L(G) \cap S) \cup (L(G) \cap T)$. $L(G) \cap T$ has measure zero, and by Proposition 7, $L(G) \cap S =$

$$L(G) \cap S = \cup_{g\in G}\, g(L(G_1) \cup L(G_2)).$$

G of course is countable, and by assumption $L(G_1)$ and $L(G_2)$ both have measure zero.

It remains only to prove conclusion 5. Let g be some element of G. Among all the conjugates of g in G, there is at least one, which we call g^*, of minimal length n. We trivially observe that either $n = 1$, or n is even. If $n = 1$, there is nothing to prove, and so we assume that n is even, and we assume for simplicity that $g^* > 0$. Then one easily sees that $g^*(B_1) \subset B_1$, and hence g^* is loxodromic.

This concludes the proof of combination theorem I.

§5. PROOF OF COMBINATION THEOREM II.

It should be remarked that, in combination theorem II, if H_1 and H_2 are both trivial, then G is the free product of G_1 and G_2. Of course, this free product need not occur in the sense of combination theorem I; the geometry is somewhat different. The basic idea in the proof below is to imitate, as far as possible, the elementary proof of combination theorem I given above. From here on, we assume that the hypotheses of combination theorem II hold.

Since G is generated by G_1 and G_2, every element $g \in G$, can be written in the form

$$7) \qquad g = f^{\alpha_n} \circ g_n \circ \cdots \circ f^{\alpha_1} \circ g_1 \ ,$$

where $\alpha_1, \ldots, \alpha_{n-1} \neq 0$, and $g_2, \ldots, g_n \neq 1$. Using the relations in hypothesis (m), we can assume that if $\alpha_{i-1} < 0$ and $g_i \in H_1$, then $\alpha_i \leq 0$; we can also assume that if $\alpha_{i-1} > 0$ and $g_i \in H_2$, then $\alpha_i \geq 0$.

An expression of the form (7) satisfying these assumptions is called a *normal form* for the element g. Let F be the complement of $\overline{B}_1 \cup \overline{B}_2$, so that F is the interior of D_2.

We make the following observations.

If $\alpha > 0$, then $f^{\alpha}(F \cup B_2 \cup \gamma_2) \subset B_2$.

If $\alpha < 0$, then $f^{\alpha}(F \cup B_1 \cup \gamma_1) \subset B_1$.

If $g \in G_1$, $g \neq 1$, then $g(D) \cap D = \emptyset$.

If $g \in G_1 - H_1$, then $g(B_1 \cup \gamma_1) \subset F - D$.

If $g \in G_1 - H_2$, then $g(B_2 \cup \gamma_2) \subset F - D$.
If $\alpha > 0$ and $g \in H_2$, then $g \circ f^{\alpha}(F \cup B_2 \cup \gamma_2) \subset B_2$.
If $\alpha < 0$ and $g \in H_1$, then $g \circ f^{\alpha}(F \cup B_1 \cup \gamma_1) \subset B_1$.

Now let g be some non-trivial element of G, expressed in the normal form (7), and let z be some point of D. We put together the above observations as follows. If $g_1 = 1$, then $\alpha_1 \neq 0$, in any case, $g_1(z) \in F \cup \gamma_1$. Then $f^{\alpha_1} \circ g_1(z)$ belongs to B_1 if $\alpha_1 < 0$, or $B_2 \cup \gamma_2$ if $\alpha_1 > 0$. In either case $g_2 \circ f^{\alpha_1} \circ g_1(z)$ belongs to $F - D$, unless $\alpha_1 < 0$ and $g_2 \in H_1$, or $\alpha_1 > 0$ and $g_2 \in H_2$. If $\alpha_1 < 0$ and $g_2 \in H_1$, then $g_2 \circ f^{\alpha_1} \circ g_1(z) \in B_1$, and $\alpha_2 \leq 0$, so that $f^{\alpha_2} \circ g_2 \circ f^{\alpha_1} \circ g_1(z) \in B_1$. Similarly if $\alpha_1 > 0$ and $g_2 \in H_2$, then $f^{\alpha_2} \circ g_2 \circ f^{\alpha_1} \circ g_1(z) \in B_2$. In any case, we can continue in this manner and observe that we have proven

PROPOSITION 12: If $g = f^{\alpha_n} \circ \cdots \circ g_1$ is in normal form, where either $g_1 \neq 1$, or $\alpha_1 \neq 0$, and if $z \in D$, then $g(z) \notin D$.

Simple modifications in the above argument, together with hypotheses (i) and (k), yield

PROPOSITION 13: If $g \in G - H_1$, then $g(\gamma_1) \cap \gamma_1 = \emptyset$; if $g \in G$, $g \neq f \circ g_1$, where $g_1 \in G_1$, then $g(\gamma_1) \cap \gamma_2 \neq \emptyset$.

By hypothesis (j), D has a non-empty interior. This statement, together with Proposition 12 trivially implies conclusions 1 and 2.

It is somewhat more convenient to restate Proposition 12 as

PROPOSITION 14: If $g \in G$, $g \neq 1$, and $z \in D$, then $g(z) \notin D$.

We next want to show that $D \subset R(G)$. Consider the set of all the translates under G_1 of $B_1 \cup B_2$, and call the complement of this set F_1. Observe that for every $z \in F_1$, there is a $g \in G_1$, so that $g(z) \in D \cup \gamma_2 \cup L(G_1)$. Hence for every $z_0 \in D$, either z_0 is an interior point of F_1, or z_0 is a point of γ_1. If z_0 is an interior point of F_1, then z_0 trivially has a neighborhood U so that every $z \in U$ is equivalent under G_1, to a point of D.

If z_0 is a point of γ_1, then using hypothesis (f), there is a neighborhood U of z_0 so that for every $z \in U$, either $z \in D$, or $f(z) \in F_1 \cap R(G_1)$.

We restate this as

PROPOSITION 15: $D \subset R(G)$.

Using hypothesis (f) again, there are neighborhoods V_1 and V_2 of γ_1 and γ_2 respectively, so that $\Delta_1 \cap V_1 \subset D_1$, and $\Delta_2 \cap V_2 \subset D_1$. Set $V = V_1 \cap f^{-1}(V_2)$. We define a new FS D_1' for G_1 as follows. Delete $f(V) \cap D_1$ from D_1; for each $z \in F(V) \cap D_1$, there is an $h \in H_1$ so that $h \circ f^{-1}(z) \in \Delta_1$; adjoin $h \circ f^{-1}(z)$ to the truncated D_1, and in this way we get a new FS D_1' for G_1. Similarly pick a FS D_2' for G_2 so that $V \subset D_2'$. Thus, using Propositions 14 and 15, there is a FS D' for G, so that $\Delta_1 \cap V \subset D'$. Hence $(G, D', H_1, \Delta_1, \gamma_1)$ is a conglomerate; this is conclusion 4.

We recall that every $g \in G$, $g \neq 1$, can be written, not uniquely, in the normal form

$$g = f^{\alpha_n} \circ g_n \circ \cdots \circ f^{\alpha_1} \circ g_1 \ , \tag{7}$$

where $g_1, \ldots, g_n \in G_1$; $g_2, \ldots, g_n \neq 1$; $\alpha_1, \ldots, \alpha_{n-1} \neq 0$; if $\alpha_i < 0$ and $g_{i+1} \in H_1$, then $\alpha_{i+1} \leq 0$; if $\alpha_i > 0$ and $g_{i+1} \in H_2$, then $\alpha_{i+1} \geq 0$.

Suppose now that g can also be written as some other normal form, say

$$g = f^{\beta_k} \circ \tilde{g}_k \circ \cdots \circ f^{\beta_1} \circ \tilde{g}_1 \ . \tag{8}$$

Proposition 12 asserts that no non-trivial normal form can be the identity. We form the expression

$$(f^{\beta_k} \circ \cdots \circ f^{\beta_1} \circ \tilde{g}_1) \circ (g_1^{-1} \circ f^{-\alpha_1} \circ \cdots \circ f^{-\alpha_n}) \tag{9}$$

and we know that (9) is not a normal form; i.e., $\tilde{g}_1 \circ g_1^{-1}$ belongs to either H_1 or H_2, and α_1 and β_1 have the same sign. Since the expression (9) must reduce to the identity, we easily see that as we reduce (9), we will always get expressions of the form

$$10)\qquad (f^{\beta_k} \circ \tilde{g}_k \circ \cdots \circ f^{\beta_i} \circ \tilde{g}_i) \circ (g_j^{-1} \circ \cdots \circ g_n^{-1} \circ f^{-\alpha_n}) .$$

Again, since (10) cannot be a normal form, we must have that $\tilde{g}_i \circ g_j^{-1}$ belongs to either H_1 or H_2, and β_i and α_j have the same sign. We conclude that

$$11)\qquad \sum_{i=1}^{n} |\alpha_i| = \sum_{j=1}^{k} |\beta_j| .$$

We call the invariant, given by (11), the *length* of g, and denote it by $|g|$. If $g \in G_1$, then $|g| = 0$; in particular, the identity has length zero. We trivially remark that if $g_1 \in G_1$, then $|gg_1| = |g_1 g| = |g|$, for every $g \in G$.

The right hand side of (7) is called a *positive* normal form if $g_1 \in H_1$ and $\alpha_1 > 0$. If we also represent the same element g by the normal form given in (8), then we again form the expression (9) and get a word which is not in normal form. It follows that $\tilde{g}_1 \in H_1$ and $\beta_1 > 0$; i.e., the right hand side of (8) is also positive. It follows that we can refer to the element g as being positive. Our major interest is in elements which are non-positive; if $g \in G$ is non-positive, then we write $g \leq 0$.

In analogy with the above, the right hand side of (7) is called *negative* if $g_1 \in H_2$ and $\alpha_1 < 0$. Again, if $g \in G$ can be expressed by a negative normal form, then every normal form for g is negative, and we say that g is *negative*. Again our major interest is in elements which are *non-negative*; if $g \in G$ is non-negative, then we write $g \geq 0$.

One easily sees that if $g \leq 0$ and $h \in H_1$, then $g \circ h \leq 0$; also if $g \geq 0$ and $h \in H_2$, then $g \circ h \geq 0$.

We decompose G into cosets modulo H_1. For a fixed coset gH_1, every element has the same length; this length is also called the *length* of the coset. Again for fixed g, either every element of gH_1 is positive, or every element is non-positive, and we say that the coset is *positive* or *non-positive* respectively. For each *non-positive* coset of length N, we pick out some coset representative and denote it by a_{NM}.

Similarly, we decompose G into cosets modulo H_2; define the *length* of the coset gH_2 as the length of g; define gH_2 to be *negative* if g is negative; and pick out a set of coset representatives b_{NM} for the *non-negative cosets of length N.*

For $N = 0, 1, 2, \ldots$, we set

$$T_N = \cup_M \{z \mid z \in a_{NM}(B_1)\} \cup \cup_M \{z \mid z \in b_{NM}(B_2)\} .$$

Let S_N be the complement of T_N; set $S = \cup_N S_N$, and let T be the complement of S.

In analogy with the preceding theorem, our goal is to show that every point of S is a translate of some point in $D \cup L(G_1) \cup L(G_2)$, and that T has measure zero and consists of limit points of G.

For $N > 0$, let z_0 be some point of T_N. We assume for simplicity that there is an M so that $z_0 \in a_{NM}(B_1)$; there is an obvious analogous treatment for the case that $z_0 \in b_{NM}(B_2)$.

Let $g = a_{NM}$ be expressed in the normal form (7). If $\alpha_1 > 0$, we set

$$g^* = f^{\alpha_n} \circ g_n \circ \cdots \circ f^{\alpha_1 - 1} \quad . \tag{12}$$

Since $a_{NM} \leq 0$, $g_1 \notin H_1$, and so $g(B_1) \subset g^*(B_2)$. We see at once that $|g^*| = N-1$, and that the right hand side of (12) is a non-negative normal form. We conclude that $z_0 \in T_{N-1}$.

Similarly, if $\alpha_1 < 0$, then we set

$$g^* = f^{\alpha_n} \circ \cdots \circ f^{\alpha_1 + 1} \quad , \tag{13}$$

and observe that $|g^*| = N-1$, $g^* \leq 0$, and $g(B_1) \subset g^*(B_1)$. We again conclude that $z_0 \in T_{N-1}$.

We summarize these arguments as

PROPOSITION 16: For $N > 0$, $T_N \subset T_{N-1}$, $S_N \supset S_{N-1}$.

Now let z_0 be some point of S_0. Recall that $T_0 = \cup_{g \epsilon G_1} g(B_1 \cup B_2)$, and so either $z_0 \epsilon L(G_1)$ or there is a $g \epsilon G_1$ with $g(z) \epsilon D_1$. In the latter case, $g(z)$ either belongs to D, or lies on γ_2, in which case there is an $h \epsilon H_1$ so that $h \circ f^{-1} \circ g(z) \epsilon D$. We have proven

PROPOSITION 17: Let $z_0 \epsilon S_0$, then there is a $g \epsilon G$ so that $g(z_0) \epsilon D \cup L(G_1)$.

PROPOSITION 18. Let $z_0 \epsilon S$, then there is a $g \epsilon G$, so that $g(z_0) \epsilon D \cup L(G_1) \cup L(G_2)$.

Proof: Using Propositions 16 and 17, there is an $N \geq 0$ so that $z_0 \epsilon S_{N+1}$, $z_0 \notin S_N$. Since $z_0 \epsilon T_N$, there is either an a_{NM} with $z_0 \epsilon a_{NM}(B_1)$, or there is a b_{NM} with $z_0 \epsilon b_{NM}(B_2)$. We assume for simplicity that the former case occurs; the latter case is treated analogously.

Let $z_1 = a_{NM}^{-1}(z_0) \epsilon B_1$. Either z_1 is a fixed point of f, or there is a $\beta > 0$ so that $z_2 = f^{\beta}(z_1) \epsilon D_2$. In the former case,

$$a_{NM}^{-1}(z_0) \epsilon L(G_2) ,$$

and we are finished; we assume the latter case occurs.

We want to show that z_2 in fact lies in S_0. Suppose not. Then there would be a $\tilde{g} \epsilon G_1$ so that $z_3 = \tilde{g}(z_2) \epsilon B_1 \cup B_2$. Set

$$g^* = a_{NM} \circ f^{-\beta} \circ \tilde{g}^{-1} .$$

Recall that $a_{NM} \leq 0$ and $\beta > 0$, so that $N^* = |g^*| = |g| + \beta > N$. If $z_3 \epsilon B_1$, then $\tilde{g} \notin H_1$, and so $g^* \leq 0$; if $z_3 \epsilon B_2$, then $\tilde{g} \notin H_2$, and so $g^* \geq 0$. Since $z_0 = g^*(z_3)$, we conclude that $z_0 \epsilon T_{N^*}$ contradicting the assumption that $z_0 \epsilon S_{N+1}$. Hence z_2 does lie in S_0, and so by Proposition 17, there is a $g \epsilon G$, with $g(z_2) \epsilon D \cup L(G_1)$.

We remark, incidentally, that it is not clear whether or not the translates of the fixed points of f actually do occur as points of S.

We have already observed that $D \subset S_0$, and so $T_N \cap D = \emptyset$ for every N. This remark, together with Proposition 13, yields

PROPOSITION 19: Every connected component T_{NM} of T_N is either a set of the form $a_{NM}(B_1)$ or a set of the form $b_{NM}(B_2)$.

Each connected component T_{NM} of T_N is bounded either by $a_{NM}(\gamma_1)$ or by $b_{NM} \circ f(\gamma_1)$.

For each T_{NM} we pick c_{NM} so that T_{NM} is bounded by $c_{NM}(\gamma_1)$, and observe that the c_{NM} all represent distinct cosets of H_1 in G.

We now normalize G so that $\infty \in D$ and so that γ_1 and all its translates are bounded. We have already shown that there is a FS D' for G so that $(G, D', H_1, \Delta_1, \gamma_1)$ is a conglomerate; hence Koebe's theorem is applicable. We observe that

$$14) \qquad \Sigma_N \Sigma_M \operatorname{dia}^2 T_{NM} = \Sigma_N \Sigma_M \operatorname{dia}^2 c_{NM}(\gamma_1) < \infty ,$$

and so

$$15) \qquad \lim_{N \to \infty} \Sigma_M \operatorname{dia}^2 c_{NM}(\gamma_1) = 0 .$$

Now if $z_0 \in T$, then $z_0 \in T_N$ for every N, and so for each N, there is an M(N) with $z_0 \in T_{N,M(N)}$. $T_{N,M(N)}$ is bounded by $c_{N,M(N)}(\gamma_1)$ and by (15) $\operatorname{dia} c_{N,M(N)}(\gamma_1) \to 0$; i.e., if z is any point of γ_1, then $c_{N,M(N)}(z) \to z_0$. It follows at once that z is a limit point of G. We restate this as

PROPOSITION 20: $T \subset L(G)$.

Putting together Propositions (18) and (20), we see that for every $z \in R(G)$, there is a $g \in G$ so that $g(z) \in D$. This completes the proof of conclusion 3; i.e., D is a FS for G.

In order to prove conclusion 6, we observe that $L(G_2)$ consists of two points, and $L(G) \cap S$ has measure zero if and only if $L(G_1)$ has measure zero.

Hence it suffices to prove that T has measure zero. Observe that

$$16)\quad \text{meas}(T) \leq \text{meas}(T_N) = \sum_M \text{meas}(T_{NM}) \leq (\pi/4) \sum_M \text{dia}^2 T_{NM}$$

$$= (\pi/4) \sum_M \text{dia}^2 c_{NM}(\gamma_1) \quad .$$

Putting together (15) and (16), we get

PROPOSITION 21: T has measure zero.

The only thing left is to prove that if $g \in G$ is elliptic or parabolic, then g is conjugate in G to an element of G_1.

Let g be some element of G, and let g^* be a conjugate of g, where $|g^*|$ is minimal among all conjugates of g. It suffices to show that if $|g^*| > 0$, then g^* is loxodromic.

We write $g^* = f^{\alpha_n} \circ \cdots \circ g_1$, in normal form, and we can assume that $\alpha_n \neq 0$. We in fact assume that $\alpha_n > 0$; the case that $\alpha_n < 0$ is treated in an analogous manner. With this assumption, the minimality of $|g^*|$ implies that $g^* \geq 0$. Using the appropriate remarks in the proof of Proposition 12, we now observe that $g^*(B_2) \subset B_2$; it follows at once that g^* is loxodromic, and conclusion 5 has been established.

REFERENCES

[1] L.V. Ahlfors, Fundamental polyhedrons and limit point sets of Kleinian groups, *Proc. Nat. Acad. Sci.*, 55 (1966), 251-254.

[2] V. Chuckrow, to appear.

[3] L.R. Ford, *Automorphic Functions*, 2nd ed. Chelsea, New York, 1951.

[4] F. Klein, Neue Beiträge zur Riemann'schen Functionentheorie, *Math. Ann.* 21 (1883), 141-218.

[5] P. Koebe, Über die Uniformisierung der algebraischen Kurven II, *Math. Ann.* 69 (1910), 1-81.

[6] ———, Über die Uniformisierung der algebraischen Kurven III, *Math. Ann.* 72 (1912), 437-516.

[7] B. Maskit, On Klein's combination theorem, *Trans. Am. Math. Soc.* 120 (1965), 499-509.

[8] ———, On Klein's combination theorem II, *Trans. Am. Math. Soc.* 131 (1968), 32-39.

Massachusetts Institute of Technology
Cambridge, Massachusetts

ON FINSLER GEOMETRY AND
APPLICATIONS TO TEICHMÜLLER SPACES

by

Brian O'Byrne*

§1. Introduction.

In this paper we investigate the following situation in Finsler geometry: (X, α) and (Y, β) are complete Finsler manifolds such that there exists a C^{1+}-foliation $f: X \to Y$ and the Finsler structure β is, in a sense to be defined later, the infimum via f of the Finsler structure α. It is possible to define two metrics on Y which are related to α. One is the metric induced by β and the other is, again in a sense to be defined later, the infimum via f of the metric induced on X by α. Our main result is the proof of the equality of these two metrics.

Earle and Eells in [2] have shown that for any Fuchsian group Γ it is possible to introduce Finsler structures $\alpha(\Gamma)$ on the space $M(\Gamma)$ of Beltrami differentials of Γ and $\beta(\Gamma)$ on $T(\Gamma)$ the Teichmüller space of Γ in such a way that $(M(\Gamma), \alpha(\Gamma))$ and $(T(\Gamma), \beta(\Gamma))$ satisfy the hypotheses of our main theorem. As a consequence of this, we are able to show that for any $\mu \in M(\Gamma)$ which is extremal in the sense of [4] $t\mu$ is also extremal for $0 \leq t \leq 1$ and that for any extremal μ its norm as a linear functional on the integrable quadratic differentials of Γ is equal to its supremum norm. The latter result was first proved by Hamilton in [4].

In §2 and §3 we describe the situation we are concerned about. In §4 we prove two preliminary lemmas, the second of which involves the lifting

---

* This work was done at Cornell University as part of the author's doctoral dissertation.

of regular curves. In §5 we prove our main theorems and in §6 we give their applications to Teichmüller spaces.

§2. Finsler Manifolds

Let E be a Banach space with norm $\|\cdot\|_E$ and X a paracompact C^{1+}-manifold modeled on E. Let $\pi_X: TX \to X$ denote the tangent vector bundle of X and $X(x) = \pi_X^{-1}(x)$ the tangent space to X at x. Then (X, E, α) is a *Finsler manifold* and α is a *Finsler structure* on X if $\alpha: TX \to R$ is a map such that for each x in X, the restriction of α to $X(x)$ is a norm and x is centered at a coordinate chart (θ, U) in which α has the following two properties:

(2.1) There is a number $A > 0$ such that

$$A^{-1}\alpha(v) \leq \|\theta_*(x)v\|_E \leq A\alpha(v) \quad \text{for all } v \text{ in } X(x).$$

Here $\theta_*(x)$ denotes the differential of θ at x.

(2.2) There is a number $K > 0$ such that

$$|\alpha(\theta_*^{-1}(\theta(z))v) - \alpha(\theta_*^{-1}(\theta(y))v)| \leq K\|\theta(z) - \theta(y)\|_E \|v\|_E$$

for all z, y in U and all v in E.

These properties imply that α is locally Lipschitz in TX and that the norms on $X(y)$ induced by α and $\|\cdot\|_E$ are equivalent uniformly for y in U.

A curve is called *regular* if it is continuously differentiable and a local homeomorphism onto its range. If $b: [a, c] \to X$ is a piecewise regular curve then its *Finsler length* $L_\alpha(b)$ is defined by

$$(2.3) \qquad L_\alpha(b) = \int_a^b \alpha(b'(t))\,dt \quad \text{where } b'(t) = b_*(t)(1) \in X(b(t)).$$

Assuming that X is connected, for any two points, $x, y \in X$ the *Finsler distance* $d_\alpha(x, y)$ is defined by

$$(2.4) \quad d_\alpha(x, y) = \inf\{L_\alpha(b) : b \text{ is a piecewise regular curve in } X \text{ joining } x \text{ to } y\}.$$

The following lemma is well known and is fairly straightforward.

LEMMA. *If* (X, E, α) *is a Finsler manifold, then the Finsler distance is a metric on* X *compatible with its topology.*

In the future, when speaking of a *complete Finsler manifold* (X, E, α) we refer to completeness with respect to Cauchy sequences in X relative to the metric d_α.

§3. Foliations

Let X and Y be differentiable manifolds modeled on Banach spaces and $f: X \to Y$ a differentiable surjective map. The map is called a *foliation* if, for each point x in X, $f_*(x)$ maps $X(x)$ onto $Y(f(x))$ and the kernel $\operatorname{Ker} f_*(x)$ of $f_*(x)$ is a direct summand of $X(x)$, (i.e., $\operatorname{Ker} f_*(x)$ is a closed subspace of $X(x)$ admitting a closed supplement).

Now let (X, E, α) and (Y, F, β) be complete Finsler manifolds subject to the following two assumptions:

(3.1) There is a C^{1+}-foliation $f: x \to y$.

(3.2) Given y in Y then for each x in $f^{-1}(y)$ and each v in $Y(y)$

$$\beta(v) = \inf\{\alpha(u) : f_*(x)u = v,\ u \text{ in } X(x)\}.$$

For any two points y, z in Y we define

$$\sigma(y, z) = \inf\{d_\alpha(x, w) : x \in f^{-1}(y),\ w \in f^{-1}(z)\}. \tag{3.3}$$

The main result of this paper is the proof that $\sigma = d_\beta$ and that therefore it is a metric on Y.

§4. Preliminary Lemmas

In this section and the next we assume that (X, E, α) and (Y, F, β) are complete Finsler manifolds satisfying (3.1) and (3.2). Here we prove two lemmas that are needed for the proof of Theorem I.

LEMMA I: *Given* y *in* Y, v *in* Y(y), x *in* $f^{-1}(y)$ *and* w *in* $(f_*(y))^{-1}(v)$ *then there exists a continuous linear map* $\hat{S}: Y(y) \to X(x)$ *such that*

$$f_*(x) \circ \hat{S} = I_{Y(y)} \text{ and } \hat{S}v = w \ .$$

The proof is quite easy and is therefore omitted.

LEMMA II: *Let* $c: [p, q] \to Y$ *be any regular curve which has no multiple points and let* $\varepsilon > 0$ *be arbitrary. Then for each point* x *in* $f^{-1}(c(p))$ *there is a regular curve* $b_x: [p, q] \to X$ *such that*

$$f \circ b_x = c \text{ and } b_x(p) = x \tag{4.1}$$

and

$$\alpha(b_x'(t)) \leq \beta(c'(t)) + \varepsilon \quad \text{for all } t \text{ in } [p, q] \ . \tag{4.2}$$

Proof: First we assume that, for the moment, y in Y, x in $f^{-1}(y)$ and v in Y(y) are fixed. By condition (3.2) on $\beta(v)$ we know there is a u in X(x) such that

$$\alpha(u) < \beta(v) + \varepsilon/2 \ , \quad f_*(x)u = v \ . \tag{4.3}$$

The conditions on f imply that it is locally a trivial fibration and therefore, there are C^{1+}-coordinate charts (θ, U) centered at x and (ψ, V) centered at y which trivialize f (see [5] Chapter II, section 2). This means that θ is a C^{1+}-homeomorphism of U onto a product of open subsets U_1 and V_1 of Banach spaces E_1 and F respectively and ψ is a C^{1+}-homeomorphism of V onto V_1 in such a way that the map

$$U_1 \times V_1 \times E_1 \times F \xrightarrow{\theta_*^{-1}} \pi_X^{-1}(U) \xrightarrow{f_*} \pi_Y^{-1}(V) \xrightarrow{\psi_*} V_1 \times F \tag{4.4}$$

is projection on the second and fourth factors.

By Lemma I there is a continuous linear map $\hat{S}: Y(y) \to X(x)$ such that

$$f_*(x) \circ \hat{S} = I_{Y(y)} \text{ and } \hat{S}v = u \ . \tag{4.5}$$

This implies that $\psi_*(y) \circ f_*(x) \circ \theta_*^{-1}(\theta(x)) \circ \theta_*^{-1}\theta(x) \circ \theta_*(x) \circ \hat{S} \circ \psi_*^{-1}(\psi(y))$ $= I_F$ and since θ and ψ trivialize f we have

(4.6) $$\psi_*(f(z)) \circ f_*(z) \circ \theta_*^{-1}(\theta(z)) \circ \theta_*(x) \circ \hat{S} \circ \psi_*^{-1}(\psi(y)) = I_F$$

for all z in U.

Let $f^*TY \to X$ denote the vector bundle over X obtained by pulling back TY via f. Since U and V are coordinate charts, the subbundle $f^*\pi_Y^{-1}(V)$ of f^*TY obtained by restricting f to U is homeomorphic to the vector bundle over U whose fibre over x is $Y(f(x))$. This homeomorphism is locally Lipschitz. Thus, for convenience, we may represent a point in $f^*\pi_Y^{-1}(V)$ as a pair (z,w) where $z \in U$ and $w \in Y(f(z))$. Also, let $L(f^*\pi_Y^{-1}(V), \pi_X^{-1}(U))$ denote the set of bundle maps which are locally Lipschitz and are continuous and linear in each fibre.

We define a map $\hat{S}$ in $L(f^*\pi_Y^{-1}(V), \pi_X^{-1}(U))$ by

(4.7) $$\tilde{S}(z)w = \theta_*^{-1}(\theta(z)) \circ \theta_*(x) \circ \hat{S} \circ \psi_*^{-1}(\psi(y)) \circ \psi_*(f(z))$$

for all $(z,w) \in f^*\pi_Y^{-1}(V)$.

This map is continuous and linear in each fibre and locally Lipschitz because it is composed of maps with these properties. Clearly

$$\tilde{S}(z): Y(f(z)) \to X(z) \text{ and } f_*(z) \circ \tilde{S}(z) = I_{Y(f(z))}$$

for all z in U.

The map $P: f^*\pi_Y^{-1}(V) \to R$ defined by

(4.8) $P(z,w) = \alpha(\tilde{S}(z)w) - \beta(w)$ for all (z,w) in $f^*\pi_Y^{-1}(V)$

is locally Lipschitz for similar reasons. Therefore, if U and V are restricted sufficiently, we know that there exists a $\delta > 0$ such that $P(z,w) < \varepsilon$ for all (z,w) in $f^*\pi_Y^{-1}(V)$ such that $|\beta(w) - \beta(v)| < \delta$ because $\tilde{S}(x) = \hat{S}$ and $\alpha(u) \leq \beta(v) + \varepsilon/2$.

The restriction that y, x and v are fixed will now be removed and all the maps, sets and constants from the above will be written with the subscripts x and v (i.e., $\tilde{S}$ will now be written as $\tilde{S}_{x,v}$).

If $y \in Y$ and $y \notin c([p,q])$ then we select any v in Y(y) and for each x in $f^{-1}(y)$ we construct the map $\tilde{S}_{x,v}$ as above with the additional restriction that $U_{x,v} \cap f^{-1}(c([p,q])) = \emptyset$. This is possible because $f^{-1}(c([p,q]))$ is closed.

If $y \in Y$ and $y = c(t')$ for some t' in $[p,q]$ then we set $v = c'(t')$ and for each x in $f^{-1}(y)$ we construct the map $\tilde{S}_{x,v}$ as above. Without loss of generality, we can assume that $V_{x,v}$ meets only a connected arc of c. Since $\beta(c'(t))$ is a continuous function of t, we may also assume that $V_{x,v}$ has been restricted sufficiently so that $|\beta(c'(t)) - \beta(c'(t'))| < \delta$ for all t in $[p,q]$ such that $c(t)$ is in $V_{x,v}$.

The sets $\{U_{x,v}\}$ which were constructed in the above two paragraphs form an open cover of X. Since X is paracompact this open cover has a neighborhood finite subcover $\{U_\tau\}$. If $\{\rho_\tau\}$ is a locally Lipschitz partition of unity subordinate to it, then we define a map $S \in L(f^*TY, TX)$ by

$$S(x)\,w = \sum_\tau \rho_\tau(x)\,\tilde{S}_\tau(x)\,w \quad \text{for all} \quad (x,w) \quad \text{in} \quad f^*TY \; . \tag{4.9}$$

This sum involves only a finite number of terms because the open cover is neighborhood finite. It is clear that the map is in $L(f^*TY, TX)$. Since f_* is linear in each fibre and $\Sigma_\tau\, \rho_\tau(x) = 1$ for all x in X, we have

$$f_*(x) \circ S(x) = I_{Y(f(x))} \quad \text{for all} \quad x \quad \text{in} \quad X \; . \tag{4.10}$$

Given a map S in $L(f^*TY, TX)$ satisfying (4.10) Earle and Eells have shown [3] that for $x \in f^{-1}(c(p))$ there exists a number r in $(p,q]$ and a curve $b_x : [p,r] \to X$ satisfying (4.1) and

$$b_x'(t) = S(b_x(t))\,c'(t) \quad \text{for all } t \text{ in } [p,q] \; . \tag{4.11}$$

Further ([3], Lemma 3B), we can take $r = q$ provided that the numbers $\alpha(S(x)\,c'(t))$ are bounded uniformly for t in $[p,q]$ and x in $f^{-1}(c(t))$. But

$$\begin{aligned}\alpha(S(x)\,c'(t)) &= \alpha\Big(\sum_\tau \rho_\tau(x)\,\tilde{S}_\tau(x)\,c'(t)\Big)\\ &\leq \sum_\tau \rho_\tau(x)\,\alpha(\tilde{S}_\tau(x)\,c'(t))\\ &\leq \beta(c'(t)) + \varepsilon\end{aligned}$$

by construction. Therefore $b_x : [p, q] \to X$ is clearly regular, satisfies (4.1) and (4.11) and

$\alpha(b_x'(t)) = \alpha(S(b_x(t))\,c'(t)) \leq \beta(c'(t)) + \varepsilon$ by the above calculation.

Q.E.D.

COROLLARY I: *Let* $c : [p, q] \to y$ *be any piecewise regular curve and let* $\varepsilon > 0$ *be arbitrary. Then for each point* x *in* $f^{-1}(c(p))$ *there is a piecewise regular curve* $b_x : [p, q] \to X$ *satisfying* (4.1) *and* (4.2).

Proof: Since c is piecewise regular it has only a finite number of zeros and points of discontinuity. Thus, there is a partition $p = t_0 < \cdots < t_n = q$ of $[p, q]$ such that $c_i = c\,|\,[t_{i-1}, t_i]$ is a regular curve which has no multiple points. By Theorem I, for each x in $f^{-1}(c(p))$ there is a regular curve $b_{x,1} : [t_0, t_1] \to X$ satisfying (4.1) and (4.2) for c_1. Similarly, for each i there is a regular curve $b_{x,i} : [t_{i-1}, t_i] \to X$ starting at $b_{x,i-1}(t_{i-1})$ satisfying (4.1) and (4.2) for c_i. Therefore, the curve $b_x : [p, q] \to X$ defined by $b_x\,|\,[t_{i-1}, t_i] = b_{x,i}$ is a piecewise regular curve which satisfies (4.1) and (4.2) for c.

§5. Main Theorems

THEOREM I: $d_\beta = \sigma$.

Proof: Let $\varepsilon > 0$ be given and let y_1 and y_2 be two points in Y. By the definition of σ there exist $x_1 \in f^{-1}(y_1)$ and $x_2 \in f^{-1}(y_2)$ such that $d_\alpha(x_1, x_2) \leq \sigma(y_1, y_2) + \varepsilon/2$. Let $b : [0, 1] \to X$ be a piecewise regular curve joining x_1 to x_2 for which $L_\alpha(b) \leq d_\alpha(x_1, x_2) + \varepsilon/2$. Then, since $\alpha(b'(t)) \leq \beta((f \circ b)'(t))$ for all t in $[0, 1]$ we have

$$\begin{aligned} d_\beta(y_1, y_2) &\le L_\beta(f \circ b) \le L_\alpha(b) \\ &\le d_\alpha(x_1, x_2) + \varepsilon/2 \\ &\le \sigma(y_1, y_2) + \varepsilon \quad . \end{aligned}$$

In the other direction, let $c: [0, 1] \to Y$ be a piecewise regular curve joining y_1 to y_2 such that $L_\beta(c) \le d_\beta(y_1, y_2) + \varepsilon/2$. By Corollary I there is a piecewise regular curve $b: [0, 1] \to X$ starting at x_1 such that $f \circ b = c$ and $\alpha(b'(t)) \le \beta(c'(t)) + \varepsilon/2$ for all t in $[0, 1]$. Thus

$$\begin{aligned} \sigma(y_1, y_2) &\le d_\alpha(b(0), b(1)) \\ &\le \int_0^1 \alpha(b'(t))\, dt \\ &\le \int_0^1 (\beta(c'(t)) + \varepsilon/2)\, dt \\ &= L_\beta(c) + \varepsilon/2 \\ &\le d_\beta(y_1, y_2) + \varepsilon \quad . \end{aligned}$$

Since ε is arbitrary we have $\sigma(y_1, y_2) = d_\beta(y_1, y_2)$.

Q.E.D.

If (Z, α) is a complete Finsler manifold then a curve $\gamma: [a, b] \to Z$ is *extremal* if it is piecewise regular and $L_\alpha(\gamma) = d_\alpha(\gamma(a), \gamma(b))$. It is easy to show that any subarc of an extremal curve is also extremal. A complete Finsler manifold in which any two points can be joined by an extremal curve is called an *extremal space*.

Let x_0 be a point in X. A point x is said to be *extremal with respect to* x_0 if $d_\alpha(x_0, x) = \inf\{d_\alpha(x_0, z): f(z) = f(x)\} = d_\beta(f(x_0), f(x))$.

THEOREM II. *Let* (X, E, α) *be an extremal space in addition to its other properties. Let* x *in* X *be extremal with respect to a point* x_0 *and* $\gamma: [0, 1] \to X$ *be an extremal curve joining* x_0 *to* x. *Then* $L_\alpha(\gamma) = L_\beta(f \circ \gamma)$,

$f \circ \gamma$ is extremal and $\alpha(\gamma'(t)) = \beta((f \circ \gamma)'(t))$. Further, $\gamma(t)$ is extremal with respect to x_0 for all $0 \leq t \leq 1$.

Proof: We have

$$\begin{aligned} d_\alpha(x_0, x) &= \int_0^1 \alpha(\gamma'(t))\,dt \\ &\geq \int_0^1 \beta((f \circ \gamma)'(t))\,dt \\ &\geq d_\beta(f(x_0), f(x)) . \end{aligned}$$

However, $d_\alpha(x_0, x) = d_\beta(f(x_0), f(x))$. Therefore we must have equality everywhere. This gives the first two parts. Since $\alpha(\gamma'(t)) \geq \beta((f \circ \gamma)'(t)) \geq 0$ for all t in $[0, 1]$ it also gives the third part.

Since any subarc of an extremal curve is extremal then for all t in $[0, 1]$ $d_\alpha(x_0, \gamma(t)) = d_\beta(f(x_0), f(\gamma(t)))$ because both sides are differentiable functions of t and their derivatives are equal by part three.

Q.E.D.

§6. Applications to Teichmüller Spaces

In this section our notation is essentially the same as that used by Earle and Eells in their paper [2]. We ask the reader to refer to that paper for the proofs of any results quoted here.

Let U denote the upper half plane of the complex plane and let M denote the unit ball of $L^\infty(U, C)$. If Γ is any Fuchsian group acting on U, we define $L^\infty(\Gamma) = \{\mu \in L^\infty(U, C) : \mu = (\mu \circ A)\,\bar{A}'/A' \text{ for all } A \text{ in } \Gamma\}$ and set $M(\Gamma) = M \cap L^\infty(\Gamma)$.

Teichmüller's metric $\sigma(\Gamma)$ on $M(\Gamma)$ is defined by

$$\sigma(\Gamma)(\mu, \nu) = \log \frac{1 + \left\| \dfrac{\mu - \nu}{1 - \bar{\mu}\nu} \right\|_\infty}{1 - \left\| \dfrac{\mu - \nu}{1 - \bar{\mu}\nu} \right\|_\infty} \quad \text{for all } \mu, \nu \in M(\Gamma) . \tag{6.1}$$

We can also define a Finsler structure $a(\Gamma)$ on $M(\Gamma)$ by

$$(6.2)\qquad a(\Gamma)(\mu,\nu) = 2\left\| \frac{\nu}{1-|\mu|^2} \right\|_\infty$$

for all (μ,ν) in $TM(\Gamma) = M(\Gamma) \times L^\infty(\Gamma)$. It turns out that $d_{a(\Gamma)} = \sigma(\Gamma)$.

Let B denote the complex Banach space of functions holomorphic in the lower half plane U^* with norm

$$(6.3)\qquad \|\phi\|_B = \sup\{|\bar{z} - z|^2\, |\phi(z)| : z \text{ in } U^*\} < \infty .$$

Then we define $B(\Gamma) = \{\phi \in B: (\phi \circ A)(A')^2 = \phi \text{ for all } A \in \Gamma\}$. Now there is a map $\Phi: M(\Gamma) \to B(\Gamma)$ which is a C^{1+}-foliation onto its range $T(\Gamma)$. $T(\Gamma)$ is called the *Teichmüller space of* Γ.

Teichmüller's metric $\tau(\Gamma)$ on $T(\Gamma)$ is defined by

$$(6.4)\quad \tau(\Gamma)(\phi,\psi) = \inf\{\sigma(\Gamma)(\mu,\nu): \Phi(\mu) = \phi, \Phi(\nu) = \psi,\ \mu,\nu \in M(\Gamma)\} .$$

In the same spirit we define a map $\beta(\Gamma)$: $TT(\Gamma) = T(\Gamma) \times B(\Gamma) \to R$ by

$$(6.5)\qquad \beta(\Gamma)(\Phi(\mu),\psi) = \inf\{a(\Gamma)(\mu,\nu): \Phi_*(\mu)\nu = \psi,\ \nu \in L^\infty(\Gamma)\} .$$

The map $\beta(\Gamma)$ satisfies (3.2) and is a Finsler structure on $T(\Gamma)$. We note here that our definition of $\beta(\Gamma)$ is not the same as that found in [2]. Howevery, the proofs that it depends only on $\Phi(\mu)$ and that it is a Finsler structure are the same. By Theorem I we have

THEOREM III: *For any Fuchsian group* Γ *acting on* U $\tau(\Gamma) = d_{\beta(\Gamma)}$.

Theorem II has applications to the study of extremal quasiconformal maps. We call μ in $M(\Gamma)$ *extremal* if the corresponding quasiconformal map $w_\mu: U \to U$ is *extremal* (see [4]). The extremal μ are exactly those μ which are *extremal with respect to the origin* in the sense of §5. Thus we obtain

THEOREM IV: *Let* Γ *be any Fuchsian group acting on* U. *If* μ *in* $M(\Gamma)$ *is extremal, then so is* $t\mu$ *for all* t *in* [0, 1].

Proof: The curve $\gamma(t) = t\mu$, $0 \leq t \leq 1$, is an extremal curve joining 0 to μ.

If $L^1(\Gamma) = \{\phi \in L^1(U/\Gamma, C): (\phi \circ A)(A')^2 = \phi$ for all A in $\Gamma\}$ then we can define a pairing (,) between $L^1(\Gamma)$, and $L^\infty(\Gamma)$ by

$$(6.6) \qquad (\phi, \mu) = \int_{U/\Gamma} \phi(z)\mu(z)\, dx\, dy$$

for all ϕ in $L^1(\Gamma)$, μ in $L^\infty(\Gamma)$. Under this pairing $L^\infty(\Gamma)$ is the dual space of $L^1(\Gamma)$ (i.e., every complex valued linear functional 1 on $L^1(\Gamma)$ is equal to $(\ ,\mu)$ for some μ in $L^\infty(\Gamma)$).

The space of integrable quadratic differentials of Γ, $A(\Gamma)$, is the subspace of $L^1(\Gamma)$ consisting of holomorphic differentials.

THEOREM V: *Let* Γ *be the Fuchsian group acting on* U. *If* μ *in* $M(\Gamma)$ *is extremal then* $\|\mu\|_\infty \leq \|\mu+\lambda\|_\infty$ *for all* λ *in* $A(\Gamma)^\perp$ *where* $A(\Gamma)^\perp = \{\lambda \in L^\infty(\Gamma): (\psi, \lambda) = 0$ *for all* $\psi \in A(\Gamma)\}$.

Proof: By Theorem II we have

$$a(\Gamma)(0,\mu) = \beta(\Gamma)(0, \Phi_*(0)\mu) \ .$$

But

$$\begin{aligned}\beta(\Gamma)(0, \Phi_*(0)\mu) &= \inf\{a(\Gamma)(0,\nu): \Phi_*(0)\nu = \Phi_*(0)\mu,\ \nu \in L^\infty(\Gamma)\} \\ &= \inf\{a(\Gamma)(0,\mu+\lambda): \Phi_*(0)\lambda = 0,\ \lambda \in L^\infty(\Gamma)\} \\ &= \inf\{a(\Gamma)(0,\mu+\lambda): \lambda \in A(\Gamma)^\perp\}\end{aligned}$$

because $\operatorname{Ker} \Phi_*(0) = A(\Gamma)^\perp$ by Bers [1].

Since $L^\infty(\Gamma)$ is the dual of $L^1(\Gamma)$, Theorem V is just a reformulation of a theorem by Hamilton [4] which says that if μ in $M(\Gamma)$ is extremal then its norm $\|\mu \mid A(\Gamma)\|$ as a linear functional on $A(\Gamma)$ equals its norm $\|\mu\|_\infty$ as a linear functional on $L^1(\Gamma)$. Clearly $\|\mu \mid A(\Gamma)\| \leq \|\mu\|_\infty$. By the Hahn-Banach theorem there is a linear functional k on $L^1(\Gamma)$ with norm equal to $\|\mu \mid A(\Gamma)\|$ such that

$$k(\phi) = (\phi, \mu) \text{ for all } \phi \in A(\Gamma).$$

Therefore $k = (\cdot, \nu)$ for some ν in $L^\infty(\Gamma)$ and $\|\nu\|_\infty = \|\mu \mid A(\Gamma)\|$. But of course

$$(\phi, \nu) = (\phi, \mu) \text{ for all } \phi \text{ in } A(\Gamma) .$$

So $\nu = \mu+\lambda$ with $\lambda \epsilon A(\Gamma)^{\perp}$.

Thus $\|\mu | A(\Gamma)\| = \|\nu\|_{\infty} = \|\mu+\lambda\|_{\infty} \geq \|\mu\|_{\infty}$ by Theorem V. Therefore $\|\mu | A(\Gamma)\| = \|\mu\|_{\infty}$.

REFERENCES

[1] L. Bers, On moduli of Riemann surfaces, lecture notes *Eidgenössische Technische Hochsule*, Zürich, 1964.

[2] C. J. Earle and J. Eells, Jr., On the differential geometry of Teichmüller spaces, *Journal D'Analyse Mathématique* (1967), 35-52.

[3] ______, Foliations and fibrations, *Journal of Differential Geometry* (1967), 33-41.

[4] R. S. Hamilton, Extremal quasiconformal mappings with prescribed boundary values, *Transactions of the American Mathematical Society* (1969), 399-406.

[5] S. Lang, *Introduction to Differentiable Manifolds*, Interscience, New York, 1962.

REPRODUCING FORMULAS FOR POINCARÉ SERIES OF DIMENSION −2 AND APPLICATIONS

K. V. Rajeswara Rao*

§1. *Introduction:* Let Γ denote a group of conformal self-maps of the open unit disc U of the complex plane, acting discontinuously and freely on U. Thus Γ is to contain no elliptic transformations. Our concern is with the Hilbert space $A(\Gamma)$ of square integrable automorphic forms of dimension −2 belonging to Γ, or, equivalently, with holomorphic differentials on the Riemann surface U/Γ which are square integrable in the sense of Ahlfors and Sario ([1], Ch. V).

We try to construct elements of $A(\Gamma)$ as Poincaré series of dimension -2: $\Sigma_{T \epsilon \Gamma} f(Tz) \cdot T'(z)$. For the existence of such Poincaré series, the natural condition to impose on Γ is that $\Sigma_{T \epsilon \Gamma} |T'(z)| < \infty$. We thus assume, throughout, that Γ is of convergence type, i.e., for one, and hence all, z in U

$$\sum_{T \epsilon \Gamma} (1 - |Tz|^2) < \infty \ . \tag{1.1}$$

It is known (Tsuji [8], p. 522) that this is equivalent to assuming that the Riemann surface U/Γ is hyperbolic.

The object of this paper is to exhibit, as *Poincaré series*, the reproducing kernels, in the Hilbert space $A(\Gamma)$, for the Taylor coefficients (§4) and periods (§5) and consider some applications (§6). The construction of these kernels as *differentials* on the surface U/Γ and their interrelations

* Supported, in part, by the National Science Foundation.

are well-known (Ahlfors and Sario [1], Ch. V, Theorems 18H and 19I). The original motivation for our work came from Bers's completeness theorem ([2], Th. 2) and his surjection theorem ([3], Th. 2) for forms of dimensions < -2. The analogue of the former is valid in our context (Corollary 2 of this paper) but there is no analogue of the latter (cf: our Th. 5).

The present short proofs of Theorem 1 and Corollary 1, showing their relation to known results on reproducing kernels and other domain functionals on Riemann surfaces, are suggested by the referee and replace longer and entirely different arguments independent of known results. The author wishes to express his thanks to the referee for this and related suggestions.

§2. *Preliminaries:* A function F holomorphic on U is said to be an automorphic form of dimension -2 belonging to Γ if $(F \circ T) \cdot T' = F$ for all T in Γ. The complex vector space of all such F is denoted by $H(\Gamma)$. Given a group Γ, we fix an open fundamental domain Ω whose boundary has Lebesgue (plane) measure zero. For any Lebesgue measurable function F defined on Ω, set $\|F\| = (\int_\Omega \int |F(z)|^2\, dx\, dy)^{1/2}$ $(z = x + iy)$. Denote by $A(\Gamma)$ the subspace of all F in $H(\Gamma)$ with $\|F\| < \infty$. It is easily verified that convergence in the norm $\| \quad \|$ in $A(\Gamma)$ implies uniform convergence on compact subsets of U and then one can conclude that $A(\Gamma)$ with the inner product $\int_\Omega \int F_1 \bar{F}_2\, dx\, dy$, is a complete (complex) inner-product space.

Let R be the quotient Riemann surface U/Γ and

$$\phi : U \to R \tag{2.1}$$

be the natural projection. The only local parameters on R that we shall use below are those given by the local homeomorphism ϕ, i.e., the variable z on U. On R, consider the space of all differentials $a(z)dx + b(z)dy$ $(z = x + iy)$ with Lebesgue measurable complex-valued coefficients a and b such that $\int_R \int (|a|^2 + |b|^2)\, dx\, dy < \infty$ (cf: Ahlfors and Sario [1], Ch. V). Denote this space by D(R). There is a natural inner product $\langle\ ,\ \rangle$ in D(R): if $\beta_i = a_i dx + b_i dy$ $(i = 1, 2)$, $\langle \beta_1, \beta_2 \rangle = \frac{1}{2} \int\int_R (a_1 \bar{a}_2 + b_1 \bar{b}_2)\, dx\, dy$.

Let $\|\beta\| = <\beta,\beta>^{1/2}$. If, as usual, we identify differentials which "differ" only on a set of measure zero, it is readily verified that D(R), with the above inner product, is a Hilbert space. Note that the correspondence $F \to \beta \equiv F(z)\,dz$ is a linear isometry of $A(\Gamma)$ into D(R). This allows us to regard $A(\Gamma)$ as a closed subspace of D(R). We shall use this identification. The standard proof of the following lemma is omitted.

LEMMA: Let $\alpha_n \equiv a_n(z)\,dx + b_n(z)\,dy \in D(R)$ $(n = 1, 2, \dots \infty)$ and $\alpha \equiv a(z)\,dx + b(z)dy$ be a differential on R. Assume that α_n converges weakly in D(R) to $\tilde{\alpha}$ and that, in any closed parametric disc, $a_n(z) \to a(z)$ and $b_n(z) \to b(z)$ uniformly as $n \to \infty$. Then $\tilde{\alpha} = \alpha$.

§3. *Poincaré series:* For a function f defined on U, we say that $\theta(f,\Gamma)$ exists if every arrangement of the series $\Sigma_{T\in\Gamma}\, f(Tz)T'(z)$ converges uniformly on compact subsets of U. It is elementary that the existence of $\theta(f,\Gamma)$ implies that $\Sigma_{T\in\Gamma}\, |f(Tz)\cdot T'(z)|$ converges for all z in U and that the sum

(3.1) $$\theta(f,\Gamma)(z) \equiv \sum_{T\in\Gamma} f(Tz)T'(z)$$

is independent of the arrangement of the series. Also, if f is holomorphic on U, $\theta(f,\Gamma) \in H(\Gamma)$.

If, for all z in U,

$$Tz = e^{it}\,\frac{z-\zeta}{1-z\bar{\zeta}} \qquad \text{(t real)}$$

then

(3.2) $$\zeta = T^{-1}0, \qquad |\zeta| = |T0| < 1$$

and

$$\frac{(1-|z|)(1-|\zeta|^2)}{2} \le 1-|Tz|^2 \le \frac{1+|z|}{1-|z|}\cdot(1-|\zeta|^2)\ .$$

Also $1-|Tz|^2 = (1-|z|^2)\,|T'(z)|$.

These remarks show that, if (as we assume throughout) (1.1) holds for one z, then it holds for all z in U and that $\Sigma_{T\epsilon\Gamma} |T'(z)|$ converges uniformly on compact subsets of U. Hence $\theta(f,\Gamma)$ exists whenever f is bounded on U.

§4. *Reproducing formulas for Taylor coefficients:* For z and w in U, set

$$(4.1) \qquad f_w(z) = \frac{1}{(1-z\bar{w})^2} \ .$$

The following is basic.

THEOREM 1. For every w in U, $\theta(f_w,\Gamma) \epsilon A(\Gamma)$ and, for all F in $A(\Gamma)$,

$$(4.2) \qquad \int_\Omega\int F(z)\,\overline{\theta(f_w,\Gamma)(z)}\,dx\,dy = \pi F(w) \ .$$

Proof: Let $g(p,q)$ be the Green's function of the Riemann surface $R = U/\Gamma$. If $\phi(z) = p$ and $\phi(w) = q$, it is well known (Tsuji [8], p. 522) that $g(p,q) = \Sigma_{T\epsilon\Gamma}\, g_U(Tz,w)$, g_U being the Green's function of U. A straightforward computation based on the equality

$$g_U(z,w) = \frac{1}{2}\log\left|\frac{1-z\bar{w}}{z-w}\right|^2$$

shows that

$$(4.3) \qquad \frac{\partial^2 g}{\partial z\partial\bar{w}}(z,w) = -\frac{1}{2}\,\theta(f_w,\Gamma)(z) \ .$$

Fix w_0 in U. It then suffices to show that the differential

$$\alpha \equiv -2\frac{\partial^2 g}{\partial z\partial\bar{w}}(z,w_0)\,dz$$

belongs to $D(R)$ and reproduces the value at $q_0 = \phi(w_0)$ for members of the subspace $A(\Gamma)$: if, for F in $A(\Gamma)$, $\beta = F(z)\,dz$, then

$$<\beta, a> = \pi F(w_0) \quad . \tag{4.4}$$

If R is the interior of a compact bordered Riemann surface, this is well-known (Schiffer and Spencer [7], p. 122). To establish the general case, choose an exhaustion $\{R_n\}$ of R such that $q_0 \in R_n \subset \bar{R}_n \subset R_{n+1}$, ($n = 1, 2, \dots \infty$) and each $\bar{R}_n$ is a compact bordered Riemann surface with R_n as its interior. Let $g_n(p, q)$ be the Green's function of R_n and, with $\phi(z) = p$ and $\phi(w) = q$, set

$$a_n = -2 \frac{\partial^2 g_n}{\partial z \partial \overline{w}} (z, w_0)\,dz \quad .$$

Then, for any F in $A(\Gamma)$, $\beta = F(z)\,dz \in D(R_n)$ and hence

$$<\beta, a_n> = \pi \cdot F(w_0) \quad ; \tag{4.5}$$

in particular, $\|a_n\|^2 = -2\pi \dfrac{\partial^2 g_n}{\partial z \partial \overline{w}} (w_0, w_0)$ (both $< \,,\, >$ and $\| \quad \|$ being computed on R_n). Since, as $n \to \infty$, $g_n(p, q) \to g(p, q)$ uniformly on compact subsets of $R \times R$, it follows readily that, on every closed parametric disc of R,

$$\frac{\partial^2 g_n}{\partial z \partial \overline{w}} (z, w_0) \text{ converges uniformly to } \frac{\partial^2 g}{\partial z \partial \overline{w}} (z, w_0) \quad .$$

Hence $\sup_n \|a_n\| < \infty$. Thus, if we set $\tilde{a}_n = a_n$ on R_n and $= 0$ on $R \setminus R_n$, $\tilde{a}_n \in D(R)$ and $\sup_n \|\tilde{a}_n\| < \infty$. Hence $\{\tilde{a}_n\}$ contains a subsequence converging weakly in $D(R)$ to, say, $\tilde{a}$. This, together with (4.5), implies that $<\beta, \tilde{a}> = \pi F(w_0)$ for all F in $A(\Gamma)$. To establish (4.4), it remains only to identify a with $\tilde{a}$ and this is readily done via the lemma in §2.

COROLLARY 1. For any fixed integer $m \geq 0$, let $f_m(z) = z^m$. Then $\theta(f_m, \Gamma) \in A(\Gamma)$ and, for any F in $A(\Gamma)$,

$$\int_{\Omega}\int F \cdot \overline{\theta(f_m, \Gamma)}\,dx\,dy = \pi \frac{B_m}{m+1} ,$$

where

$$(4.6) \qquad F(z) = \sum_{n=0}^{\infty} B_n z^n .$$

Proof: Set $k_0(z, w) = \theta(f_w, \Gamma)(z)$. It follows at once from our Theorem 1 and Theorem 18H of Ahlfors and Sario [1] that

$$k_m(z, w) \equiv \frac{1}{(m+1)!} \frac{\partial^m k_0}{\partial \bar{w}^m}(z, w) \in A(\Gamma)$$

and that, for all F in $A(\Gamma)$ as in (4.6),

$$\int_{\Omega}\int F(z)\,\overline{k_m(z, 0)}\,dx\,dy = \frac{\pi}{(m+1)!} F^{(m)}(0) .$$

From the definition of k_0, it follows that

$$\frac{\partial^m k_0}{\partial \bar{w}^m} = (m+1)! \sum_{T \in \Gamma} T'(z) \cdot \frac{T(z)^m}{(1 - \bar{w} \cdot Tz)^{m+1}}$$

and the result is now immediate.

COROLLARY 2. $\{\theta(f, \Gamma) \mid f(z)$ a polynomial in $z\}$ is a dense subspace of $A(\Gamma)$.

Remark. This is an analogue of Bers' completeness theorem ([2], Th. 2) for dimensions < -2.

§5. *Reproducing kernels for periods:* Let g be an antiderivative of G in $H(\Gamma)$ so that, for each S in Γ, there exists a constant $C(S, G)$ such that $g \circ S = g + C(S, G)$. Clearly

$$C(S,G) = \int_0^{S0} G(u)\,du \tag{5.1}$$

and it is the period of G "along" S.

THEOREM 2. For any S in Γ, let $S0 = \eta$ and $f_S(z) = \bar{\eta}/1 - z\bar{\eta}$. Then $\theta(f_S, \Gamma) \in A(\Gamma)$ and, for all G in $A(\Gamma)$,

$$\pi C(S,G) = \iint_\Omega G \cdot \overline{\theta(f_S,\Gamma)}\,dx\,dy \;.$$

Proof: Theorem 1 implies that, for all u in U,

$$\|\theta(f_u,\Gamma)\|^2 = \pi\theta(f_u,\Gamma)(u) = \pi \cdot \sum_{T\in\Gamma} \frac{T'(u)}{(1-\bar{u}\,T u)^2}$$

$$\leq \frac{\pi}{(1-|u|)^2} \sum_{T\in\Gamma} |T'(u)| \leq \frac{\pi}{(1-|u|)^4} \sum_{T\in\Gamma} (1-|\zeta|^2)$$

(cf: (3.2)). This and Schwarz's inequality imply that

$$\int_\Omega\int |G \cdot \theta(f_u,\Gamma)|\,dx\,dy \leq \frac{\|G\|}{(1-|u|)^2}\left[\pi \sum_{T\in\Gamma} (1-|\zeta|^2)\right]^{1/2} < \infty \;. \tag{5.2}$$

Now (5.1) and Theorem 1 imply that, for G in $A(\Gamma)$,

$$\pi C(S,G) = \int_0^\eta \left(\int_\Omega\int G(z)\overline{\theta(f_u,\Gamma)}\,dx\,dy\right)du$$

$$= \iint_\Omega G(z)\left(\int_0^\eta \overline{\theta(f_u,\Gamma)(z)}\,du\right)dx\,dy \;,$$

the interchange being justified by (5.2) and Fubini's theorem. Substituting the defining series for $\theta(\;,\;)(z)$ we can interchange summation and integration in the inner integral because the series converges uniformly on compact subsets of U. If we recall the definitions of f_u and f_S, this completes the proof of Theorem 2.

Remarks: For the groups Γ under consideration, Theorem 2 is equivalent to Theorem 19I of Ahlfors and Sario [1]. Also, in the special case where U/Γ is the interior of a compact bordered Riemann surface, Theorem 2 was proved also by Earle and Marden ([4], p. 207).

COROLLARY 3: If S is a parabolic transformation in Γ then

$$\theta(f_S, \Gamma) \equiv 0 \ .$$

Proof: It is well-known (see, for instance, [6], p. 635) that, if S is parabolic, then $C(S, G) = 0$ for all G in $A(\Gamma)$. Since reproducing kernels are unique in a complete inner product space, the result follows from Theorem 2.

§6. *Applications:* Combining Corollary 1 with results of [6] we shall prove the following.

THEOREM 3. The linear map $f \to \theta(f, \Gamma)$ is one-to-one on the space of polynomials. In particular, $A(\Gamma)$ is infinite dimensional.

Proof: Since θ is a linear map, it suffices to show that, if p is a non-zero polynomial, then $\theta(p, \Gamma)$ is not identically zero. Assume, if possible, that, with $p(z) = a_0 + a_1 z + \cdots + a_n z^n$, $(a_n \neq 0)$, $\theta(p, \Gamma)(z) \equiv 0$. Corollary 1 then implies that, for all F (as in (4.6)) in $A(\Gamma)$,

$$\sum_{k=0}^{n} \frac{\bar{a}_k B_k}{k+1} = 0 \ .$$

We arrive at a contradiction by exhibiting a Blaschke product b such that $F \equiv \theta(b, \Gamma)$ is in $A(\Gamma)$ and has a zero of (exact) order n at the origin. Let $a(z)$ be "the" Blaschke product with simple zeros at points of $\Gamma 0$ but no other zeros. Set $b(z) = a(z)^n$ and $F = \theta(b, \Gamma)$. Clearly $|b(Tz)| = |b(z)|$ for all T in Γ and all z. Hence, for each T in Γ, there exists a complex constant $\nu(T)$ such that $b \circ T = \nu(T) \cdot b$. It is readily verified that

$T \to \nu(T)$ is a homomorphism of Γ into the circle group. Thus ν is a character of Γ and

$$F(z) = b(z)F_1(z)\,, \tag{6.1}$$

where $F_1(z) = \Sigma_{T\epsilon\Gamma}\,\nu(T)\,T'(z)$. By Theorem 4 of this paper (proved below) $F \epsilon A(\Gamma)$. Also, by Theorem 5 of [6], $F_1(0) > 0$. Since b has a zero of order n at the origin, this, together with (6.1), completes the proof of Theorem 3.

Recall that $H^2(U)$ is the Hilbert space of functions $f(z) = \Sigma_{n=0}^{\infty} a_n z^n$ holomorphic in U with norm given by:

$$\|f\|^2 = \frac{1}{2\pi}\int_0^{2\pi} |f(e^{i\theta})|^2 d\theta = \sum_{n=0}^{\infty} |a_n|^2 < \infty\,.$$

Here $f(e^{i\theta})$ is the radial limit function of f (see Hoffman [5]). As another application of Corollary 1 we can prove the following.

THEOREM 4. $f \to \theta(f, \Gamma)$ is a bounded linear map from the Hardy space $H^2(U)$ into $A(\Gamma)$.

Proof: Let $F \epsilon A(\Gamma)$. Then the measure $d\mu = |F|^2\,dx\,dy$ is invariant under the action of Γ, and, if ψ is any non-negative measurable function on U,

$$\int_U\int \psi\, d\mu = \sum_{T\epsilon\Gamma} \int_{T\Omega}\int \psi\, d\mu = \sum_{T\epsilon\Gamma} \int_{\Omega}\int \psi(Tz)\,d\mu(z)$$
$$= \int_{\Omega}\int \left(\sum_{T\epsilon\Gamma} \psi(Tz)\right) d\mu(z)\,.$$

If we choose $\psi(z) = 1-|z|^2$ and recall (see, for instance, [6], p. 636; also Tsuji [8], p. 517) that $M \equiv \sup_{z\epsilon U} \Sigma_{T\epsilon\Gamma}\,\psi(Tz) < \infty$, we obtain the inequality

$$\int_U\int (1-|z|^2)|F(z)|^2\,dx\,dy \le M\|F\|^2 .$$

If $F(z) = \Sigma_{n=0}^{\infty} B_n z^n$, this can be rewritten as

$$\pi \sum_{n=0}^{\infty} \frac{|B_n|^2}{(n+1)(n+2)} \le M\|F\|^2 .$$

Thus, if $\Lambda(F)(z) \equiv \frac{\pi}{z}\int_0^z F(u)\,du = \pi\,\Sigma_{n=0}^{\infty} \frac{B_n}{n+1} z^n$, then $F \to \Lambda(F)$

is a bounded linear map from the hilbert space $A(\Gamma)$ into the Hilbert space $H^2(U)$. Hence the adjoint map $\Lambda^*\colon H^2(U) \to A(\Gamma)$ is also a bounded linear map. We compute $\Lambda^*(f_k)$ where $f_k(z) = z^k$ $(k = 0, 1, 2, \dots \infty)$. We have for all F in $A(\Gamma)$ (as in (4.6))

$$\int_\Omega\int F\cdot\overline{\Lambda^*(f_k)}\,dx\,dy = \frac{1}{2\pi}\int_0^{2\pi} \Lambda(F)(e^{i\theta})\cdot\overline{f_k(e^{i\theta})}\,d\theta = \frac{\pi B_k}{k+1} .$$

This, together with the uniqueness of reproducing kernels in $A(\Gamma)$ and Corollary 1, implies that $\Lambda^*(f_k) = \theta(f_k, \Gamma)$. Also we know ([6], Th. 1) that $\theta(g, \Gamma)$ exists if $g \in H^2(U)$ and that, if g_n tends to g in $H^2(U)$ then, $\theta(g_n, \Gamma)$ tends to $\theta(g, \Gamma)$ for all z in U (see Th. 1 and inequality (3.6) of [6]). Since the f_k $(k = 0, 1, \dots, \infty)$ span $H^2(U)$, this shows that $\Lambda^*(f) = \theta(f, \Gamma)$ for all f in $H^2(U)$ thus completing the proof.

The following result should be contrasted with Bers' surjection theorem for A_q, $q \ge 2$ ([3], Th. 2).

THEOREM 5. There exists no Banach space B of functions holomorphic on U such that the following hold for every Γ under consideration: (i) $\theta(f, \Gamma)$ exists for every f in B, and (ii) $f \to \theta(f, \Gamma)$ is a bounded linear map of B onto $A(\Gamma)$.

Proof: If Γ is the trivial group write A for $A(\Gamma)$. Clearly, if $g(z) = \Sigma_{n=0}^{\infty} b_n z^n \in A$, then

$$(6.2)\qquad \|g\|^2 = \int_U\int |g|^2 dx\,dy = \pi \sum_{n=0}^{\infty} \frac{|b_n|^2}{n+1} .$$

Suppose, if possible, a Banach space B of holomorphic functions exists as in the statement of the theorem. Since, for $\Gamma = \{id\}$, $\theta(f, \Gamma)$ is the identity map, we can, by invoking the closed graph theorem, assume that $B = A$ with norm as in (6.2). Thus, for any group Γ and any f in A, $\theta(f, \Gamma)$ exists and $f \to \theta(f, \Gamma)$ is a bounded linear map of the Hilbert space A into the Hilbert space $A(\Gamma)$. Hence the adjoint map $\theta^*: A(\Gamma) \to A$ is a bounded linear map too. For $F(z) = \Sigma_{n=0}^{\infty} B_n z^n$, let $\theta^*(F)(z) = \Sigma_{n=0}^{\infty} B_n^* z^n$. We have, with $f_k(z) = z^k$, for all $k = 0, 1, 2, \ldots$, $\int_\Omega \int F \cdot \overline{\theta(f_k)}\, dx\,dy = \int_U \int \theta^*(G) \overline{f}_k\, dx\,dy$, or (by Corollary 1) $B_k/k+1 = B_k^*/k+1$. Thus $\theta^* =$ identity and hence $\int_U \int |F|^2 dx\,dy < \infty$ for all F in $A(\Gamma)$, i.e., $\Sigma_{T \epsilon \Gamma} \int\int |F^2|\, dx\,dy < \infty$ for all F in $A(\Gamma)$. But this is impossible for a non-zero F and a non-trivial Γ, Q.E.D.

Note: (added in proof). The references (in the proof of Corollary 1 and the remarks after Theorem 2) to Theorems 18H and 19I of Ahlfors and Sario [1] are to those results in Chapter V of that book.

REFERENCES

[1] L. V. Ahlfors and L. Sario: *Riemann Surfaces*, Princeton University Press, Princeton, N.J., (1960).

[2] L. Bers: Completeness theorems for Poincaré series in one variable, *Proc. International Symposium on Linear Spaces*, Jerusalem (1960), 88-100.

[3] L. Bers: Automorphic forms and Poincaré series for infinitely generated Fuchsian groups, *Amer. Jour. Math.*, 87 (1965), 196-214.

[4] C. J. Earle and A. Marden: On Poincaré series with application to H^p spaces on bordered Riemann surfaces, *Illinois Jour. Math.*, 13 (1969), 202-219.

[5] K. Hoffman: *Banach Spaces of Analytic Functions*, Prentice-Hall, Inc., Englewood Cliffs, N. J., (1962).

[6] K. V. Rajeswara Rao: Fuchsian groups of convergence type and Poincaré series of dimension -2, *Jour. Math. and Mech.*, 18 (1969), 629-644.

[7] M. Schiffer and D. C. Spencer: *Functionals of Finite Riemann Surfaces*, Princeton University Press, Princeton, N. J., (1954).

[8] M. Tsuji: *Potential Theory in Modern Function Theory*, Maruzen, Tokyo (1959).

PERIOD RELATIONS ON RIEMANN SURFACES

Harry E. Rauch*

1. Introduction.

In recent joint work, [4], Farkas and I established explicit period relations on a compact Riemann surface of genus $g \geq 4$, relations which include Schottky's relation, [11] and [10], for $g = 4$ (the first such relation known), and, indeed, give what we regard as the first satisfactory proof of it, and which for $g > 4$ are the direct structural generalizations of it first conjectured rather imprecisely by Schottky and Jung, [12]. For convenience I call such period relations *relations of Schottky type*. Very succinctly, given a Riemann surface S of genus $g \geq 4$ and on it a canonical homology basis $\gamma_1, \ldots, \gamma_g$; $\delta_1, \ldots, \delta_g$ (thus, in effect, a Torelli surface, [7], over S), in order to write down a relation of Schottky type among the periods $\pi_{ij} = \int_{\delta_j} d\zeta_i$, where $d\zeta$ is the vector of normal differentials of first kind on S determined by $\delta_{ij} = \int_{\gamma_j} d\zeta_i$, $i, j = 1, \ldots, g$, one has only to use

Prescription A. Take any identity (actually only certain identities are useful—see below) among the $(g-1)$-theta constants on the generalized upper half-plane $\mathfrak{S}_{g-1}$, of $(g-1) \times (g-1)$ symmetric complex matrices with positive-definite imaginary part and replace each such constant by the square root of the product of the two g-theta constants whose characteristics consist of the $(g-1)$-characteristic of that constant bordered on the left by the columns $\left[\begin{smallmatrix}0\\0\end{smallmatrix}\right]$ and $\left[\begin{smallmatrix}0\\1\end{smallmatrix}\right]$, respectively, and whose period matrix arguments

* Research partially sponsored by the Air Force Office of Scientific Research, Office of Aerospace Research, U.S. Air Force, under AFOSR Grant No. AFOSR 1873-70.

are precisely $\pi = (\pi_{ij})$. (In [9] and [4], to which I also refer for definitions, these are called Riemann theta constants on S with respect to the given homology basis.)

In [4] this strikingly simple prescription is used to derive (one form of) Schottky's relation for $g = 4$ and a typical relation of Schottky type for $g = 5$.

The results of which I have spoken are the outgrowth of a several year period of intensive collaboration between Professor Farkas and me, and the final outcome is joint work. During the course of our work, Farkas, on his own, had a brilliant insight that provided us with a fundamental breakthrough and which constitutes a beautiful theorem in its own right. All of this is recorded in three PNAS notes [9], [3], [4] of which the first and last are joint and the middle is by Farkas.

Farkas will speak about his work in his lecture. Since the note [4] has quite a readable outline and a fair amount of detail it seems pointless to me to rehash it here in view of time limitations. I prefer here, instead, to illustrate by a concrete example the application of Prescription A. This will serve the additional purpose of introducing the listener and reader to the marvelously fascinating calculus–perhaps I should say algebra–of the theta constants and their characteristics. This has a game-like quality to it, particularly the manipulation of the zeros and ones in the characteristics, and, indeed, many of the rules of the game, although quite recondite in origin and derivation, could probably be taught to grammar-school children.

Since the key to note [4] is the profound study of hyperelliptic surfaces I choose to illustrate a phenomenon on such surfaces for $g = 4$.

To explain the origin of the problem let me backtrack a little and say that the significance of Prescription A and the relations of Schottky type lies in the attempt to make quantitatively explicit the qualitative fact, observed heuristically by Riemann in 1857 and proved in sharper form by me in 1955 (cf. [7]), that the $g(g+1)/2$ periods π_{ij} depend holomorphically on $3g-3$ complex parameters, "the" moduli, which parametrize the set of (conformal equivalence classes of) Riemann surfaces of genus $g \geq 2$ and hence are subject to $g(g+1)/2 - (3g-3) = (g-2)(g-3)/2$ holomorphic relations. The relations of Schottky type are such relations—one may view them alternatively as necessary conditions for a matrix in $\mathfrak{S}_g$ to come from a Riemann surface as π does. There is, at present, the question whether the totality of relations of Schottky type imply all the required relations for given g, so to say whether they are ample. But it is clear even when they do as for $g = 4$ that they are superabundant in the sense that one derives algebraically a great many more relations than can possibly be functionally independent.

An analogous but considerably simpler situation has long existed for the set of hyperelliptic surfaces. It was recognized certainly by Rosenhain in 1850, if not earlier, that the hyperelliptic surfaces of genus g depend on $2g-1$ moduli, the $2g+2-3$ unnormalized branch points of a two sheeted representation. Again this was proved by me in sharper form in 1955 (cf. [7]). On the other hand Riemann and Weierstrass both recognized by 1865 at the latest that a necessary condition for a surface to be hyperelliptic or for a period matrix to come from such a surface is the vanishing of an algorithmically determined large set of theta constants and partial derivatives of theta *functions* for zero values of the argument (cf. [5], Chapter X and [7]). Thus for $g = 4$ one has 10 theta constants vanishing as compared with $g(g+1)/2 - (2g-1) = (g-1)(g-2)/2 = 3$ period relations, while for $g = 5$ one has $4.3/2 = 6$ period relations while 66 theta constants and the first partials of an odd theta function at the origin vanish (cf. [5], Chapter X). Max Noether accounted (cf. [5], Chapter X) for the discrepancy in these two cases by showing that for $g = 4$ the 10 vanishing conditions are "equivalent" to the vanishing of a subset of 3 theta constants, i.e., the vanishing of the remaining 7 imposes no new conditions and, similarly, the residual 61 vanishing conditions are "dependent" on the vanishing of a certain set of 6 theta constants for $g = 5$. Here, "dependent" means the existence of a linear relation among the theta constants and derivatives in question whose coefficients are monomials in theta constants none of which vanishes for the particular values of the periods in question. "Equivalent" then has the obvious meaning: mutually dependent. Here, however, the reverse dependence usually means logical inclusion.

But in view of the paragraph before the last the additional question arises: if one already knows a minimal functionally independent set of $(g-2)(g-3)/2$ *necessary* period relations (that cut down $g(g+1)/2$ to $(3g-3)$) on a surface of genus g, is it true that the *necessary* conditions for hyperellipticity are "dependent" on the preceding *general* necessary period relations and a *subset* of $g-2 = 3g-3-(2g-1)$ members of the hyperelliptic conditions?

I call attention to the careful formulation of the question. I do not ask for *sufficient* conditions that a surface be hyperelliptic—like the "$g-2$ conjecture" (cf. [2] and [1]). Nor do I demand that the vanishing of certain theta constants or period relations *imply* the vanishing of others—that would require an excursion into the mysteries of theta algebra more extensive than I am now prepared to make. I do not even wish to discuss here the question of precise sense of the *functional* independence of the minimal set of hyperelliptic necessary conditions since there are also mysteries there (cf. [8]).

With that *caveat* I turn to showing that the answer to the question posed above is affirmative when $g = 4$. More precisely I show that on a hyperelliptic surface with $g = 4$ one can choose a canonical homology basis and a form of Schottky's relation such that the vanishing of a minimal set of three (out of the 10 necessary hyperelliptic conditions) is "equivalent" to Schottky's relation and the vanishing of two of them. Here, "dependence" and "equivalence" must be generalized to include the case that there is a polynomial (not necessarily linear) relation with nonvanishing coefficients in the preceding sense and with the "dependent" theta constant occurring in at least one monomial which does not contain the "independent" theta constants. It is important to observe that Schottky's relation although written in irrational form is really a polynomial relation obtained by multiplication of all conjugates (plus and minus combinations). This comes about by exhibiting in each term of the three term Schottky relation in the special form chosen exactly one theta constant from the special list of 10 which vanish. It seems clear to me, although I have not yet checked the eighth roots of unity which occur, that linear transformation of the theta constants leads to the conclusion that this is always true, i.e., for any choice of canonical homology basis on a hypereilliptic surface there is a form of Schottky's relation such that exactly one of the corresponding list of 10 vanishing constants occurs in each term. What I am unable to foresee now is whether or not *any* three of the 10 occur in *some* form of Schottky's relation. This contrasts with Noether's reduction (for $g = 4$) in which *any* three can serve as a minimal set.

To finish off I shall exhibit Noether's reduction for the particular three constants occurring in Schottky's relation. Thus I shall show that 10 is really 3 is really 2, i.e., Schottky plus 2.

2. Hyperelliptic surfaces of genus four.

Let S be a hyperelliptic surface of genus $g = 4$. One may assume S represented as a two-sheeted surface over the z-plane with branch points at $0, 1, \lambda_1, \ldots, \lambda_7, \infty$, where no λ_i lies on the real axis between $-\infty$ and 1 and a simple (rectilinear) polygon joins the branch points in the order indicated. With these conventions I may represent S without loss of topological generality as shown by the conventional diagram in Figure 1, and I may draw a canonical homology basis as shown there.

I now compute the table of 9 half-periods obtained by integrating the vector of differentials of first kind from the branch point 0 to the 9 other branch points. (See Table 1 on the following page.)

The expressions on the left are defined by those in the middle.

The sum of the half-periods of Table 1 with odd (or even) symbols is

$$(\chi) = \binom{0\ 1\ 1\ 1}{0\ 1\ 1\ 0} ,$$

Table 1.

$$(4) = \binom{1\,0\,0\,0}{0\,0\,0\,0} = \int_0^1$$

$$(5) = \binom{1\,1\,0\,0}{1\,0\,0\,0} = \int_0^{\lambda_1}$$

$$(6) = \binom{1\,1\,0\,0}{1\,1\,0\,0} = \int_0^{\lambda_2}$$

$$(7) = \binom{1\,0\,0\,0}{1\,1\,0\,0} = \int_0^{\lambda_3}$$

$$(1) = \binom{0\,0\,0\,0}{1\,0\,1\,1} = \int_0^{\lambda_4}$$

$$(2) = \binom{0\,0\,1\,0}{1\,0\,1\,1} = \int_0^{\lambda_5}$$

$$(3) = \binom{0\,0\,1\,0}{1\,0\,0\,1} = \int_0^{\lambda_6}$$

$$(8) = \binom{0\,0\,0\,1}{1\,0\,0\,1} = \int_0^{\lambda_7}$$

$$(9) = \binom{0\,0\,0\,1}{1\,0\,0\,0} = \int_0^{\infty} .$$

which is the vector of Riemann constants $K(0)$ with respect to 0 (cf. [5], Chapter X). I denote by

$$[\chi] = \begin{bmatrix} 0\,1\,1\,1 \\ 0\,1\,1\,0 \end{bmatrix}$$

the characteristic which is the symbol of (χ), and I compose $[\chi]$ with the half-periods of Table 1 to obtain the following table of 10 even characteristics. (See Table 2.)

It is a theorem (cf. [5], Chapter X, and [6]) that these are precisely (no others) the characteristics of the 10 vanishing theta constants (in particular that they are all *even*). The half-periods of Table 1 form a fundamental set, and the last nine characteristics of Table 2 are a principal set.

It will be observed that [1], [2], [3], [8], [9], $[\chi]$ are of the form to appear in a relation of Schottky type according to Prescription A, since they

Table 2.

$$[\chi] = \left[\begin{smallmatrix} 0 & 1 & 1 & 1 \\ 0 & 1 & 1 & 0 \end{smallmatrix}\right]$$

$$[4] = \left[\begin{smallmatrix} 1 & 1 & 1 & 1 \\ 0 & 1 & 1 & 0 \end{smallmatrix}\right] = [\chi] + (4)$$

$$[5] = \left[\begin{smallmatrix} 1 & 0 & 1 & 1 \\ 1 & 1 & 1 & 0 \end{smallmatrix}\right] = [\chi] + (5)$$

$$[6] = \left[\begin{smallmatrix} 1 & 0 & 1 & 1 \\ 1 & 0 & 1 & 0 \end{smallmatrix}\right] = [\chi] + (6)$$

$$[7] = \left[\begin{smallmatrix} 1 & 1 & 1 & 1 \\ 1 & 0 & 1 & 0 \end{smallmatrix}\right] = [\chi] + (7)$$

$$[1] = \left[\begin{smallmatrix} 0 & 1 & 1 & 1 \\ 1 & 1 & 0 & 1 \end{smallmatrix}\right] = [\chi] + (1)$$

$$[2] = \left[\begin{smallmatrix} 0 & 1 & 0 & 1 \\ 1 & 1 & 0 & 1 \end{smallmatrix}\right] = [\chi] + (2)$$

$$[3] = \left[\begin{smallmatrix} 0 & 1 & 0 & 1 \\ 1 & 1 & 1 & 1 \end{smallmatrix}\right] = [\chi] + (3)$$

$$[8] = \left[\begin{smallmatrix} 0 & 1 & 1 & 0 \\ 1 & 1 & 1 & 1 \end{smallmatrix}\right] = [\chi] + (8)$$

$$[9] = \left[\begin{smallmatrix} 0 & 1 & 1 & 0 \\ 1 & 1 & 1 & 0 \end{smallmatrix}\right] = [\chi] + (9) \quad .$$

have $\left[\begin{smallmatrix} 0 \\ 0 \end{smallmatrix}\right]$ or $\left[\begin{smallmatrix} 0 \\ 1 \end{smallmatrix}\right]$ in the first column. This asymmetry of Prescription A can probably be cured by using a different 2-sheeted covering in the proof and by linear transformation of the thetas. The first three and last three have a particular symmetry but especially the first three, hence their labels. I want a Schottky relation with one each of $\theta[1], \theta[2], \theta[3]$ in each of its three separate terms, To create one I first need a three term biquadratic identity among 12 distinct 3-theta constants with $\theta\left[\begin{smallmatrix} 1 & 1 & 1 \\ 1 & 0 & 1 \end{smallmatrix}\right]$, $\theta\left[\begin{smallmatrix} 1 & 0 & 1 \\ 1 & 0 & 1 \end{smallmatrix}\right]$, $\theta\left[\begin{smallmatrix} 1 & 0 & 1 \\ 1 & 1 & 1 \end{smallmatrix}\right]$ in distinct terms, the characteristics being obtained from [1], [2], [3] by deletion of their first columns. According to Schottky, [10], one needs a syzygetic group of 3-half-periods G of degree 2 and three even 3-characteristics which when composed with G give 12 distinct even 3-characteristics. I observe that

$$\left[\begin{smallmatrix} 1 & 1 & 1 \\ 1 & 0 & 1 \end{smallmatrix}\right] = \left[\begin{smallmatrix} 0 & 1 & 0 \\ 0 & 0 & 0 \end{smallmatrix}\right] + \left(\begin{smallmatrix} 1 & 0 & 1 \\ 1 & 0 & 1 \end{smallmatrix}\right)$$

$$[\begin{smallmatrix}1&0&1\\1&0&1\end{smallmatrix}] = [\begin{smallmatrix}0&0&0\\0&0&0\end{smallmatrix}] + (\begin{smallmatrix}1&0&1\\1&0&1\end{smallmatrix})$$

$$[\begin{smallmatrix}1&0&1\\1&1&1\end{smallmatrix}] = [\begin{smallmatrix}0&0&0\\0&1&0\end{smallmatrix}] + (\begin{smallmatrix}1&0&1\\1&0&1\end{smallmatrix})$$

and

$$(\begin{smallmatrix}1&0&1\\1&0&1\end{smallmatrix}) = (\begin{smallmatrix}1&0&1\\0&0&0\end{smallmatrix}) + (\begin{smallmatrix}0&0&0\\1&0&1\end{smallmatrix}) .$$

Hence, I take

$$G = \{(\begin{smallmatrix}0&0&0\\0&0&0\end{smallmatrix}), (\begin{smallmatrix}1&0&1\\0&0&0\end{smallmatrix}), (\begin{smallmatrix}0&0&0\\1&0&1\end{smallmatrix}), (\begin{smallmatrix}1&0&1\\1&0&1\end{smallmatrix})\}$$

and the three characteristics to be

$$[\begin{smallmatrix}0&1&0\\0&0&0\end{smallmatrix}]$$

$$[\begin{smallmatrix}0&0&0\\0&0&0\end{smallmatrix}]$$

$$[\begin{smallmatrix}0&0&0\\0&1&0\end{smallmatrix}]$$

to get (the signs are not important here)

$$\theta[\begin{smallmatrix}0&1&0\\0&0&0\end{smallmatrix}]\theta[\begin{smallmatrix}1&1&1\\0&0&0\end{smallmatrix}]\theta[\begin{smallmatrix}0&1&0\\1&0&1\end{smallmatrix}]\theta[\begin{smallmatrix}1&1&1\\1&0&1\end{smallmatrix}] \pm$$

$$\theta[\begin{smallmatrix}0&0&0\\0&0&0\end{smallmatrix}]\theta[\begin{smallmatrix}1&0&1\\0&0&0\end{smallmatrix}]\theta[\begin{smallmatrix}0&0&0\\1&0&1\end{smallmatrix}]\theta[\begin{smallmatrix}1&0&1\\1&0&1\end{smallmatrix}] \pm$$

$$\theta[\begin{smallmatrix}0&0&0\\0&1&0\end{smallmatrix}]\theta[\begin{smallmatrix}1&0&1\\0&1&0\end{smallmatrix}]\theta[\begin{smallmatrix}0&0&0\\1&1&1\end{smallmatrix}]\theta[\begin{smallmatrix}1&0&1\\1&1&1\end{smallmatrix}] = 0$$

By Prescription A I get the desired Schottky relation

1)

$$\sqrt{\theta[\begin{smallmatrix}0&0&1&0\\0&0&0&0\end{smallmatrix}]\theta[\begin{smallmatrix}0&0&1&0\\1&0&0&0\end{smallmatrix}]\theta[\begin{smallmatrix}0&1&1&1\\0&0&0&0\end{smallmatrix}]\theta[\begin{smallmatrix}0&1&1&1\\1&0&0&0\end{smallmatrix}] \times}$$

$$\sqrt{\theta[\begin{smallmatrix}0&0&1&0\\0&1&0&1\end{smallmatrix}]\theta[\begin{smallmatrix}0&0&1&0\\1&1&0&1\end{smallmatrix}]\theta[\begin{smallmatrix}0&1&1&1\\0&1&0&1\end{smallmatrix}]\theta[\begin{smallmatrix}0&1&1&1\\1&1&0&1\end{smallmatrix}]} \pm$$

$$\sqrt{\theta[\begin{smallmatrix}0&0&0&0\\0&0&0&0\end{smallmatrix}]\theta[\begin{smallmatrix}0&0&0&0\\1&0&0&0\end{smallmatrix}]\theta[\begin{smallmatrix}0&1&0&1\\0&0&0&0\end{smallmatrix}]\theta[\begin{smallmatrix}0&1&0&1\\1&0&0&0\end{smallmatrix}] \times}$$

$$\sqrt{\theta[\begin{smallmatrix}0&0&0&0\\0&1&0&1\end{smallmatrix}]\theta[\begin{smallmatrix}0&0&0&0\\1&1&0&1\end{smallmatrix}]\theta[\begin{smallmatrix}0&1&0&1\\0&1&0&1\end{smallmatrix}]\theta[\begin{smallmatrix}0&1&0&1\\1&1&0&1\end{smallmatrix}]} \pm$$

$$\sqrt{\theta\left[\begin{smallmatrix}0&0&0&0\\0&0&1&0\end{smallmatrix}\right]\theta\left[\begin{smallmatrix}0&0&0&0\\1&0&1&0\end{smallmatrix}\right]\theta\left[\begin{smallmatrix}0&1&0&1\\0&0&1&0\end{smallmatrix}\right]\theta\left[\begin{smallmatrix}0&1&0&1\\1&0&1&0\end{smallmatrix}\right]}\times$$

$$\sqrt{\theta\left[\begin{smallmatrix}0&0&0&0\\0&1&1&1\end{smallmatrix}\right]\theta\left[\begin{smallmatrix}0&0&0&0\\1&1&1&1\end{smallmatrix}\right]\theta\left[\begin{smallmatrix}0&1&0&1\\0&1&1&1\end{smallmatrix}\right]\theta\left[\begin{smallmatrix}0&1&0&1\\1&1&1&1\end{smallmatrix}\right]} = 0$$

I observe that $\theta[1], \theta[2], \theta[3]$ are, respectively, the last members of the three terms of 1) and that Table 2 shows that no other theta constant in 1) vanishes, so that $\theta[3] = 0$, say, is dependent on $\theta[1] = \theta[2] = 0$. Thus 3 is really 2 plus Schottky.

Now, following Noether, with an assist from [5], Chapter X, p. 462, I shall show that $\theta[\chi] = \theta[4] = \cdots = \theta[9] = 0$ are dependent on $\theta[1] = \theta[2] = \theta[3] = 0$ so that 10 is really 3 is really 2 plus Schottky.

Following Schottky, [10], to get a four term biquadratic relation among 16 distinct 4-theta constants one takes four 4-characteristics forming an azygetic set and again a syzygetic group G, this time of 4-half-periods, of degree 2, which when composed with the four 4-characteristics gives 16 even 4-characteristics. Now as remarked before, the half-periods of Table 1 are a fundamental set and thus the 10 characteristics of Table 2 are a fundamental (maximal azygetic) set so that any subset of four, in particular, is also an azygetic set, hence, [1], [2], [3], and any other. To obtain the generators of G I use Krazer's device, which is designed to ensure that no other constants in the identity than the original four are from Table 2. This ensures that the others do not vanish and the identities furnish non-trivial relations. That the device in question delivers syzygetic half-periods is seen empirically but can be proved easily. That it gives the right G follows from the properties of a fundamental set of half-periods and related facts.

Accordingly, I set

$$G = \left\{\left(\begin{smallmatrix}0&0&0&0\\0&0&0&0\end{smallmatrix}\right), (4)+(5)+(6)+(7), (4)+(5)+(8)+(9),\right.$$
$$\left.(6)+(7)+(8)+(9)\right\} =$$
$$= \left\{\left(\begin{smallmatrix}0&0&0&0\\0&0&0&0\end{smallmatrix}\right), \left(\begin{smallmatrix}0&0&0&0\\1&0&0&0\end{smallmatrix}\right), \left(\begin{smallmatrix}0&1&0&0\\1&0&0&1\end{smallmatrix}\right), \left(\begin{smallmatrix}0&1&0&0\\0&0&0&1\end{smallmatrix}\right)\right\}$$

for an identity among $\theta[\chi]$, $\theta[1]$, $\theta[2]$, $\theta[3]$ and obtain (with these as first members of the respective terms)

$$2)\qquad \theta\left[\begin{smallmatrix}0&1&1&1\\0&1&1&0\end{smallmatrix}\right]\theta\left[\begin{smallmatrix}0&1&1&1\\1&1&1&0\end{smallmatrix}\right]\theta\left[\begin{smallmatrix}0&0&1&1\\1&1&1&1\end{smallmatrix}\right]\theta\left[\begin{smallmatrix}0&0&1&1\\0&1&1&1\end{smallmatrix}\right] \pm$$
$$\theta\left[\begin{smallmatrix}0&1&1&1\\1&1&0&1\end{smallmatrix}\right]\theta\left[\begin{smallmatrix}0&1&1&1\\0&1&0&1\end{smallmatrix}\right]\theta\left[\begin{smallmatrix}0&0&1&1\\0&1&0&0\end{smallmatrix}\right]\theta\left[\begin{smallmatrix}0&0&1&1\\1&1&0&0\end{smallmatrix}\right] \pm$$
$$\theta\left[\begin{smallmatrix}0&1&0&1\\1&1&0&1\end{smallmatrix}\right]\theta\left[\begin{smallmatrix}0&1&0&1\\0&1&0&1\end{smallmatrix}\right]\theta\left[\begin{smallmatrix}0&0&0&1\\0&1&0&0\end{smallmatrix}\right]\theta\left[\begin{smallmatrix}0&0&0&1\\1&1&0&0\end{smallmatrix}\right] \pm$$
$$\theta\left[\begin{smallmatrix}0&1&0&1\\1&1&1&1\end{smallmatrix}\right]\theta\left[\begin{smallmatrix}0&1&0&1\\0&1&1&1\end{smallmatrix}\right]\theta\left[\begin{smallmatrix}0&0&0&1\\0&1&1&0\end{smallmatrix}\right]\theta\left[\begin{smallmatrix}0&0&0&1\\1&1&1&0\end{smallmatrix}\right] = 0\ .$$

It is not immediately evident in Krazer how to replace $\theta[\chi]$ by $\theta[4]$, say, but I find the following device works. Define

$$(4') = (4),\quad (5') = (4)+(5),\ldots,(9') = (4)+(9)\ .$$

Then

$$\begin{aligned}(4')+(5')+(6')+(7') &= (4)+(4)+(5)+(6)+(4)+(7)\\ &= (4)+(5)+(6)+(7)+(4)\\ &= \left(\begin{smallmatrix}0&0&0&0\\1&0&0&0\end{smallmatrix}\right)+\left(\begin{smallmatrix}1&0&0&0\\0&0&0&0\end{smallmatrix}\right) = \left(\begin{smallmatrix}1&0&0&0\\1&0&0&0\end{smallmatrix}\right)\ ,\end{aligned}$$

and $(4')+(5')+(8')+(9') = (4)+(4)+(5)+(4)+(8)+(4)+(9) =$

$$(4)+(5)+(8)+(9)+(4) = \left(\begin{smallmatrix}1&1&0&0\\1&0&0&0\end{smallmatrix}\right)\ .$$

One then uses

$$G' = \left\{\left(\begin{smallmatrix}0&0&0&0\\0&0&0&0\end{smallmatrix}\right),\left(\begin{smallmatrix}1&0&0&0\\1&0&0&0\end{smallmatrix}\right),\left(\begin{smallmatrix}1&1&0&0\\1&0&0&0\end{smallmatrix}\right),\left(\begin{smallmatrix}0&1&0&0\\0&0&0&0\end{smallmatrix}\right)\right\}$$

to obtain

$$3)\qquad \theta\left[\begin{smallmatrix}1&1&1&1\\0&1&1&0\end{smallmatrix}\right]\theta\left[\begin{smallmatrix}0&1&1&1\\1&1&1&0\end{smallmatrix}\right]\theta\left[\begin{smallmatrix}0&0&1&1\\1&1&1&1\end{smallmatrix}\right]\theta\left[\begin{smallmatrix}1&0&1&1\\0&1&1&1\end{smallmatrix}\right] \pm$$
$$\theta\left[\begin{smallmatrix}0&1&1&1\\1&1&0&1\end{smallmatrix}\right]\theta\left[\begin{smallmatrix}1&1&1&1\\0&1&0&1\end{smallmatrix}\right]\theta\left[\begin{smallmatrix}1&0&1&1\\0&1&0&0\end{smallmatrix}\right]\theta\left[\begin{smallmatrix}0&0&1&1\\1&1&0&0\end{smallmatrix}\right] \pm$$
$$\theta\left[\begin{smallmatrix}0&1&0&1\\1&1&0&1\end{smallmatrix}\right]\theta\left[\begin{smallmatrix}1&1&0&1\\0&1&0&1\end{smallmatrix}\right]\theta\left[\begin{smallmatrix}1&0&0&1\\0&1&0&0\end{smallmatrix}\right]\theta\left[\begin{smallmatrix}0&0&0&1\\1&1&0&0\end{smallmatrix}\right] \pm$$
$$\theta\left[\begin{smallmatrix}0&1&0&1\\1&1&1&1\end{smallmatrix}\right]\theta\left[\begin{smallmatrix}1&1&0&1\\0&1&1&1\end{smallmatrix}\right]\theta\left[\begin{smallmatrix}1&0&0&1\\0&1&1&0\end{smallmatrix}\right]\theta\left[\begin{smallmatrix}0&0&0&1\\1&1&1&0\end{smallmatrix}\right] = 0\ .$$

Proceeding in this way I derive the following additional 5 identities, which complete my story.

4)

$$\begin{aligned}
&\theta\begin{bmatrix}1&0&1&1\\1&1&1&0\end{bmatrix}\theta\begin{bmatrix}0&1&1&1\\1&1&1&0\end{bmatrix}\theta\begin{bmatrix}0&0&1&1\\1&1&1&1\end{bmatrix}\theta\begin{bmatrix}1&1&1&1\\1&1&1&1\end{bmatrix}\pm\\
&\theta\begin{bmatrix}0&1&1&1\\1&1&0&1\end{bmatrix}\theta\begin{bmatrix}1&0&1&1\\1&1&0&1\end{bmatrix}\theta\begin{bmatrix}1&1&1&1\\1&1&0&0\end{bmatrix}\theta\begin{bmatrix}0&0&1&1\\1&1&0&0\end{bmatrix}\pm\\
&\theta\begin{bmatrix}0&1&0&1\\1&1&0&1\end{bmatrix}\theta\begin{bmatrix}1&0&0&1\\1&1&0&1\end{bmatrix}\theta\begin{bmatrix}1&1&0&1\\1&1&0&0\end{bmatrix}\theta\begin{bmatrix}0&0&0&1\\1&1&0&0\end{bmatrix}\pm\\
&\theta\begin{bmatrix}0&1&0&1\\1&1&1&1\end{bmatrix}\theta\begin{bmatrix}1&0&0&1\\1&1&1&1\end{bmatrix}\theta\begin{bmatrix}1&1&0&1\\1&1&1&0\end{bmatrix}\theta\begin{bmatrix}0&0&0&1\\1&1&1&0\end{bmatrix} = 0 \ .
\end{aligned}$$

5)

$$\begin{aligned}
&\theta\begin{bmatrix}1&0&1&1\\1&0&1&0\end{bmatrix}\theta\begin{bmatrix}0&1&1&1\\1&1&1&0\end{bmatrix}\theta\begin{bmatrix}1&1&1&1\\0&0&1&1\end{bmatrix}\theta\begin{bmatrix}0&0&1&1\\0&1&1&1\end{bmatrix}\pm\\
&\theta\begin{bmatrix}0&1&1&1\\1&1&0&1\end{bmatrix}\theta\begin{bmatrix}1&0&1&1\\1&0&0&1\end{bmatrix}\theta\begin{bmatrix}0&0&1&1\\0&1&0&0\end{bmatrix}\theta\begin{bmatrix}1&1&1&1\\0&0&0&0\end{bmatrix}\pm\\
&\theta\begin{bmatrix}0&1&0&1\\1&1&0&1\end{bmatrix}\theta\begin{bmatrix}1&0&0&1\\1&0&0&1\end{bmatrix}\theta\begin{bmatrix}0&0&0&1\\0&1&0&0\end{bmatrix}\theta\begin{bmatrix}1&1&0&1\\0&0&0&0\end{bmatrix}\pm\\
&\theta\begin{bmatrix}0&1&0&1\\1&1&1&1\end{bmatrix}\theta\begin{bmatrix}1&0&0&1\\1&0&1&1\end{bmatrix}\theta\begin{bmatrix}0&0&0&1\\0&1&1&0\end{bmatrix}\theta\begin{bmatrix}1&1&0&1\\0&0&1&0\end{bmatrix} = 0 \ .
\end{aligned}$$

6)

$$\begin{aligned}
&\theta\begin{bmatrix}1&1&1&1\\1&0&1&0\end{bmatrix}\theta\begin{bmatrix}0&1&1&1\\1&1&1&0\end{bmatrix}\theta\begin{bmatrix}1&0&1&1\\0&0&1&1\end{bmatrix}\theta\begin{bmatrix}0&0&1&1\\0&1&1&1\end{bmatrix}\pm\\
&\theta\begin{bmatrix}0&1&1&1\\1&1&0&1\end{bmatrix}\theta\begin{bmatrix}1&1&1&1\\1&0&0&1\end{bmatrix}\theta\begin{bmatrix}0&0&1&1\\0&1&0&0\end{bmatrix}\theta\begin{bmatrix}1&0&1&1\\0&0&0&0\end{bmatrix}\pm\\
&\theta\begin{bmatrix}0&1&0&1\\1&1&0&1\end{bmatrix}\theta\begin{bmatrix}1&1&0&1\\1&0&0&1\end{bmatrix}\theta\begin{bmatrix}0&0&0&1\\0&1&0&0\end{bmatrix}\theta\begin{bmatrix}1&0&0&1\\0&0&0&0\end{bmatrix}\pm\\
&\theta\begin{bmatrix}0&1&0&1\\1&1&1&1\end{bmatrix}\theta\begin{bmatrix}1&1&0&1\\1&0&1&1\end{bmatrix}\theta\begin{bmatrix}0&0&0&1\\0&1&1&0\end{bmatrix}\theta\begin{bmatrix}1&0&0&1\\0&0&1&0\end{bmatrix} = 0 \ .
\end{aligned}$$

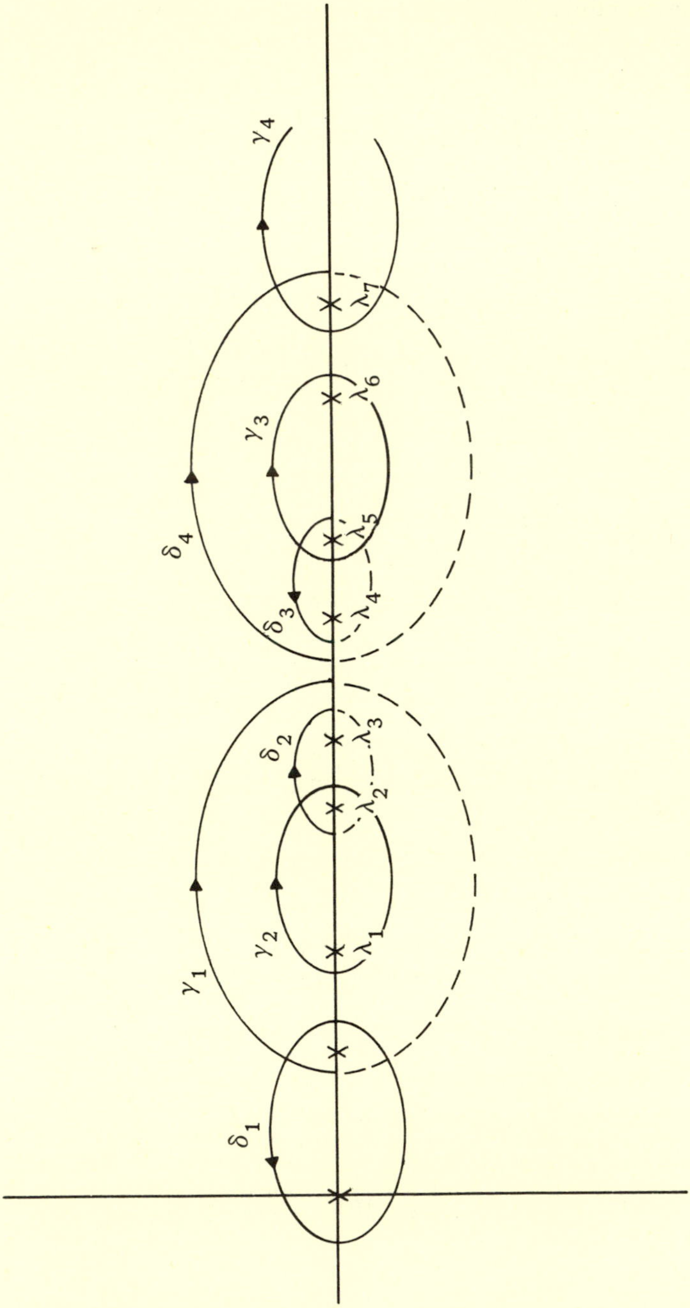

γ_4
λ_7
λ_6
γ_3
δ_4
λ_5
λ_4
δ_3
λ_3
δ_2
λ_2
γ_2
λ_1
γ_1
δ_1

7)

$$\theta\begin{bmatrix}0\,1\,1\,0\\1\,1\,1\,1\end{bmatrix}\theta\begin{bmatrix}0\,1\,1\,0\\0\,1\,1\,1\end{bmatrix}\theta\begin{bmatrix}0\,0\,1\,1\\1\,1\,1\,1\end{bmatrix}\theta\begin{bmatrix}0\,0\,1\,1\\0\,1\,1\,1\end{bmatrix}\pm$$

$$\theta\begin{bmatrix}0\,1\,1\,1\\1\,1\,0\,1\end{bmatrix}\theta\begin{bmatrix}0\,1\,1\,1\\0\,1\,0\,1\end{bmatrix}\theta\begin{bmatrix}0\,0\,1\,0\\1\,1\,0\,1\end{bmatrix}\theta\begin{bmatrix}0\,0\,1\,0\\0\,1\,0\,1\end{bmatrix}\pm$$

$$\theta\begin{bmatrix}0\,1\,0\,1\\1\,1\,0\,1\end{bmatrix}\theta\begin{bmatrix}0\,1\,0\,1\\0\,1\,0\,1\end{bmatrix}\theta\begin{bmatrix}0\,0\,0\,0\\1\,1\,0\,1\end{bmatrix}\theta\begin{bmatrix}0\,0\,0\,0\\0\,1\,0\,1\end{bmatrix}\pm$$

$$\theta\begin{bmatrix}0\,1\,0\,1\\1\,1\,1\,1\end{bmatrix}\theta\begin{bmatrix}0\,1\,0\,1\\0\,1\,1\,1\end{bmatrix}\theta\begin{bmatrix}0\,0\,0\,0\\1\,1\,1\,1\end{bmatrix}\theta\begin{bmatrix}0\,0\,0\,0\\0\,1\,1\,1\end{bmatrix} = 0 \ .$$

8)

$$\theta\begin{bmatrix}0\,1\,1\,0\\1\,1\,1\,0\end{bmatrix}\theta\begin{bmatrix}0\,1\,1\,0\\0\,1\,1\,0\end{bmatrix}\theta\begin{bmatrix}0\,0\,1\,1\\1\,1\,1\,1\end{bmatrix}\theta\begin{bmatrix}0\,0\,1\,1\\0\,1\,1\,1\end{bmatrix}\pm$$

$$\theta\begin{bmatrix}0\,1\,1\,1\\1\,1\,0\,1\end{bmatrix}\theta\begin{bmatrix}0\,1\,1\,1\\0\,1\,0\,1\end{bmatrix}\theta\begin{bmatrix}0\,0\,1\,0\\1\,1\,0\,0\end{bmatrix}\theta\begin{bmatrix}0\,0\,1\,0\\0\,0\,0\,1\end{bmatrix}\pm$$

$$\theta\begin{bmatrix}0\,1\,0\,1\\1\,1\,0\,1\end{bmatrix}\theta\begin{bmatrix}0\,1\,0\,1\\0\,1\,0\,1\end{bmatrix}\theta\begin{bmatrix}0\,0\,0\,0\\1\,1\,0\,0\end{bmatrix}\theta\begin{bmatrix}0\,0\,0\,0\\0\,1\,0\,0\end{bmatrix}\pm$$

$$\theta\begin{bmatrix}0\,1\,0\,1\\1\,1\,1\,1\end{bmatrix}\theta\begin{bmatrix}0\,1\,0\,1\\0\,1\,1\,1\end{bmatrix}\theta\begin{bmatrix}0\,0\,0\,0\\1\,1\,1\,0\end{bmatrix}\theta\begin{bmatrix}0\,0\,0\,0\\0\,1\,1\,0\end{bmatrix} = 0 \ .$$

REFERENCES

[1] Accola, R. D. M., On the problem of characterizing hyperelliptic surfaces in terms of vanishing theta nulls, unpublished.

[2] Farkas, H. M., Special divisors and analytic subloci of Teichmueller space, *Am. J. Math.* 88 (1966), 881-901.

[3] ______, Theta constants, moduli, and compact Riemann surfaces, *Proc. Nat. Acad. Sci.* 62 (1969), 320-325.

[4] Farkas H. M. and Rauch, Harry E., Two kinds of theta constants and period relations on a Riemann surface, *ibid.* 62 (1969), 679-686.

[5] Krazer, A., *Lehrbuch der Thetafunktionen,* Leipzig, Teubner, 1903.

[6] Lewittes, J., Riemann surfaces and the theta function, *Acta Math.* 111 (1964), 37-61.

[[7] Rauch, Harry E., A transcendental view of the space of algebraic Riemann surfaces, *Bull. Am. Math. Soc.* 71 (1965), 1-39.

[8] ______ , The vanishing of a theta constant is a peculiar phenomenon, *Bull. Am. Math. Soc.*73 (1967), 339-342.

[9] Rauch, Harry E. and Farkas, H. M. Relations between two kinds of theta constants on a Riemann surface, *Proc. Nat. Acad. Sci.* 59 (1968), 52-55.

[10] Schottky, F., Über die Moduln der Thetafunktionen, *Acta Math.* 27 (1903), 235-288.

[11] ______ , Zur Theorie der Abelschen Functionen von vier Variabeln, *J. Reine Ang. Math.* 102 (1888), 304-352.

[12] Schottky, F. and Jung, H., Neue Sätze Über Symmetralfunktionen und die Abelschen Funktionen der Riemann'sche Theorie, S.-B, Preuss. Akad. Wiss. (Berlin), *Phys. Math. Cl.* 1909, 282-297.

Yeshiva University
New York, New York

SCHOTTKY IMPLIES POINCARÉ

Harry E. Rauch*

1. Introduction.

In this note I show that (one form of) Schottky's period relation for Riemann surfaces of genus four, [7] and [6], implies Poincaré'a approximate period relation, [3] (cf. also [2]), for surfaces of genus four whose period matrices are close to diagonal form.

Namely:

1)

$$\sqrt{\theta\begin{bmatrix}1&1&0&0\\1&1&0&0\end{bmatrix}\theta\begin{bmatrix}0&1&1&1\\1&1&1&0\end{bmatrix}\theta\begin{bmatrix}1&0&1&1\\0&0&1&1\end{bmatrix}\theta\begin{bmatrix}1&1&1&1\\1&0&0&1\end{bmatrix}} \times$$

$$\sqrt{\theta\begin{bmatrix}0&0&0&0\\0&0&0&1\end{bmatrix}\theta\begin{bmatrix}0&1&0&0\\1&0&1&1\end{bmatrix}\theta\begin{bmatrix}1&0&0&0\\0&1&1&0\end{bmatrix}\theta\begin{bmatrix}0&0&1&1\\0&1&0&0\end{bmatrix}} \pm$$

$$\sqrt{\theta\begin{bmatrix}1&1&1&0\\0&0&0&0\end{bmatrix}\theta\begin{bmatrix}0&1&0&1\\0&0&1&0\end{bmatrix}\theta\begin{bmatrix}1&0&0&1\\1&1&1&1\end{bmatrix}\theta\begin{bmatrix}1&1&0&1\\0&1&0&1\end{bmatrix}} \times$$

$$\sqrt{\theta\begin{bmatrix}0&0&1&0\\1&1&0&1\end{bmatrix}\theta\begin{bmatrix}0&1&1&0\\0&1&1&1\end{bmatrix}\theta\begin{bmatrix}1&0&1&0\\1&0&1&0\end{bmatrix}\theta\begin{bmatrix}0&0&0&1\\1&0&0&0\end{bmatrix}} \pm$$

$$\sqrt{\theta\begin{bmatrix}0&0&0&0\\0&0&1&0\end{bmatrix}\theta\begin{bmatrix}1&0&1&1\\0&0&0&0\end{bmatrix}\theta\begin{bmatrix}0&1&1&1\\1&1&0&1\end{bmatrix}\theta\begin{bmatrix}0&0&1&1\\0&1&1&1\end{bmatrix}} \times$$

$$\sqrt{\theta\begin{bmatrix}1&1&0&0\\1&1&1&1\end{bmatrix}\theta\begin{bmatrix}1&0&0&0\\0&1&0&1\end{bmatrix}\theta\begin{bmatrix}0&1&0&0\\1&0&0&0\end{bmatrix}\theta\begin{bmatrix}1&1&1&1\\1&0&1&0\end{bmatrix}} = 0$$

---

* Research partially sponsored by the Air Force Office of Scientific Research, Office of Aerospace Research, U.S. Air Force, under AFOSR Grant No. AFOSR - 1873-70.

implies, I claim,

$$2)\qquad \sqrt{\pi_{12}\,\pi_{23}\,\pi_{34}\,\pi_{14} + 0(\epsilon^{10})} \pm \sqrt{\pi_{14}\,\pi_{24}\,\pi_{23}\,\pi_{13} + 0(\epsilon^{10})} \pm \sqrt{\pi_{13}\,\pi_{34}\,\pi_{24}\,\pi_{12} + 0(\epsilon^{10})} = 0 ,$$

where $\epsilon = \max\sqrt{|\pi_{ij}|}$ over all $1 \leq i < j \leq 4$. It must be understood that I am dealing here with a Riemann surface S of genus four and on it a canonical homology basis $\gamma_1,\ldots,\gamma_4;\ \delta_1,\ldots,\delta_4$ and the set $d\zeta_1,\ldots,d\zeta_4$ of normal differentials of first kind with respect to that basis, i.e., such that $\delta_{ij} = \int_{\gamma_j} d\zeta_i$, $i, j = 1,\ldots,4$. One then has $\pi_{ij} = \int_{\delta_j} d\zeta_i$ and the theta constants in 1) are computed with the theta matrix $\pi = (\pi_{ij})$ (these are called Riemann theta constants in [5]).

In fact, in section 2 I shall show

$$3)\qquad \mathrm{Sch}(\pi) = K\,P_0(\pi) ,$$

where $\mathrm{Sch}(\pi)$ and $P_0(\pi)$ are the left sides of 1) and 2), respectively, and

$$4)\qquad K = \frac{\pi^2}{4}\,\Pi_{j=1}^{4}\,(\theta^2[{}^0_0]\theta^2[{}^0_1]\theta^2[{}^1_0])_j \neq 0 ,$$

where the subscript j indicates that all three 1-theta constants in parenthesis are to be evaluated at π_{jj} and the π in 4) is the area of the unit circle. From 3) the implication 1) ==> 2) is immediate.

2) is strictly speaking not Poincaré's identity. Poincaré makes the (dubious) assumption that

$$\pi_{12} = t^2\pi^0_{12}, \ldots, \pi_{34} = t^2\pi^0_{34} .$$

If one inserts these in 2), assumes that no π^0_{ij} $(i < j)$ is zero, divides by t^4, and passes to the limit $t = 0$, one obtains Poincaré's relation for what one may call the infinitesimal periods $\pi^0_{12},\ldots,\pi^0_{34}$. If one then multiplies through by t^4, one obtains Poincaré's approximate relation:

$$\sqrt{\pi_{12}\pi_{23}\pi_{34}\pi_{14}} \pm \sqrt{\pi_{14}\pi_{24}\pi_{23}\pi_{13}} \pm \sqrt{\pi_{13}\pi_{24}\pi_{34}\pi_{12}} = 0 .$$

A more realistic assumption such as $\pi_{12} = t^2\pi^0_{12} + \cdots, \cdots, \pi_{34} = t^2\pi^0_{34} + \cdots$, the dots representing terms with higher powers of t, under the non-vanishing hypothesis on the infinitesimal periods results in the preceding with an $O(t^6)$ added on *outside the radicals,* when one substitutes in 2).

In section 2 I shall prove 3) and 4). In section 3 I shall deduce 1) from recent work of Farkas and me, [1], and a result of Schottky.

In section 4 I attempt to indicate my reasoning in arriving at the particular form 1) of Schottky's relation. As a first clue let me say that the technique for deriving 3) from 1) is already indicated in my note [4], where the crucial role played by columns of the form $\left[{1 \atop 1}\right]$ in a characteristic is noted. In fact, if one looks carefully at 1), one will discover that each of the three radicands has as factors exactly four theta constants whose characteristics contain precisely two such columns while the other four have no such columns (indeed for a 4-characteristic to be even it must have no, two, or four such columns). Furthermore one will observe that the positions of these columns in each term of 1) correspond precisely to the indices of the π's in the corresponding term of 2). Thus the first four factors of the first term have $\left[{1 \atop 1}\right]$ columns in the (1,2), (2,3), (3,4), and (1,4) positions, the next term has such columns in the (1,4), (2,4), (2,3), and (1,3) positions in the third, fourth, sixth, and seventh factors, and similarly for the third terms of both relations. This fact was evidently my heuristic *Ansatz*. However, the real trick is how to juggle the intricate theta algebra to get Schottky's relation in such a form. I must warn the reader that, in addition to the indications in section 4, a lot of informed guesswork also played a role.

Before proceeding I make the important remark that 3) shows in particular that 1) is a non-trivial relation and not an identity (Schottky also proved this in a different way). Indeed, choose values $\pi^0_{12}, \ldots, \pi^0_{34}$ such that

$$\sqrt{\pi^0_{12}\pi^0_{23}\pi^0_{34}\pi^0_{14}} \pm \sqrt{\pi^0_{14}\pi^0_{24}\pi^0_{23}\pi^0_{13}} \pm$$

$$\sqrt{\pi^0_{13}\pi^0_{34}\pi^0_{24}\pi^0_{12}} \neq 0 .$$

For example pick $\pi^0_{12} = 0$ and all others not zero. Set $\pi_{12} = t^2\pi^0_{12}, \ldots, \pi_{34} = t^2\pi^0_{34}$. Then for $0 < |t|$ sufficiently small 3) implies that the left side of Schottky's relation 1) is not zero.

2. Proof of 3).

I consider each radicand in 1) as a function of the 6 off-diagonal variables $\pi_{12}, \ldots, \pi_{34}$ for fixed values of the diagonal variables $\pi_{11}, \ldots, \pi_{44}$ of π, and I expand it in a Maclaurin series about the "origin" $\pi_{12} = \cdots = \pi_{34} = 0$. To do this I expand each theta constant appearing as a factor in a Maclaurin series and formally multiply the eight series thus obtained. Now the theta constants and their first partials with respect to the π's have a particularly simple appearance when evaluated at the "origin," i.e., on the diagonal as I point out in [4], to which the reader is referred for definitions and more detail.

Namely, *on the diagonal*

5) $$\theta\begin{bmatrix}\epsilon_1 \epsilon_2 \epsilon_3 \epsilon_4 \\ \epsilon'_1 \epsilon'_2 \epsilon'_3 \epsilon'_4\end{bmatrix} = \theta\begin{bmatrix}\epsilon_1 \\ \epsilon'_1\end{bmatrix}_1 \theta\begin{bmatrix}\epsilon_2 \\ \epsilon'_2\end{bmatrix}_2 \theta\begin{bmatrix}\epsilon_3 \\ \epsilon'_3\end{bmatrix}_3 \theta\begin{bmatrix}\epsilon_4 \\ \epsilon'_4\end{bmatrix}_4$$

and

6) $$\frac{\partial}{\partial \pi_{ij}} \theta\begin{bmatrix}\epsilon_1 & \epsilon_2 & \epsilon_3 & \epsilon_4 \\ \epsilon'_1 & \epsilon'_2 & \epsilon'_3 & \epsilon'_4\end{bmatrix} = \frac{1}{2\pi i} \prod_{k \neq i,j} \theta\begin{bmatrix}\epsilon_k \\ \epsilon'_k\end{bmatrix}_k \theta'\begin{bmatrix}\epsilon_i \\ \epsilon'_i\end{bmatrix}_i \theta'\begin{bmatrix}\epsilon_j \\ \epsilon'_j\end{bmatrix}_j ,$$

where $i < j$,

$$\theta\begin{bmatrix}\epsilon_\ell \\ \epsilon'_\ell\end{bmatrix}_\ell = \theta\begin{bmatrix}\epsilon_\ell \\ \epsilon'_\ell\end{bmatrix}(0, \pi_{\ell\ell}) ,$$

and

$$\theta'\begin{bmatrix}\epsilon_\ell \\ \epsilon'_\ell\end{bmatrix}_\ell = \frac{\partial}{\partial u} \theta\begin{bmatrix}\epsilon_\ell \\ \epsilon'_\ell\end{bmatrix}(u, \pi_{\ell\ell})\Big|_{u=0} .$$

Here, $\theta[{}^{\mu}_{\mu'}](u,\tau)$ is the classical (elliptic) 1-theta function, which as a function of u is even for $[{}^{\mu}_{\mu'}] = [{}^0_0], [{}^0_1], [{}^1_0]$ and odd for $[{}^{\mu}_{\mu'}] = [{}^1_1]$ so that $\theta[{}^1_1]_\ell = \theta'[{}^0_0]_\ell = \theta'[{}^0_1]_\ell = \theta'[{}^1_0]_\ell = 0$, $\ell = 1,2,3,4$. On the other hand it is known that the 1-theta constants $\theta[{}^0_0]$, $\theta[{}^0_1]$, $\theta[{}^1_0]$, as well as $\theta'[{}^1_1]_\ell$ are not zero (for period in the upper half-plane, as is the case here). Finally one knows

$$\theta'[{}^1_1]_\ell = -\pi\theta[{}^0_0]_\ell\,\theta[{}^0_1]_\ell\,\theta[{}^1_0]_\ell \,. \tag{7}$$

Putting all this together with 5) and 6) one obtains the following expansions for the eight factors of the first radicand, where I observe that 6) and the preceding imply that the first partials of a theta constant with no $[{}^1_1]$ columns all vanish on the diagonal:

$$\theta[{}^{1\,1\,0\,0}_{1\,1\,0\,0}] = \{\tfrac{\pi}{2i}\,\theta[{}^0_0]_3\,\theta[{}^0_0]_4 \prod_{i=1,2} (\theta[{}^0_0]\,\theta[{}^0_1]\,\theta[{}^1_0])_i\}\,\pi_{12} + 0(\epsilon^4) \;,$$

$$\theta[{}^{0\,1\,1\,1}_{1\,1\,1\,0}] = \{\tfrac{\pi}{2i}\,\theta[{}^0_1]_1\theta[{}^1_0]_4 \prod_{i=2,3} (\theta[{}^0_0]\,\theta[{}^0_1]\,\theta[{}^1_0])_i\}\,\pi_{23} + 0(\epsilon^4) \;,$$

$$\theta[{}^{1\,0\,1\,1}_{0\,0\,1\,1}] = \{\tfrac{\pi}{2i}\,\theta[{}^1_0]_1\theta[{}^0_0]_2 \prod_{i=3,4} (\theta[{}^0_0]\,\theta[{}^0_1]\,\theta[{}^1_0])_i\}\,\pi_{34} + 0(\epsilon^4) \;,$$

$$\theta[{}^{1\,1\,1\,1}_{1\,0\,0\,1}] = \{\tfrac{\pi}{2i}\,\theta[{}^1_0]_2\theta[{}^1_0]_3 \prod_{i=1,4} (\theta[{}^0_0]\,\theta[{}^0_1]\,\theta[{}^1_0])_i\}\,\pi_{14} + 0(\epsilon^4) \;,$$

$$\theta[{}^{0\,0\,0\,0}_{0\,0\,0\,1}] = \theta[{}^0_0]_1\theta[{}^0_0]_2\theta[{}^0_0]_3\theta[{}^0_1]_4 + 0(\epsilon^4) \;,$$

$$\theta[{}^{0\,1\,0\,0}_{1\,0\,1\,1}] = \theta[{}^0_1]_1\,\theta[{}^1_0]_2\theta[{}^0_1]_3\theta[{}^0_1]_4 + 0(\epsilon^4) \;,$$

$$\theta[{}^{1\,0\,0\,0}_{0\,1\,1\,0}] = \theta[{}^1_0]_1\,\theta[{}^0_1]_2\,\theta[{}^0_1]_3\theta[{}^0_0]_4 + 0(\epsilon^4) \;,$$

$$\theta[{}^{0\,0\,1\,1}_{0\,1\,0\,0}] = \theta[{}^0_0]_1\,\theta[{}^0_1]_2\,\theta[{}^1_0]_3\theta[{}^1_0]_4 + 0(\epsilon^4) \,.$$

Multiplying the terms of 8), one obtains the first radicand of 1), the term $\pi_{12}\pi_{23}\pi_{34}\pi_{14}$ being $0(\epsilon^8)$, and the two remaining radicands are handled similarly.

3. Schottky's relation.

In [7] and [6] Schottky asserts that to find any one form of his relation one takes a syzygetic group of 4-half-periods G of degree 3, i.e., order $2^3 = 8$ and composes it with the uniquely determined three 4-characteristics such that each of them composed with G gives 8 even 4-characteristics. The theta constants with each of these three sets when multiplied together give a radicand, let them be r_1, r_2, r_3, and his relation is then

$$\sqrt{r_1} \pm \sqrt{r_2} \pm \sqrt{r_3} = 0 .$$

He proves correctly, since the proof involves only formal algebra, that one gets the same relation no matter what G one starts with in the precise sense that the rationalized left side, namely,

$$r_1^2 + r_2^2 + r_3^2 - 2(r_1 r_2 + r_2 r_3 + r_1 r_3)$$

has the same numerical value for all G. In order to prove his relation then, it is sufficient to prove that the left side vanishes on the Torelli sublocus of S_4 for some G. Here Schottky's argument in [7] is clouded by some assertions about generic linear independence of certain multiplicative functions and non-vanishing of certain determinants which can probably be established with considerable labor. Also it is not clear why the particular determinantal relation he uses should be the only relation, whether, in other words, there may not be other additional relations automatically implied, although it is clear from dimensional considerations that there cannot be such. In [6] he relies instead on a "well-known algebraic theorem" on the vanishing of certain quadratic forms subject to linear side conditions, but this theorem seems to be false as stated. Hence, I appeal instead to the recent joint note [1] of Farkas and mine where we prove the vanishing of the left side with the three 4-characteristics

$$\begin{bmatrix}0&0&0&0\\0&0&0&0\end{bmatrix}, \begin{bmatrix}0&0&0&1\\0&0&0&0\end{bmatrix}, \begin{bmatrix}0&0&0&0\\0&0&0&1\end{bmatrix},$$

and the syzygetic group

$$G = \{\begin{pmatrix}0&0&0&0\\0&0&0&0\end{pmatrix}, \begin{pmatrix}0&0&0&0\\1&0&0&0\end{pmatrix}, \begin{pmatrix}0&1&0&0\\0&0&0&0\end{pmatrix}, \begin{pmatrix}0&0&1&0\\0&0&0&0\end{pmatrix},$$

$$\begin{pmatrix}0&1&0&0\\1&0&0&0\end{pmatrix}, \begin{pmatrix}0&0&1&0\\1&0&0&0\end{pmatrix}, \begin{pmatrix}0&1&1&0\\0&0&0&0\end{pmatrix}, \begin{pmatrix}0&1&1&0\\1&0&0&0\end{pmatrix}\}.$$

By the preceding then, Schottky's relation is established for any G and the corresponding three 4-characteristics.

But the reader will see immediately that the three 4-characteristics

$$\begin{bmatrix}1&1&0&0\\1&1&0&0\end{bmatrix}, \begin{bmatrix}1&1&1&0\\0&0&0&0\end{bmatrix}, \begin{bmatrix}0&0&0&0\\0&0&1&0\end{bmatrix},$$

and the syzygetic group

$$G = \{\begin{pmatrix}0&0&0&0\\0&0&0&0\end{pmatrix}, \begin{pmatrix}1&0&1&1\\0&0&1&0\end{pmatrix}, \begin{pmatrix}0&1&1&1\\1&1&1&1\end{pmatrix}, \begin{pmatrix}0&0&1&1\\0&1&0&1\end{pmatrix},$$

$$\begin{pmatrix}1&1&0&0\\1&1&0&1\end{pmatrix}, \begin{pmatrix}1&0&0&0\\0&1&1&1\end{pmatrix}, \begin{pmatrix}0&1&0&0\\1&0&1&0\end{pmatrix}, \begin{pmatrix}1&1&1&1\\1&0&0&0\end{pmatrix}\}$$

result in 1).

4. Procedure for discovering G and the three characteristics.

I have already mentioned the heuristic *Ansatz* in section 1. Thus to get the radicand $\pi_{12}\,\pi_{23}\,\pi_{34}\,\pi_{14}$ in Poincaré I need four 4-theta constants whose characteristics have $\begin{bmatrix}1\\1\end{bmatrix}$ columns in the (1, 2), (2, 3), (3, 4), (1, 4) positions, respectively, while the remaining four constants have no $\begin{bmatrix}1\\1\end{bmatrix}$ columns in their characteristics. Thus I set up

$$\begin{bmatrix}1&1&\alpha&\beta\\1&1&\alpha'&\beta'\end{bmatrix}, \begin{bmatrix}\delta&1&1&\epsilon\\\delta'&1&1&\epsilon'\end{bmatrix}, \begin{bmatrix}\mu&\lambda&1&1\\\mu'&\lambda'&1&1\end{bmatrix}, \begin{bmatrix}1&\nu&\tau&1\\1&\nu'&\tau'&1\end{bmatrix}.$$

I assume that these four come from composing the first with the first four elements of some G, thus, the identity and three generators. To find these I add the first elementwise to the last three, and then I complete G by composition to get the remaining four elements:

$$G = \{\begin{pmatrix} 0\,0\,0\,0 \\ 0\,0\,0\,0 \end{pmatrix}, \begin{pmatrix} 1+\delta & 0 & 1+\alpha & \beta+\epsilon \\ 1+\delta' & 0 & 1+\alpha' & \beta+\epsilon \end{pmatrix},$$

$$\begin{pmatrix} 1+\mu & 1+\lambda & 1+\alpha & 1+\beta \\ 1+\mu' & 1+\lambda' & 1+\alpha' & 1+\beta' \end{pmatrix}, \begin{pmatrix} 0 & 1+\nu & \alpha+\tau & 1+\beta \\ 0 & 1+\nu' & \alpha'+\tau' & 1+\beta' \end{pmatrix}$$

$$\begin{pmatrix} \delta+\mu & 1+\lambda & 0 & 1+\epsilon \\ \delta'+\mu' & 1+\lambda' & 0 & 1+\epsilon' \end{pmatrix}, \begin{pmatrix} 1+\delta & 1+\nu & 1+\tau & 1+\epsilon \\ 1+\delta' & 1+\nu' & 1+\tau' & 1+\epsilon' \end{pmatrix},$$

$$\begin{pmatrix} 1+\mu & \lambda+\nu & 1+\tau & 0 \\ 1+\mu' & \lambda'+\nu & 1+\tau' & 0 \end{pmatrix}, \begin{pmatrix} \delta+\mu & \lambda+\nu & \alpha+\tau & \beta+\epsilon \\ \delta'+\mu' & \lambda'+\nu' & \alpha'+\tau' & \beta'+\epsilon' \end{pmatrix}$$

I then compose the last four elements of G with $\begin{bmatrix} 1 & 1 & \alpha & \beta \\ 1 & 1 & \alpha' & \beta' \end{bmatrix}$ to get

$$\begin{bmatrix} 1+\delta+\mu & \lambda & \alpha & 1+\epsilon+\beta \\ 1+\delta'+\mu' & \lambda' & \alpha' & 1+\epsilon'+\beta' \end{bmatrix}, \begin{bmatrix} \delta & \nu & 1+\alpha+\tau & 1+\epsilon+\beta \\ \delta' & \nu' & 1+\alpha'+\tau' & 1+\epsilon'+\beta' \end{bmatrix},$$

$$\begin{bmatrix} \mu & 1+\lambda+\nu & 1+\alpha+\tau & \beta \\ \mu' & 1+\lambda'+\gamma' & 1+\alpha'+\tau' & \beta' \end{bmatrix}, \begin{bmatrix} 1+\delta+\mu & 1+\lambda+\nu & \tau & \epsilon \\ 1+\delta'+\mu' & 1+\lambda'+\gamma' & \tau' & \epsilon' \end{bmatrix}.$$

Since none of these should have a $\begin{bmatrix} 1 \\ 1 \end{bmatrix}$ column I take $\delta+\mu = 1$, $\epsilon+\beta = 1$, $\lambda+\nu = 1$, $\alpha+\tau = 1$ and then $\alpha = \beta = \lambda = \delta = 0$ so that $\mu = \epsilon = \nu = \tau = 1$. This immediately gives me the top rows of the symbols of the elements of G as shown in section 3.

I then look for the top row of the first characteristic of the second radicand. Because the $\begin{bmatrix} 1 \\ 1 \end{bmatrix}$ columns of some four of the eight terms obtained should correspond to the indices of $\pi_{14}\,\pi_{24}\,\pi_{23}\,\pi_{13}$ I find almost by inspection that I get an acceptable distribution of 1's by taking

$$\begin{bmatrix} 1 & 1 & 1 & 0 \\ a_1 & a_2 & a_3 & a_4 \end{bmatrix}.$$

This turns out to be a little deceptive but correct since not (0 1 0 1) but (1 1 0 1) turns out to be the top row in the characteristic corresponding to π_{24}.

Again, looking at the top rows of the elements of G, I decide, purely on hunch to try

$$\begin{bmatrix} 0 & 0 & 0 & 0 \\ b_1 & b_2 & b_3 & b_4 \end{bmatrix}$$

as the characteristic of the leading factor of the third radicand.

Going back now that the top rows of all eight characteristics of the first radicand are determined one sees that one must have $\epsilon' = \mu' = \nu' = \tau' = 0$ to avoid any additional $\left[\begin{smallmatrix}1\\1\end{smallmatrix}\right]$ columns.

Similarly, composing G with $\begin{bmatrix} 1 & 1 & 1 & 0 \\ a_1 & a_2 & a_3 & a_4 \end{bmatrix}$ one finds it expedient to take $a_1 = a_2 = a_3 = 0$ to avoid having characteristics with only one $\left[\begin{smallmatrix}1\\1\end{smallmatrix}\right]$ column. On making this substitution in the eight characteristics one finds that in order to make the third, fourth, sixth and seventh factors correspond to (1,4), (2,4), (2,3), and (1,3) $\left[\begin{smallmatrix}1\\1\end{smallmatrix}\right]$ columns and no others one needs $\beta' + \epsilon' + a_4 \equiv 0$ but since $\epsilon' = 0$ this means $a_4 \equiv -\beta'$ (all mod 2).

Now working with $\begin{bmatrix} 0 & 0 & 0 & 0 \\ b_1 & b_2 & b_3 & b_4 \end{bmatrix}$ similarly, in order to make the third radicand contain the third, fourth, seventh, and eighth factors corresponding to (2,3), (3,4), (1,3), (1,2), respectively, one first needs $1 + \alpha' + c_3 \equiv 0$, $\beta' + c_4 \equiv 0$, $c_1 = 0$, $\lambda' + c_2 \equiv 0$ and $\delta' = 1$. Then one needs $\lambda' = 0$ so that $c_2 = 0$, and $c_4 = 0$ so that $\beta' = 0$ and thus $a_4 = 0$, and $c_3 = 1$ so that $\alpha' = 0$.

Checking through then one finds the G of section 3 and

$$\begin{bmatrix} 1\,1\,0\,0 \\ 1\,1\,0\,0 \end{bmatrix}, \begin{bmatrix} 1\,1\,1\,0 \\ 0\,0\,0\,0 \end{bmatrix}, \begin{bmatrix} 0\,0\,0\,0 \\ 0\,0\,1\,0 \end{bmatrix}$$

as the three initial characteristics in section 3.

REFERENCES

[1] Farkas, H. M. and Rauch, Harry E., Two kinds of theta constants and period relations on a Riemann surface, *Proc. Nat. Acad. Sci.* 62 (1969), 679-686.

[2] Garabedian, P., Asymptotic identities among periods of integrals of the first kind, *Amer. J. Math.* 73 (1951), 107-121.

[3] Poincaré, H., Remarques diverses sur les fonctions abéliennes, *J. de Math.*, ser. 5, 1 (1895), 219-314.

[4] Rauch, Harry E., Functional independence of theta constants, *Bull. Am. Math. Soc.* 74 (1968), 633-638.

[4] Rauch, Harry E. and Farkas, H. M., Relations between two kinds of theta constants on a Riemann surface, *Proc. Nat. Acad. Sci.* 59 (1968), 52-55.

[6] Schottky, F., Über die Moduln der Thetafunktionen, *Acta Math.* 27 (1903), 235-288.

[7] Schottky, F. Zur Theorie der Abelschen Funktionen von vier Variabeln, *J. Reine Ang. Math.* 102 (1888), 304-352.

Yeshiva University
New York, New York

TEICHMÜLLER MAPPINGS WHICH KEEP THE BOUNDARY POINTWISE FIXED

Edgar Reich and Kurt Strebel

Given the measurable function $\mu(z)$, $U: \{|z| < 1\}$,

$$\operatorname*{ess\,sup}_U |\mu(z)| < 1$$

let f^μ denote the quasiconformal homeomorphism of U onto U with complex dilatation μ, normalized such that $f^\mu(1) = 1$, $f^\mu(i) = i$, $f^\mu(-1) = -1$. Our purpose is to study conditions on μ for which the boundary correspondence induced by f^μ is the identity. This abstract contains a summary of some of the results. For proofs, further results, applications, and bibliography the reader is referred to [1] and [2].

Let $\mathfrak{F} = \{\mu \mid f^\mu(e^{i\theta}) \equiv e^{i\theta},\ 0 \leq \theta < 2\pi\}$.

THEOREM 1. *If* $\mu \in \mathfrak{F}$ *the following condition is necessary:*

$$\left| \iint_U \frac{\mu(z)\,g(z)}{1-|\mu(z)|^2}\,dx\,dy \right| \leq \iint_U \frac{|\mu(z)|^2\,|g(z)|}{1-|\mu(z)|^2}\,dx\,dy$$

for all functions $g(z)$ *holomorphic in* U *with norm*

$$\|g\| = \iint_U |g(z)|\,dx\,dy < \infty .$$

From now on we assume that $\mu(z)$ is of Teichmüller type:

$$\mu(z) = k\frac{\overline{\phi(z)}}{|\phi(z)|} = k\frac{\overline{\Phi'(z)}}{\Phi'(z)} = k\nu(z) ,$$

where $\Phi(z)$ is single-valued analytic in U, $0 \leq k < 1$. In this case

$$f^{\mu} = \Psi^{-1} \cdot F_k \cdot \Phi \ ,$$

where Ψ is single-valued analytic in U, and F_k is the affine mapping

$$F_k(\zeta) = \zeta + k\bar{\zeta} \ .$$

We shall say that ν satisfies the strong orthogonality condition if $\iint \nu(z)\, g(z) dx\, dy = 0$ for all functions g holomorphic in U except possibly for poles which may be located at the zeroes of $\Phi'(z)$, $\|g\| < \infty$.

THEOREM 2. *Suppose* Φ *is a rational function. Then the following conditions are equivalent.*

(a) $k\nu \in \mathfrak{F}$ *for all* k, $0 \leq k < 1$;

(b) ν *satisfies the strong orthogonality condition;*

(c) Φ *is holomorphic in the extended plane except possibly for poles on* $\{|z| = 1\}$,

$$\Phi^{(j)}(z_0) = 0, \ |z_0| < 1, j = 1, 2, \ldots, m \Longrightarrow$$

$$\Phi\left(\frac{1}{\bar{z}_0}\right) = \Phi(z_0), \ \Phi^{(j)}\left(\frac{1}{\bar{z}_0}\right) = 0, \ j = 1, 2, \ldots, m.$$

The assumption of rationality of Φ is a special case of a more general growth condition in U. Let

$$\tilde{\Psi}(k; z) = \Phi(z) + k[\overline{\Phi(z)} - \Phi'(z)\, p(z)] \ ,$$

where

$$p(z) = \frac{1}{-\pi} \iint_U \nu(\zeta) \frac{d\xi\, d\eta}{\zeta - z} \ ,$$

and put

$$\lambda(z) = \frac{\Phi'(z) - \tilde{\Psi}'(k; z)}{k\Phi'(z)} \ .$$

(λ is meromorphic in U, and completely determined by ν.) If Φ is rational and any one of the conditions (a), (b), (c), of Theorem 2 is satisfied λ turns out to be bounded in U. More generally, for arbitrary Φ, holomorphic in U, we have the following.

THEOREM 3. *Suppose* ν *satisfies the strong orthogonality condition, and* λ *is bounded. Then* $k\nu \in \mathfrak{F}$. *Moreover,* $w = f^{t\nu}(z)$ *is the unique solution of the differential equation*

$$\begin{cases} \dfrac{dw}{dt} = \dfrac{1}{1-t^2} \dfrac{p(w) - t\nu(w)\overline{p(w)}}{1 - t\lambda(w)} \\ w(0) = z \end{cases},$$

and $f^{t\nu}(z) = \Psi^{-1} \cdot F_t \cdot \Phi(z)$, *where* $\Psi \equiv \tilde{\Psi}(t; w)$, $0 \leq t < 1$.

REFERENCES

[1] Edgar Reich and Kurt Strebel, "On quasiconformal mappings which keep the boundary pointwise fixed," *Trans. Am. Math. Soc.*, 138 (1969), 211-222.

[2] ______, "Einige Klassen Teichmüllerscher Abbildungen die die Randpunkte festhalten," *Ann. Acad. Sci. Fenn.*, Ser A. I., 457 (1970), 1-19.

AUTOMORPHISMS AND ISOMETRIES OF TEICHMÜLLER SPACE

H. L. Royden*

1. **The Teichmüller space** T^g. Let W be a compact surface of genus $g \geq 2$, and let W_0 be W with a fixed complex analytic structure. Then every complex analytic structure on W is specified by giving a Beltrami differential μ on W_0. The Riemann surface given by μ is denoted by W_μ. Two structures μ and ν are said to be conformally equivalent if there is a conformal mapping of W_μ onto W_ν, and they are said to be equivalent if this mapping can be taken to be homotopic to the identity. The Teichmüller space T^g of Riemann surfaces of genus g is the space of analytic structures modulo equivalence. The Teichmüller distance between $[W_\mu]$ and $[W_\nu]$ is defined to be ½ log K where K is the infimum of the maximal dilations of mappings $\phi: W_\mu \to W_\nu$ with ϕ homotopic to the identity. It is clearly a metric on Teichmüller space. Equivalently, the distance between W_0 and W_μ is given by

$$\tau(W_0, W_\mu) = \inf \tfrac{1}{2} \log \frac{1 + |\nu|}{1 - |\nu|}$$

as ν ranges over all structures equivalent to μ. According to the Ahlfors-Bers-Teichmüller theory ([2], [3]), this has a minimum which is given by a ν of the form $k\bar{\eta} / |\eta|$ for some analytic quadratic differential η on W_0, and ν's of this form are the unique minima in their equivalence classes.

The complex linear structure on the space of Beltrami differentials induces a complex analytic structure on T^g which makes T^g a $3g-3$ -

* This work was supported in part by the U.S. Army Research Office (Durham) and in part by the National Science Foundation.

dimensional complex manifold homeomorphic to a ball in R^{6g-6}. The cotangent space at a point $[W_\mu]$ of T^g is given by the space Q_μ of analytic quadratic differentials on W_μ. If μ is small, the Teichmüller distance from $[W_0]$ to $[W_\mu]$ is

$$(1) \qquad \sup_{\|\eta\|=1} \int_{W_0} \mu\eta + 0(|\mu|^2)$$

where $\eta \epsilon Q_0$ and $\|\eta\| = \int_{W_0} |\eta|$.

2. **Metrics and cometrics on** T^g. If $<x,\eta>$ with $x \epsilon T^g$, $\eta \epsilon Q_x$ is a point in the cotangent bundle of Teichmüller space, we define $G(x, \eta)$ by

$$(2) \qquad G(x, \eta) = \|\eta\| = \int_x |\eta| .$$

Then G is a function on the cotangent bundle, and we shall show that it is of class C^1 on the cotangent bundle. For each x, the function G is a homogeneous strictly convex function of η.

On the tangent bundle of T^g, we define a function F by setting

$$(3) \qquad F(x, \xi) = \sup_{\|\eta\|=1} [\xi, \eta] .$$

Thus for each x, the function F is the dual norm to that given by G and is a positive, homogeneous, convex function of ξ. The fact that G is C^1 and strictly convex implies that F is C^1 and strictly convex.

Thus F induces a Finsler metric on T^g. It follows from (1), (3) and the continuity of F that the Finsler metric induced by F is greater than or equal to the Teichmüller metric. If μ is a Beltrami differential of the form $\bar{\eta}/|\eta|$, then the mapping of $[0,k]$ into T^g given by $t \to t_\mu$ is an arc of length $\frac{1}{2}\log(1+k)/(1-k)$ joining W_0 to $W_{k\mu}$. Hence this arc is a geodesic, and since each element of Teichmüller space is given by a Beltrami differential of the form $k\bar{\eta}/|\eta|$, we see that the Finsler metric induced by F agrees with the Teichmüller metric.

The following lemmas give information about the smoothness of the functions F and G. They assert that F and G are of class C^1, that G is not of class C^2, and that in the cotangent space at a point, the Hölder continuity of the first derivative of G along rays from a quadratic differential η_0 depends on the order of the largest zero of η_0. I do not know whether F is of class C^2 or not, although I can show that for a fixed x, $F(x, \xi)$ is of class C^2 in ξ. It cannot be C^3 on an open set of the tangent space, however.

LEMMA 1. *Let η_0 be an analytic quadratic differential on W whose largest zero is of order* n, *and let η_1 be a smooth (not necessarily analytic) quadratic differential. Then the function* f *defined by*

$$f(t) = \int_W |\eta_0 + t\eta_1|$$

is differentiable as a function of t, *and its derivative at* $t = 0$ *is*

$$f'(0) = \int_W \operatorname{Re} \frac{\eta_1 \overline{\eta_0}}{|\eta_0|} .$$

If $n = 1$, f *has a second derivative at* $t = 0$ *given by*

$$f''(0) = \int_W \frac{1}{|\eta_0|^3} \cdot (\operatorname{Im} \eta_1 \overline{\eta_0})^2 .$$

For $n = 2$,

$$f(t) = f(0) + tf'(0) + c|t|^2 \log \frac{1}{|t|} + o(|t|^2 \log \frac{1}{|t|}) ,$$

and for $n > 2$,

$$f(t) = f(0) + tf'(0) + c|t|^{1+2/n} + o(|t|^{1+2/n})$$

where $c > 0$ *if* $\eta_1 \neq 0$ *at one of the largest zeros of* η_0.

Proof: If K is any compact set on which $|\eta_0| > 0$, then $\int_K |\eta_0 + t\eta_1|$ is of class C^∞ in t, and its first and second derivatives at $t = 0$ are the integrals over K of $\operatorname{Re} \eta_1 \overline{\eta}_0 / |\eta_0|$ and of $(\operatorname{Im} \eta_1 \overline{\eta}_0)^2 / |\eta_0|^3$. Thus any non-smoothness of f comes from the contribution of integrals over neighborhoods of the zeros of η_0.

At a zero of η_0 of order ν, we may choose a uniformizor z so that so that $\eta = z^\nu dz^2$, and we let $M = \sup_\Delta |\phi(z)|$, and suppose $|\phi(z) - \phi(0)| < \epsilon$ in Δ.

For complex numbers α and β, $\alpha \neq 0$, we have

$$\frac{1}{2(|\alpha| + |\beta|)} \frac{(\operatorname{Im} \alpha\overline{\beta})^2}{|\alpha|^2} \leq |\alpha + \beta| - |\alpha| - \operatorname{Re} \frac{\alpha\overline{\beta}}{|\alpha|} ,$$

and if $|\alpha| > |\beta|$, then

$$|\alpha + \beta| - |\alpha| - \operatorname{Re} \frac{\alpha\overline{\beta}}{|\alpha|} \leq \frac{1}{2(|\alpha| - |\beta|)} \left(\operatorname{Im} \frac{\alpha\overline{\beta}}{|\alpha|} \right)^2 .$$

We also have for all α and β.

$$|\alpha + \beta| - |\alpha| - \operatorname{Re} \frac{\alpha\overline{\beta}}{|\alpha|} \leq 2|\beta| .$$

Let

$$I(t) = \int_\Delta \left(|z^\nu + t\phi(z)| - |z^\nu| - t \operatorname{Re} \frac{z^\nu \overline{\phi}}{|z|^\nu} \right) dx\, dy .$$

For $\nu = 1$ and any $\delta < 1$, let I_0 be this integral over the disk $|z| \leq |t| M/\delta$ and I_1 the integral over $|t| M/\delta < |z| < a$, where we assume $|t| < \delta a/M$. Then I_0 is $O(t^3)$, and

$$I_1 \leq \frac{t^2}{2(1-\delta)} \int \frac{1}{|z|} \left(\operatorname{Im} \frac{z\overline{\phi}}{|z|} \right)^2 dx\, dy .$$

The integrand on the right is dominated by $M^2/|z|$, which is integrable over Δ. By the Lebesgue convergence theorem,

$$\lim t^{-2} I(t) = \int_\Delta \frac{1}{|z|}\left(\operatorname{Im} \frac{z\bar{\phi}}{|z|}\right)^2 dx\,dy \;,$$

and the contribution to f from Δ is of class C^2.

For $\nu = 2$ and any δ with $0 < \delta < 1$, let I_0 be that part of I taken over the disk $|z| \le \rho = (|t|M/\delta)^{1/2}$ and I_1 the integral over $\rho < |z| < a$. Then I_0 is $0(t^2)$, and

$$\begin{aligned} I_1 &\le \frac{t^2}{2(1-\delta)} \int_\Delta \frac{1}{|z|^2}\left(\operatorname{Im} \frac{z^2\bar{\phi}}{|z|^2}\right)^2 dx\,dy \\ &\le \frac{t^2}{2(1-\delta)} \int \left(\frac{1}{|z|^2} \left[\left(\operatorname{Im} \frac{z^2\overline{\phi(0)}}{|z|^2}\right)^2 \right] + 2\,\epsilon\, M\right) dx\,dy \\ &\le \frac{\pi t^2}{2(1-\delta)} \left[\left(|\phi(0)|^2 + 4\,\epsilon\, M\right) \log \frac{1}{|t|} + 0(1)\right] \end{aligned}$$

Thus

$$\overline{\lim}\; \frac{I(t)}{|t|^2 \log \frac{1}{|t|}} \le \frac{\pi}{2(1-\delta)} \left(|\phi(0)|^2 + 4\,\epsilon\, M\right) .$$

Similarly,

$$\underline{\lim}\; \frac{I(t)}{t^2 \log \frac{1}{|t|}} \ge \frac{\pi}{2(1+\delta)} \left(|\phi(0)|^2 - 4\,\epsilon\, M\right) .$$

Since δ is arbitrary and ε can be made small by taking a small, we see that the contribution to f from a double zero of η_0 is a term of the form

$$ct^2 \log \frac{1}{|t|} + o\left(t^2 \log \frac{1}{|t|}\right)$$

with $c \ge 0$ and positive if $\phi(0) \ne 0$.

For $\nu > 2$, let $z = t^{1/\nu}\zeta$. Then

$$I(t) = |t|^{1+\frac{2}{\nu}} \int_{|\zeta| < \rho} \left(|\zeta^\nu + \phi(t^{\frac{1}{\nu}}\zeta)| - |\zeta^\nu| - \operatorname{Re} \frac{\bar{\zeta}^\nu \phi(t^{\frac{1}{\nu}}\zeta)}{|\zeta^\nu|}\right) dA$$

where dA is the element of area in the ζ plane, and $\rho = a|t|^{-\frac{1}{\nu}}$. The integrand is dominated by 2M for all ζ, and by

$$\frac{M^2}{2(1-\delta)} |\zeta|^{-\nu}$$

for $|\zeta|^\nu > M\delta^{-1}$. The Lebesgue convergence theorem implies that

$$\lim |t|^{-(1+\frac{2}{\nu})} I(t) = \int_{|\zeta|<\infty} \left(|\zeta^\nu + \phi(0)| - |\zeta|^\nu - \operatorname{Re} \frac{\zeta^\nu \phi(0)}{|\zeta|^\nu} \right) dA$$

$$= |\phi(0)|^{1+\frac{2}{\nu}} \int \left(|\zeta^\nu + 1| - |\zeta|^\nu - \operatorname{Re} \frac{\zeta^\nu}{|\zeta|^\nu} \right) dA \quad .$$

This last integral is positive since its integrand is. Thus such a zero of η_0 gives a contribution to f of the form

$$c|t|^{1+\frac{2}{\nu}}$$

with $c \geq 0$, and c is positive if $\phi(0) \neq 0$. This proves the lemma.

LEMMA 2. *The function* G *is of class* C^1 *on the cotangent bundle (except on the zero section* $\eta = 0$*).*

Proof: Given a point $x_0 \in T^g$, we take our base surface W_0 to be in the equivalence class x_0. For some suitable neighborhood U of x_0 in T^g, we can represent the part of the tangent bundle over U as a product $U \times C^{3g-3}$ as follows: Let w_0 be an Abelian differential (of the first kind) on W_0 with simple zeros, and define w_0^μ to be the Abelian differential on W_μ with the same A-periods as w_0. For a suitably small U, the differentials w_0^μ will have simple zeros lying in prescribed non-overlapping neighborhoods of the zeros of w_0. Thus we can associate each of the $2g-2$ zeros of w_0^μ with one of the zeros of w_0 so that each zero depends continuously on μ. For $1 \leq i \leq g$, let w_i^μ be the normalized Abelian differential whose A_k period is δ_{ik}, and for $g+1 \leq i \leq 3g-3$, let γ_i^μ be the Abelian differential of the third kind which has a simple pole of residue 1 at the $(i-g)$-th zero of w_0^μ, a simple pole of residue -1 at the $(g-2)$-nd zero of w_0^μ, is regular elsewhere and has zero periods around a prescribed set of A-cycles. Map $U \times C^{3g-3}$ into the cotangent bundle over U by

$$\eta(\mu,t) = \left(\sum_{i=1}^{g} t_i w_i^\mu + \sum_{g+1}^{3g-3} t_j \gamma_j^\mu \right) w_0^\mu \quad .$$

This mapping depends smoothly on μ and depends only on the equivalence class of μ, and for a fixed μ, it is an isomorphism of C^{3g-3} onto the cotangent space at $[\mu]$. Thus this mapping is a diffeomorphism of $U \times C^{3g-3}$ onto the part of the cotangent bundle over U.

The quadratic differential $\eta(\mu,t)$ can be expressed locally by

$$\eta(\mu,t) = \phi(\mu,t)(dz + \mu\, d\bar{z})^2 \quad ,$$

with ϕ depending differentiably on μ and t. Then

$$|\eta| = |\phi(\mu,t)|\,(1-|\mu|^2)\,|dz|^2 \quad ,$$

and Lemma 1 implies that $\|\eta\|$ is of class C^1 in μ and t.

LEMMA 3. *The function* F *is of class* C^1 *on the tangent bundle (except on the zero section* $\xi = 0$*).*

This lemma is an immediate consequence of the following standard lemma from the calculus of variations:

LEMMA. *Let* U *be an open subset of* R^m *and* $G(x, \eta)$ *a continuous function on* $U \times R^n$, *which for each* x *is a positive convex homogeneous function of* η. *Define* $F(x, \xi)$ *on* $U \times R^n$ *by*

$$F(x,\xi) = \sup_{G(\eta,x)=1} (\xi,\eta) \quad .$$

Then F *is continuous and, for each* x, F *is a positive convex homogeneous function of* ξ. *If* G *is of class* C^1 *in* η *for* $\eta \neq 0$, *then* F *is strictly convex. If* G *is strictly convex and has continuous first derivatives with respect to* x, *then* F *is* C^1 *(except at* $\xi = 0$*).*

3. The bicanonical and dual bicanonical maps. The space of non-zero quadratic differentials on a Riemann surface W of genus g becomes a (3g − 4)-dimensional complex projective space P if we identify quadratic differentials which are multiples of one another. Let P^* be the (3g − 4)-dimensional projective space of hyperplanes in P. The bicanonical map $\phi: W \to P^*$ is the map that takes each point p of W into the hyperplane $\phi(p)$ of quadratic differentials which vanish at p. If $\eta_1, \ldots, \eta_{3g-3}$ is a basis for Q_W, then homogeneous coordinates $t_1, \ldots, t_{3g-3}$ of an $[\eta]$ in P are given when $\eta = \Sigma\, t_i \eta_i$. The image of $p \in W$ under the bicanonical map is the hyperplane whose equation is $\Sigma\, t_i \eta_i(p) = 0$. Thus if $\eta_i = \phi_i\, dz^2$ in terms of a local uniformizer near P_0, the map $p \to \langle \phi_i(p) \rangle$ gives the bicanonical map in terms of homogeneous coordinates in P^*. If $g \geq 3$, the bicanonical map is a one-to-one regular analytic map of W onto a curve in P^*, while for $g = 2$, it is two-to-one and regular except at the Weierstrass points of W, where it is branched. The image of a Riemann surface under its bicanonical map is called its bicanonical curve. Two Riemann surfaces of genus greater than 2 are conformally equivalent if and only if there is a linear map between their bicanonical curves.

If p is a point on a Riemann surface W of genus $g \geq 2$, then there is always a non-zero quadratic differential η on W which has a zero at p of order at least 3g − 4, and if $g > 2$, such a differential can have order at most $g < 3g - 4$ at any other point. The mapping ψ which assigns to each $p \in W$ that quadratic differential which has a zero of the highest possible order at p is a mapping of W into P which is one-to-one if $g > 2$, while for $g = 2$, it is two-to-one, identifying those points which are identified by the hyperelliptic self-mapping of W. The map ψ results from following the bicanonical map by the map which sends each point of the bicanonical curve into the osculating hyperplane at that point. It is an analytic map of W into P, and is regular at $p \in W$ if and only if there is no quadratic differential with a zero at p of order greater than 3g − 4; i.e., iff p is not a quadratic Weierstrass point. Let us call ψ the dual bicanonical map and the image of W under ψ the dual bicanonical curve. Then every conformal map between two Riemann surfaces of genus $g > 2$ induces and is induced by a linear map between their bicanonical curves. We shall need the following lemma:

LEMMA 4. *The dual bicanonical curve is not contained in any hyperplane of* P.

Proof: Let L be the linear subspace of Q corresponding to the hyperplane. Since $\dim L = 3g-4$, there are only a finite number of points of W at which there is an $\eta \in L$ with a zero of order at least $3g-4$. Thus the dual bicanonical curve is not contained in L.

Let $S = S(W)$ be the subset of P (considered as a quotient of the quadratic differentials on W) consisting of all quadratic differentials which have a zero of order at least $3g-4$. Then S contains the dual bicanonical curve together with linear subspaces of quadratic differentials for those quadratic Weierstrass points where there is more than one quadratic differential with zeros of order at least $3g-4$. Thus S is an analytic variety which is the union of the dual bicanonical curve and a number of linear manifolds. For $g \geq 3$, let θ be the map of S onto W which takes each quadratic differential in S onto the unique point where it has a zero of order at least $3g-4$, then θ is an analytic map, and $\theta \circ \psi$ is the identity on W. If p is a point on W, then $\theta^{-1}[p]$ is a point or linear subspace of P.

The set S, considered as a subset of the space Q of quadratic differentials on W, can be characterized in terms of the norm on Q by Lemma 1. We state this characterization in the form of a lemma.

LEMMA 5. *The set* S *is the set of those quadratic differentials* η *for which there is a quadratic differential* η_1 *so that the function* $f(t) = \|\eta + t\eta_1\|$ *is such that* $f(t) - tf'(0)$ *is not* $0(|t|^a)$ *for any* $a > (3g-2)/(3g-4)$, *if* $g \geq 3$, *and not* $0(t^2)$ *for* $g = 2$.

4. **Local isometries of T^g.** In this section, we use the smoothness properties of the norm G on the space of quadratic differentials to characterize local isometries of T^g and its cotangent space. We begin with the following theorem:

THEOREM 1. *Let* W *and* V *be Riemann surfaces of genus* $g \geq 2$, *and let* $\Phi: Q_W \to Q_V$ *be a complex linear isometry between the spaces of quadratic differentials on them. Then there is a conformal map* $\phi: V \to W$ *and a complex constant* a *with* $|a| = 1$ *such that* $\Phi(\eta) = a\eta \circ \phi$.

Proof: We first consider the case $g \geq 3$. Let S(V) and S(W) be the sets of those quadratic differentials on V and W which have a zero of order at least $3g-4$. Lemma 5 implies that Φ^{-1} maps S(V) onto S(W). Since S(V) and S(W) are homogeneous, we may consider them to be subspaces of the projective spaces P_V and P_W obtained by identifying quadratic differentials which are non-zero multiples of each other. Since Φ is homogeneous, it induces an isomorphism $\tilde{\Phi}$ of P_W onto P_V. Let $\psi: V \to S(V)$ be the dual bicanonical map taking each point of V into the multiples of the quadratic differential which has a zero of the highest order at the point, and let $\theta: S(W) \to W$ be the map taking a quadratic differential into the point where it has a zero of order at least $3g-4$. Set $\phi = \theta \circ (\tilde{\Phi})^{-1} \circ \psi$. Then ϕ is an analytic map of V into W. Since the inverse image of a point under θ is contained in a hyperplane, while by Lemma 4 we cannot have $\psi[V]$ contained in a hyperplane, the mapping ϕ cannot be constant. But a nonconstant analytic map of a compact Riemann surface of genus $g > 1$ into a Riemann surface of the same genus must be one-to-one and onto.

Define $\phi^*: Q_W \to Q_V$ by $\phi^*(\eta) = \eta \circ \phi$. Then ϕ^* is a linear isomorphism, and induces a linear map $\tilde{\phi}^*$ of P_W onto P_V. But $\tilde{\phi}^* = \tilde{\Phi}$, and hence $\Phi = a\phi^*$ for some complex constant a. Since Φ and ϕ^* are isometries, $|a| = 1$.

If $g = 2$, W and V are hyperelliptic surfaces, and each is obtained as a two-sheeted covering surface of the sphere branched over six points. The holomorphic quadratic differentials on W are the lifting to W of those meromorphic quadratic differentials on the sphere which may have simple poles at the branch points of W. Since Φ takes quadratic differentials with double zeros into quadratic differentials by Lemma 5, a simple calculation shows that Φ is induced by a conformal map of the sphere onto itself which takes the branch points of V into those of W. Thus V and W are conformally equivalent, and the mapping of the sphere can be lifted into a map of V onto W, proving the theorem for this case.

Since a partial isometry ϕ of a region in T^g into T^g must induce an isometry of the cotangent space at x with the cotangent space at $\phi(x)$, we see that x and $\phi(x)$ must be conformally equivalent. Actually, we can say a little more. Each orientation-preserving homeomorphism ψ of W onto itself induces a map ψ^* of the space of Beltrami differentials on W_0 onto itself, and since it preserves equivalence, it induces a conformal isometry of T^g onto itself. The isometry of T^g depends only on the homotopy class of ψ. We define the Teichmüller modular group Γ to be the group of orientation-preserving homeomorphisms of W modulo those homotopic to the identity. Thus Γ acts as a group of complex analytic isometries on T^g. If we have a partial isometry ϕ in T^g, then the induced conformal maps of x onto $\phi(x)$ depend continuously on x and hence are in the same homotopy class, giving us the following theorem:

THEOREM 2. *Let ϕ be a complex analytic mapping of a domain in T^g into T^g which is isometric in the Teichmüller metric. Then there is an element γ of the Teichmüller modular group such that $\phi(x) = \gamma x$ for all x.*

5. **Invariance of the Teichmüller metric.** The preceding theorem acquires added significance when coupled with the following theorem stating the invariance of the Teichmüller metric under biholomorphic maps.

THEOREM 3. *The Teichmüller metric τ is invariant under biholomorphic maps of T^g into itself. If Δ is the unit disc and ρ the Poincaré non-*

Euclidean distance on Δ, *then*

$$\tau(x, y) = \inf \rho(\phi^{-1}(x), \phi^{-1}(y))$$

for all holomorphic maps $\phi : \Delta \to T^g$. *Thus* τ *is the Kobayashi metric on* T^g *(see* [4]*)*.

Proof: Let x be the base structure W_0, and let the extremal Beltrami differential for y be $\mu = k\bar{\eta}/|\eta|$. Then the mapping ϕ given by $\zeta \to \zeta\mu$ is a one-to-one holomorphic map of Δ into T^g, and the Teichmüller metric pulled back to Δ is just the Poincaré metric for Δ; i.e., ϕ is an isometric imbedding of Δ into T^g. This shows that τ is greater than or equal to the infimum in the theorem.

To establish the inequality in the opposite direction, it suffices to show that if ψ is any holomorphic map of Δ into T^g, then the Riemann metric on Δ induced by the Finsler metric on T^g is less than or equal to the differential form of the Poincaré metric. This follows from the following two lemmas. The first is Ahlfors' version of the Schwarz-Pick lemma [1]. If $\lambda|dz|$ is a conformal metric on the disk, we define its curvature to be $-(\Delta \log \lambda)/\lambda^2$. A metric $\lambda'|dz|$ is said to be a supporting metric at p for $\lambda|dz|$ if $\lambda'(p) = \lambda(p)$ and $\lambda' \le \lambda$ in a neighborhood of p. If λ and λ' are both of class C^2, then the curvature at p of λ is less than or equal to that of λ'.

LEMMA (Ahlfors). *Let* $\lambda|dz|$ *be a conformal metric in the unit disk* Δ *such that at each* $p \in \Delta$, *there is a smooth supporting metric for* λ *with curvature at* p *which is at most* -4. *Then*

$$\lambda(z) \le \frac{1}{1-|z|^2} \quad ;$$

i.e., $\lambda|dz|$ *is at most as large as the Poincaré metric for the disk.*

The following lemma is a generalization to complex Finsler space of the well-known result that the curvature at a point p of an analytically embedded disk in a Kähler manifold is less than or equal to the sectional curvature for the tangent plane to the disk at p.

LEMMA 6. *Let* T *be a complex Finsler space whose metric form* F *is of class* C^1, *and let* ϕ *and* ψ *be holomorphic maps of the disks* Δ *and* Δ' *into* T *with* $\phi(0) = \psi(0)$ *and* $\phi'(0) = \psi'(0) \neq 0$. *Suppose that* $\phi[\Delta]$ *is totally geodesic and that the conformal metric* λ *on* Δ *induced by* F *on* $\phi[\Delta]$ *is of class* C^2 *at* 0. *Then the conformal metric* λ' *on* Δ' *induced by* F *on* $\psi[\Delta]$ *has a supporting metric at* 0 *whose curvature is equal to that of* λ *at* 0.

Proof: Let us take a holomorphic coordinate ζ at the origin in Δ so that λ has the form

$$\lambda = 1 - \kappa|\zeta|^2 + 0(|\zeta|^2) \ .$$

This implies that the lines $\arg\zeta = \text{const.}$ are geodesics for the metric $\lambda|dz|$. Choose holomorphic coordinates $x_1, \ldots, x_n$ near $\phi(0)$ so that the components ϕ_i of ϕ are given by $\phi_1(\zeta) = \zeta$, $\phi_i(\zeta) = 0$, $2 \leq i \leq n$. Since the x_1-coordinate is distinguished, we adopt the convention that i and j range from 2 to n, and write the Finsler from F as

$$F(x, \xi) = F(x_1, x_i; \xi_1, \xi_i) \ .$$

Since F is of class C^1, we have

$$F(x_1, x_i; 1, \xi_i) = F(x_1, 0; 1, \xi_i) + \operatorname{Re}\sum_i C_i(x_1, \xi_i)x_i + o(|x_i|)$$

and

$$F(x_1, 0; 1, \xi_i) = F(x_1, 0; 1, 0) + \operatorname{Re}\sum B_i(x_1)\xi_i + o(|\xi_i|)$$

where B_i and C_i are continuous functions of their arguments. The convexity of F in ξ implies that

$$F(x_1, 0; 1, \xi_i) \geq F(x_1, 0; 1, 0) + \sum \operatorname{Re} B_i(x_1)\xi_i \ .$$

Since the curves $x_1 = \alpha t$, $x_i = 0$ are geodesics, we must have

$$B_i(x_1) = B_i(0) + x_1 C_i(0, 0) + o(|x_1|) \ .$$

By a linear change of the coordinates,

$$x_1 = y_1 - \Sigma\, B_i(0) y_i \quad ,$$
$$x_i = y_i \quad ,$$

we may take $B_i(0) = 0$. By a quadratic change

$$x_1 = y_1 - \Sigma\, C_i(0, 0) y_1 y_i \quad ,$$
$$x_i = y_i \quad ,$$

we may take $C_i(0, 0) = 0$. Then

$$F(x_1, x_i; 1, \xi_i) \geq F(x_1, 0; 1, 0) + \operatorname{Re} \Sigma\, (\varepsilon_i x_i + \varepsilon_i' x_1 \xi_i) + o(|x_i|)$$

where ε_i, $\varepsilon_i' \to 0$ as x_1 and $\xi_i \to 0$. Note that $\lambda(\zeta) = F(\zeta, 0; 1, 0)$.

We may choose our coordinate ζ at 0 in Δ' so that ψ has the form $\psi_1(\zeta) = \zeta$, $\psi_i(\zeta) = a_i \zeta^2 + 0(\zeta^3)$. Then the conformal metric $\lambda' |d\zeta|$ induced by F on Δ' is given by

$$\begin{aligned} \lambda'(\zeta) &= F(\zeta, \psi_i(\zeta); 1, \psi_i'(\zeta)) \geq F(\zeta, 0; 1, 0) \\ &\quad + \operatorname{Re} \Sigma \left(\varepsilon_i a_i \zeta^2 + 2\varepsilon_i' a_i \zeta^2\right) + o(|\zeta|^2) \\ &= F(\zeta, 0; 1, 0) + o(\zeta^2) = \lambda(\zeta) + o(\zeta^2) \ . \end{aligned}$$

Thus λ is a supporting metric except for a term $o(\zeta^2)$, and the proposition is proved since such a term does not affect the curvature at 0.

As a result of Theorems 2 and 3, we have the following theorem:

THEOREM 4. *Let ϕ be a biholomorphic map of T^g into itself. Then x and $\phi(x)$ are conformally equivalent, and ϕ is the action on T^g of an element of the Teichmüller modular group Γ. Thus for $g \geq 3$, the action of Γ on T^g is an isomorphism of Γ with the group of biholomorphic automorphisms of T^g.*

A complex space T is said to be symmetric at x if there is a complex analytic involution ϕ of T onto itself with $\phi(x) = x$ and which induces the negative of the identity on the tangent space at x (and hence on the cotangent space at x). If x is fixed under ϕ, then the $\gamma \in \Gamma$ which induces ϕ must contain a conformal map of x onto itself. But there is no conformal map of any surface of genus greater than one which induces the negative identity on Q_x. Thus there is no point about which T^g is symmetric. This gives us the following corollary:

COROLLARY. *For* $g \geq 2$, *the Teichmüller space* T^g *is not homogeneous, and there is no point about which it is symmetric.*

BIBLIOGRAPHY

[1] Ahlfors, L. V., "An extension of Schwarz's lemma," *Trans. Amer. Math. Soc.*, 43 (1938), 359-364.

[2] Ahlfors, L. V., "On quasiconformal mappings," *J. d'Analyse Math.* 3 (1953), 1-58.

[3] Bers, L., "Quasiconformal mappings and Teichmüller's theorem," *Analytic Functions*, Princeton Univ. Press, Princeton, N.J., 1960, 89-119.

[4] Kobayashi, S.. "Invariant distances on complex manifolds and holomorphic mappings," *J. Math. Soc. Japan* 19 (1967), 460-480.

DEFORMATIONS OF EMBEDDED RIEMANN SURFACES*

Reto A. Rüedy

§1. *Embedded Surfaces*

It was Felix Klein who realized that each surface embedded in 3-space which is sufficiently smooth (we always assume C^∞) can be viewed as a Riemann surface in a natural way, namely by declaring a parameter admissible if and only if the corresponding mapping is conformal. The existence of such parameters is a deep but well-known fact (see [2], pp. 15-41). We will call Riemann surfaces of this kind *classical* Riemann surfaces.

Polyhedrons embedded in 3-space can be viewed as Riemann surfaces in the same way, but here the existence of admissible parameters is almost trivial: We triangulate the polyhedron and map two adjacent triangles by sense-preserving congruence mappings into the plane, such that the mappings agree on the common edge. This defines admissible parameters for all points except the vertices. If the sum of the angles at a vertex is $2\pi a$, then suitable congruence mappings of the corresponding triangles into the plane, followed by the mapping

$$z \longmapsto z^{1/a}$$

give us a topological mapping of a neighborhood of this vertex onto a neighborhood of $z = 0$. These local coordinate systems are clearly a basis for a conformal structure, the natural structure.

---

* This research was supported by the Air Force Office of Scientific Research under Contract AF 49 (638)-1591.

§2. *Deformations in the direction of the normals*

If $X: S_0 \to \mathbf{R}^3$ is an embedding function of the classical surface S_0, we call the surface described by the function

$$X + hN: S_0 \to \mathbf{R}^3$$

an ε*-deformation* of S_0, if N is the positive unit normal vector at each point of S_0 and

$$h: S_0 \mapsto (-\varepsilon, \varepsilon)$$

a differentiable (C^∞) real-valued function. We call the corresponding classical surface S_h and we denote by π_h the mapping $X \to X + hN$.

Deformations of a polyhedron S_0 are defined analogously, except that h shall be a piecewise linear map on S_0 which is zero in a neighborhood of the edges.

§3. *Embedding theorems*

Let S_0 be a fixed Riemann surface. A pair (R, f) is called a topologically marked Riemann surface, if f is a topological, orientation-preserving mapping of S_0 onto R. Two marked surfaces (R, f) and (R*, f*) are conformally equivalent if $f^* \circ f^{-1}$ is homotopic to a conformal mapping. The equivalence classes are the elements of Teichmüller space.

THEOREM A. *Let* S_0 *be any Riemann surface obtained from a compact classical surface by deleting a finite number of points and let* ε *be any positive number. Then every topologically marked Riemann surface* $(R, f: S_0 \to R)$ *is conformally equivalent to a suitable* ε*-deformation* (S_h, π_h) of S_0.

Remarks. 1. Note that R and S_0 may be quasiconformally inequivalent.
2. If ε is sufficiently small, all the S_h are classical surfaces without self-intersections.
3. In the proof of Theorem A we only make use of the fact that S_0 is

immersed rather than embedded in 3-space. Thus Theorem A remains valid even if S_0 is merely immersed in 3-space, except that then, of course, Remark 2 is no longer applicable.

4. As a corollary of the theorem and its proof we can obtain the following result:

THEOREM B. *Let* S_0 *be a compact polyhedron without self-intersections and* ε *a positive number. Then every topologically marked Riemann surface* $(R, f: S_0 \to R)$ *is conformally equivalent to a suitable* ε*-deformation* (S_h, π_h) of S_0, *such that* S_h *is a polyhedron without self-intersections.*

§4. *Proof of Theorem A*

In the special case where S_0 is a compact classical surface, the proof is given in [5]. The main ideas will be sketched in the following sections.

Another special case was solved by H. Huber in [4]. Applying a suitable Möbius transformation, his result can be stated in the following way:

There are ε*-deformations of a punctured sphere which are discs (up to conformal equivalence).*

By a standard method, the two results can be combined to obtain a proof of Theorem A. (See [5], pp. 438-441.)

The remaining sections give a proof for Theorem B. Parts which can be found in [5] are only sketched.

§5. *The continuity argument*

By explicit constructions we can only show, that the set of moduli corresponding to ε-deformations of S_0 is dense in Teichmüller space. To obtain the whole result, we have to combine these constructions with the following theorem (Brouwer, see [5], p. 423):

If the continuous function

$$f: \bar{B}(p_0, r) = \{p \in \mathbf{R}^n \mid |p - p_0| \le r\} \to \mathbf{R}^n$$

has the property

$$|f(p) - p| < r \quad \textit{for all} \quad p \in \bar{B},$$

then $p_0 \in f(\bar{B})$.

We can apply this theorem, because the Teichmüller space T_g of compact Riemann surfaces of genus g is topologically a finite-dimensional Euclidean space (the space of moduli); we denote the topological correspondence by mod. (See [1].)

Therefore Theorem B is proved, if we can find to each marked surface (R, f) a ball $\bar{B}(\mathrm{mod}\,(R, f), r)$ and a family of deformations of S_0 depending on a parameter $p \in \bar{B}$ so that

(A) $\quad |\mathrm{mod}\,(S_{h(\cdot\,,\,p)}, \pi_{h(\cdot\,,\,p)}) - p| < r \quad$ for all $p \in \bar{B}$,

and

(B) $\quad p \longmapsto \mathrm{mod}\,(S_{h(\cdot\,,\,p)}, \pi_{h(\cdot\,,\,p)})$ is continuous in $\bar{B}$,

where h is now a function from $S_0 \times \bar{B}$ into $(-\varepsilon, \varepsilon)$.

§6. *A direct approach*

In order to obtain a deformation S_h with $\mathrm{mod}\,(S_h, \pi_h) = p_0$, we define on the manifold S_0 a new conformal structure, such that the corresponding Riemann surface marked by the identity has modulus p_0. If z is a uniformizing parameter for S_0, the most convenient way to do this is to take the structure induced by the metric $|dz + \mu(z) d\bar{z}|^2$, where $\mu(z)$ is the uniquely determined Teichmüller differential on S_0 corresponding to p_0. We denote this new Riemann surface by $S_0(\mu)$, so that we have

$$\mathrm{mod}\,(S_0(\mu), \mathrm{id}) = p_0 \quad .$$

(See [1].)

Now we try to determine h in such a way that

$$\pi_h : S_0(\mu) \to S_h$$

is a conformal mapping. This leads to the differential equation

$$|d(X(z)+h(z)N(z))| = \lambda(z)\,|dz+\mu(z)d\bar{z}| \quad ,$$

(where λ has to be a function on the universal covering surface of S_0), because the dilation of π_h is obviously given by

$$D\pi_h = \sup\left(\frac{|d(X+hN)|}{|dz+\mu(z)d\bar{z}|}\right) \Big/ \inf\left(\frac{|d(X+hN)|}{|dz+\mu(z)d\bar{z}|}\right) \quad ,$$

where sup and inf are taken over all directions. But there will be no suitable solutions (h, λ), for h will certainly not be small. Therefore we have to use the procedure described in §5, which allows us to replace the differential equation by differential inequalities (see §7, (A′) and (B′)).

§7. *The deformation lemma* (Teichmüller, Garsia)

To obtain conditions for h which imply (A) and (B) in §5, we need the following lemma (see [3] p. 100, [5] pp. 422 and 432):

If (S_1, f_1) is contained in a compact subset $\bar{B}$ of T_g, then we have uniform estimates for

$$|\mathrm{mod}(S_1, f_1) - \mathrm{mod}(S_2, f_2)| \quad ,$$

if we know a K_0-quasiconformal mapping from (S_1, f_1) onto (S_2, f_2) with dilation $\leq 1+\delta$ except on a part of S_1 of areal measure $\leq \eta$. These estimates tend to zero, if $\bar{B}$ and K_0 are fixed and (δ, η) tends to $(0,0)$.

As a consequence of this lemma, the conditions (A) and (B) follow from

(A′) $$\frac{1}{K(z)} \cdot \lambda(z)\,|dz+\mu(z)d\bar{z}| \leq |d(X+hN)| \leq K(z)\cdot\lambda(z)\,|dz+\mu(z)d\bar{z}|$$

a.e., where $1 \leq K(z) \leq K_0$ and $K(z) \leq 1+\delta$ outside a set of measure $\leq \eta$, and

(B′) $|d(X(z)+h(z,p)N(z))| \longmapsto |d(X(z)+h(z,p_0)N(z))|$ uniformly on $(S_0 - M)\times\bar{B}$ if $p \to p_0$, where M is a set of areal measure zero.

§8. *The actual construction*

For the construction of differentiable functions h satisfying (A′) and (B′) and which are zero in a neighborhood of all edges, we refer to [5] pp. 425/6 and 433-437. The main ideas are sketched in the appendix.

In order to obtain piecewise linear functions on S_0 with the same properties, we approximate the above functions and apply again the deformation lemma, in order to check that (A′) and (B′) are still satisfied. The approximation may be constructed in the following way:

We triangulate S_0 and call the corresponding complex P_1. We construct P_2 by subdividing each side of P_1 according to the figure, P_3 by subdividing all sides of P_2 etc.

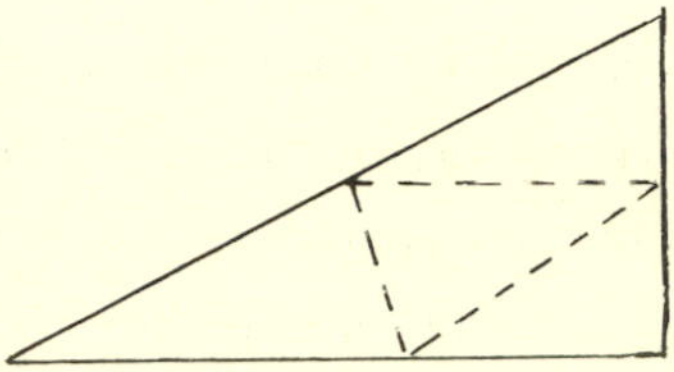

We obtain S_{h_n} by replacing each topological triangle of S_h lying over a side of P_n by the Euclidean triangle with the same vertices.

Because the angles of the approximating triangles are greater than a positive number which does not depend on n, these triangles converge towards tangent planes of S_h. Therefore the derivatives of h_n converge uniformly to thederivatives of h outside the edges. So (A′) remains true, if we replace h by h_n for a sufficiently large n. If we keep n fixed, (B′) is satisfied automatically, if we identify M with the set of points corresponding to the edges of P_n, because the vertices of S_{h_n} and therefore also dh_n vary continuously with $p \in \overline{B}$.

Self-intersections can be avoided by the following procedure: Choose a compact neighborhood U′ of the edges of P_1 which lies in the complement U of the support of h. Take n_0 so large that all triangles of P_n, $n \geq n_0$, which intersect U′ are contained in U. If we look at the construction of h (see Appendix), we realize that U′ can be fixed before we choose $\sup |h|$. Therefore, if d is the minimum of the distances d(p, q), where p and q lie in different triangles of P_1 and outside U′, we choose $\sup |h| \leq d/4$. If $n > n_0$ and if n is so large that $|h - h_n| < d/4$, S_{h_n} cannot have self-intersections.

Appendix: A solution of (A′) and (B′)

The function h and its derivatives on S_0 will be continuous on $S_0 \times \overline{B}$. Therefore (B′) will be satisfied automatically.

In (A′) we realize that $|\mu|$ is constant, $|\mu| < 1$, because μ is a Teichmüller differential, and $|dX|/|dz| = \lambda_1$ is a function, because $X(z)$ is conformal.

If we put $h^* = h/(\lambda_1 \lambda_2)$ and drop small terms, we have

$$|d(X + hN)|^2 = \lambda_1^2(|dz|^2 + \lambda_2^2(h_x^* dx + h_y^* dy)^2) .$$

On the other hand, a simple computation shows that

$$(1 - |\mu|)^2 |dz + \mu d\overline{z}|^2 = |dz|^2 + (\alpha_\mu dx + \beta_\mu dy)^2 ,$$

where α_μ and β_μ are suitable real-valued functions. We have to smooth them in a neighborhood of certain analytic curves in order to obtain continuous functions α_μ^* and β_μ^*. Taking $\lambda = \lambda_1/(1 - |\mu|)$ we see that (A′) can be replaced by

$$\text{(A″)} \quad (1/K)\,|\alpha_\mu^* dx + \beta_\mu^* dy| \leq \lambda_2\, |h_x^* dx + h_y^* dy| \leq K\, |\alpha_\mu^* dx + \beta_\mu^* dy| .$$

Choosing λ_2 in a suitable way, we can force

$$\omega = \frac{1}{\lambda_2} (\alpha_\mu^* dx + \beta_\mu^* dy)$$

to be exact, such that there is a solution h^* of the equation $dh^* = \omega$ on the universal covering surface $\tilde{S}_0$ of S_0. $h = \lambda_1 \lambda_2 h^*$ satisfies (A′), but it is neither small nor an automorphic function.

The second property is satisfied, if we multiply the restriction of h to a fundamental domain F in S_0 by a function which is identically one outside a small neighborhood of $\partial F \cup M$ and identically zero in a neighborhood of this union.

Finally, in order to obtain a small solution, we realize that only $|dh|$ is interesting to us. Therefore we take a saw-shaped periodic function ν

with slopes ± 1 and replace h by $\frac{1}{m} \cdot \nu(m \cdot h)$. Because

$$|dh| = |d(\frac{1}{m} \cdot \nu(m \cdot h))| ,$$

we only have to take a smooth version of ν and a large number m, in order

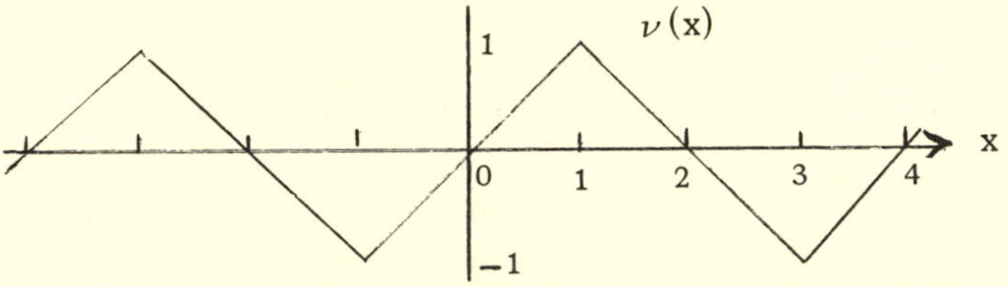

to obtain a function which satisfies all our conditions.

REFERENCES

[1] Ahlfors, L., On Quasiconformal Mappings, *J. d'Analyse Math.*, 3, 1 (1953/4).

[2] Bers, L., Lectures on Riemann Surfaces

[3] Garsuam A., An Embedding of Closed Riemann surfaces in Euclidean Space, *Comment. Math. Helv.* 35 (1961).

[4] Huber, H., Riemannsche Flächen von hyperbolischem Typus im euklidischen Raum, *Math. Ann.*, 139 (1959).

[5] Rüedy, R., Einbettungen Riemannscher Flächen in den dreidimensionalen euklidischen Raum, *Comment. Math. Helv.*, 43 (1968).

FOCK REPRESENTATIONS AND THETA-FUNCTIONS

Ichiro Satake

This lecture is mostly of expository nature. Our purpose is to explain some basic facts on Fock representations along with their connection to theta-functions. A particular emphasis will be placed on the analogy between Fock representations and discrete series representations of Harish-Chandra. We will observe Fock spaces as members of a family of (mutually equivalent) representation-spaces parametrized by a Siegel upper half-space and give an explicit form of the integral operator intertwining the representations on these spaces. In this way, we will obtain simple proofs to a result of Shale and Weil ([7], [8.2]) on the existence of a projective unitary representation of the (real) symplectic group as well as to a weaker form of the transformation formula of theta-functions.

1. Automorphy factor and kernel function. We start defining a group $\tilde{G}$ that will play a principal role in the following. Let V be a 2n-dimensional real vector-space, on which a non-degenerate alternating bilinear form A is defined. We denote by $T = R/Z$ the 1-dimensional torus, and by ε the canonical homomorphism $R \to T$, i.e., $\varepsilon(\xi) = \varepsilon^{2\pi i \xi}$ for $\xi \in R$. Since $\varepsilon(\frac{1}{2} A(u, v))$ $(u, v \in V)$ satisfies the 2-cocycle condition, we can define a central extension $\tilde{V}$ of V by T with respect to this factor set. $\tilde{V}$ is then a nilpotent Lie group with the center T such that $\tilde{V}/T \cong V$. The symplectic group $G = Sp(V, A)$ acts on $\tilde{V}$ in a natural manner as a group of automorphisms, so that one can construct a splitting extension $\tilde{G} = \tilde{V} \cdot G$ of G by $\tilde{V}$. Thus $\tilde{G}$ may be viewed as a group of triples $\tilde{g} = (t, u, g)$ $(t \in T,\ u \in V,\ g \in G)$ with the law of composition defined as follows:

$$(s, u, g)(t, v, h) = (st\ \varepsilon(\tfrac{1}{2} A(u, gv)),\ u + gv,\ gh)$$

for $s, t \in T$, $u, v \in V$, $g, h \in G$.

Let K be a maximal compact subgroup of G and put $\tilde{K} = T \cdot K\ (\cong T \times K)$. As is well-known, the homogeneous space G/K may be identified with a Siegel upper half-space $\mathfrak{H}_n$ of degree n, i.e., the space consisting of all $n \times n$ complex symmetric matrices $z = x + iy$ with the imaginary part $y > 0$. To be more precise, we fix a canonical basis $(e_1, \ldots, e_{2n})$ of V such that the corresponding matrix of A is of the form

$$\begin{bmatrix} 0 & -1_n \\ 1_n & 0 \end{bmatrix},$$

put $W = \{e_1, \ldots, e_n\}_{\mathbf{R}}$, $W^* = \{e_{n+1}, \ldots, e_{2n}\}_{\mathbf{R}}$, and consider $z \in \mathfrak{H}_n$ as a $\mathbf{C}$-linear map from $W^*_{\mathbf{C}}$ into $W_{\mathbf{C}}$ given in a matrix form with respect to these bases. If one writes $g \in G$ in the corresponding matrix form as $g = \begin{pmatrix} a & b \\ c & d \end{pmatrix}$, then the action of g on $\mathfrak{H}_n$ is given by $z \longmapsto g(z) = (az + b)(cz + d)^{-1}$. We now identify the homogeneous space $\tilde{G}/\tilde{K}$ with $W_{\mathbf{C}} \times \mathfrak{H}_n$ by the following one-to-one correspondence:

$$\tilde{G}/\tilde{K} \ni (t, u, g)\tilde{K} \longmapsto (u_z, z) \in W_{\mathbf{C}} \times \mathfrak{H}_n ,$$

where z is the point in $\mathfrak{H}_n$ corresponding to $gK \in G/K$ and, for $u = u_1 + u_2$, $u_1 \in W$, $u_2 \in W^*$, one puts $u_z = u_1 - zu_2$. [As is well-known, each point z in $\mathfrak{H}_n$ corresponds in one-to-one way to a complex structure I on V such that AI is symmetric and > 0. When one regards $\tilde{G}/\tilde{K}$ as a fibre space over $G/K \approx \mathfrak{H}_n$ in a natural manner, the above correspondence induces on the fibre over z an isomorphism $V \cong W_{\mathbf{C}}$ which precisely defines the complex structure I on V corresponding to z.] The action of $\tilde{G}$ on $W_{\mathbf{C}} \times \mathfrak{H}_n$ is then given as follows. For $\tilde{g} = (t, u, g) \in \tilde{G}$ and $\tilde{z} = (w, z) \in W_{\mathbf{C}} \times \mathfrak{H}_n$, one has

$$\tilde{g}(\tilde{z}) = (u_{g(z)} + J(g, z)w, g(z)) ,$$

where $J(g, z) = {}^t(cz + d)^{-1}$ ($\in GL(W_{\mathbf{C}})$) is the canonical automorphy factor. Thus, the action of T being trivial, $\tilde{G}/T \cong V \cdot G$ may be viewed as a group of complex analytic automorphisms of $W_{\mathbf{C}} \times \mathfrak{H}_n$ preserving the structure of the fibre space $W_C \times \mathfrak{H}_n \to \mathfrak{H}_n$. If z_0 is the origin of $\mathfrak{H}_n$ corresponding to K, then $\tilde{z}_0 = (0, z_0)$ is the origin of $W_C \times \mathfrak{H}_n$ corresponding to $\tilde{K}$.

A C-valued function $\eta(\tilde{g}, \tilde{z})$ $(\tilde{g} \in \tilde{G},\ \tilde{z} \in W_C \times \mathfrak{H}_n)$ is called a (holomorphic) *automorphy factor,* if it satisfies the following conditions:

(1a) η is C^∞ in $\tilde{g}$ and holomorphic in $\tilde{z}$,

(1b) $\eta(\tilde{g}_1\tilde{g}_2, \tilde{z}) = \eta(\tilde{g}_1, \tilde{g}_2(\tilde{z})) \cdot \eta(\tilde{g}_2, \tilde{z})$ for all $\tilde{g}_1, \tilde{g}_2 \in \tilde{G},\ \tilde{z} \in W_C \times \mathfrak{H}_n$,

It follows from the condition (1b) that $\eta(\tilde{k}, \tilde{z}_0)$ is a character of $\tilde{K}$. In the following we will exclusively be interested in the case where this character is given by $\tilde{k} = (t, k) \longmapsto t$, i.e.,

(1c) $$\eta((t, 0, k), (0, z_0)) = t \quad \text{for all } t \in T,\ k \in K.$$

Two such automorphy factors η, η' are said to be *equivalent* if there exists a holomorphic function ϕ on $W_C \times \mathfrak{H}_n$ such that one has

$$\eta'(\tilde{g}, \tilde{z}) = \phi(\tilde{g}(\tilde{z})) \cdot \eta(\tilde{g}, \tilde{z}) \cdot \phi(\tilde{z})^{-1} \quad \text{for all } \tilde{g} \in \tilde{G},\ \tilde{z} \in W_C \times \mathfrak{H}_n .$$

LEMMA 1. *There exists an automorphy factor* η *on* $(\tilde{G}, W_C \times \mathfrak{H}_n)$ *satisfying the conditions (1a) - (1c). Moreover such an automorphy factor is unique up to the equivalence.*

In fact, one can obtain a "canonical" automorphy factor

$$\tilde{J} : \tilde{G} \times (W_C \times \mathfrak{H}_n) \to C^\times \times GL(W_C)$$

by extending from G to $\tilde{G}$ the notions of Harish-Chandra imbedding and Cayley transformation and then imitating the process of defining the canonical automorphy factor $J : G \times \mathfrak{H}_n \to GL(W_C)$ (see [6]). The sought-for automorphy factor η is then obtained by taking the $C^\times$-component of $\tilde{J}$. The explicit form of η is as follows

(2) $$\eta(\tilde{g}, \tilde{z}) = t \cdot \varepsilon(\tfrac{1}{2}A(u, u_{g(z)}) + A(u, J(g, z)w) + \tfrac{1}{2}A(gw, J(g, z)w))$$

for $\tilde{g} = (t, u, g) \in \tilde{G}$ and $\tilde{z} = (w, z) \in W_C \times \mathfrak{H}_n$. The conditions (1a) -(1c) can easily be verified either by its construction, or by a straightforward calculation from (2). The uniqueness of η follows from a general criterion given by Murakami ([5.1], Appendix, or [5.2]).

LEMMA 2. *For an automorphy factor η satisfying (1a)-(1c), there exists a* $\mathbf{C}$*-valued function* $\kappa(\tilde{z}', \tilde{z})$ $(\tilde{z}, \tilde{z}' \in W_{\mathbf{C}} \times \mathfrak{H}_n)$ *satisfying the following conditions:*

(3a) $\kappa(\tilde{z}', \tilde{z})$ *is holomorphic in* $\tilde{z}'$, *and one has* $\kappa(\tilde{z}, \tilde{z}') = \overline{\kappa(\tilde{z}', \tilde{z})}$,

(3b) $\kappa(\tilde{g}(\tilde{z}'), \tilde{g}(\tilde{z})) = \eta(\tilde{g}, \tilde{z}')\kappa(\tilde{z}', \tilde{z})\overline{\eta(\tilde{g}, \tilde{z})}$ *for all* $\tilde{g} \in \tilde{G}$, $\tilde{z}, \tilde{z}' \in W_{\mathbf{C}} \times \mathfrak{H}_n$,

(3c) $\kappa(\tilde{z}_0, \tilde{z}_0) = 1$.

Moreover, such a function κ *is uniquely determined by* η.

κ is called the *kernel function* associated with η. The existence of κ can again be proved by pursuing the analogy with the case of symmetric domains. The explicit form of κ associated with the automorphy factor η given by (2) is as follows:

$$\kappa(\tilde{z}', \tilde{z}) = \varepsilon(-\tfrac{1}{2}(z' - \bar{z})^{-1}[w' - \bar{w}]) \tag{4}$$

for $\tilde{z} = (w, z)$, $\tilde{z}' = (w', z') \in W_{\mathbf{C}} \times \mathfrak{H}_n$, where [ ] is the Siegel's notation, i.e., for a symmetric matrix S, $S[u] = {}^t uSu$ is the corresponding quadratic form. To prove the uniqueness, let κ and κ' be two kernel functions associated with the same automorphy factor η, and put $\psi(\tilde{z}', \tilde{z}) = \kappa'(\tilde{z}', \tilde{z})\kappa(\tilde{z}', \tilde{z})^{-1}$. Then, from (3b), one has $\psi(\tilde{g}(\tilde{z}'), \tilde{g}(\tilde{z})) = \psi(\tilde{z}', \tilde{z})$. It follows, in particular, that $\psi(\tilde{z}, \tilde{z}) = \psi(\tilde{z}_0, \tilde{z}_0) = 1$ by (3c). Since $\psi(\tilde{z}', \tilde{z})$ is holomorphic in $\tilde{z}'$ and antiholomorphic in $\tilde{z}$, this implies that $\psi(\tilde{z}', \tilde{z})$ is identically equal to 1.

In the following, we fix η and κ given by (2) and (4), respectively. For each $z \in \mathfrak{H}_n$, we put

$$\kappa_z(w', w) = \kappa((w', z), (w, z)) = e^{-\frac{\pi}{2} y^{-1}[w' - \bar{w}]} .$$

In particular, one has

$$\kappa_z(w, w) = e^{2\pi y^{-1}[\mathrm{Im}(w)]} > 0 .$$

2. **Fock representations** (cf. [1], [2]). For each $z \in \mathfrak{H}_n$, one considers a pre-Hilbert space $\mathfrak{F}_z$ defined as follows.* Namely, $\mathfrak{F}_z$ is the space formed of all holomorphic functions ϕ on W_C such that

$$\|\phi\|_z^2 = \int_{W_C} |\phi(w)|^2 \kappa_z(w, w)^{-1} d_z w < \infty , \tag{5}$$

where $d_z w = \det(y)^{-1} dw$, dw denoting the usual Lebesgue measure on $W_C \cong C^n$ with respect to the basis $(e_1, \ldots, e_n)$. (Note that the measure $d_z w$ is actually independent of the choice of the canonical basis of V.) $\mathfrak{F}_z$ is not $= \{0\}$, for one has $e^{-\frac{\pi}{2} y^{-1}[w]} \in \mathfrak{F}_z$. Since every $\phi \in \mathfrak{F}_z$ is holomorphic, there exists, for any compact set C in W_C, a positive constant c depending only on C such that one has $|\phi(w)| \leq c\|\phi\|_z$ for all $\phi \in \mathfrak{F}_z$ and $w \in C$. It follows that a sequence in $\mathfrak{F}_z$ convergent in the norm $\| \ \|_z$ converges uniformly and absolutely on any compact set in W_C. $\mathfrak{F}_z$ is therefore complete and becomes a Hilbert space with respect to the norm $\| \ \|_z$. Moreover, since the correspondence $\phi \longmapsto \phi(w)$ is a continuous linear functional on $\mathfrak{F}_z$, there exists an element $e_w^z \in \mathfrak{F}_z$ such that

---

* In [1], the Fock space $\mathfrak{F}$ is defined as the space of holomorphic functions f on C^n satisfying the condition

$$\|f\|^2 = \pi^{-n} \int_{C^n} |f(w)|^2 e^{-{}^t\overline{w}w} dw < \infty .$$

A unitary equivalence from $\mathfrak{F}_{\sqrt{-1}1_n}$ onto $\mathfrak{F}$ is given by the correspondence

$$\mathfrak{F}_{\sqrt{-1}1_n} \ni \phi \longmapsto f(w) = e^{\frac{1}{2}{}^t ww} \cdot \phi(\sqrt{\pi}^{-1} w) \in \mathfrak{F}.$$

$$(*) \qquad \phi(w) = \int_{W_C} \overline{e_w^z(w')}\phi(w')\kappa_z(w',w')^{-1} d_z w'$$

for all $\phi \in \mathfrak{F}_z$. We shall see later on that one actually has $e_w^z(w') = \kappa_z(w',w)$.

Now, for $\tilde{g} \in \tilde{G}$ and $\phi \in \mathfrak{F}_{g(z)}$, one puts

$$(6) \qquad (T^{g(z)}_{\tilde{g}^{-1}} \phi)(w) = \eta(\tilde{g}, \tilde{z})^{-1} \phi((\tilde{g}(\tilde{z}))_w) ,$$

where $\tilde{g} = (t, u, g)$, $\tilde{z} = (w, z)$ and $(\tilde{g}(\tilde{z}))_w$ denotes the w-component of $\tilde{g}(\tilde{z})$, i.e., $(\tilde{g}(\tilde{z}))_w = u_{g(z)} + J(g, z)w$. Then, from the properties (1), (3) of η and κ, and from the invariance of the measure $d_z w$, it is clear that one has $T^{g(z)}_{\tilde{g}^{-1}}\phi \in \mathfrak{F}_z$ and $\|T^{g(z)}_{\tilde{g}^{-1}} \phi\|_z = \|\phi\|_{g(z)}$. In other words, $T^{g(z)}_{\tilde{g}^{-1}}$ is an isometry of $\mathfrak{F}_{g(z)}$ into $\mathfrak{F}_z$, or what amounts to the same, $T^z_{\tilde{g}}$ is an isometry of $\mathfrak{F}_z$ into $\mathfrak{F}_{g(z)}$. Moreover, it is also clear that one has

$$(7) \qquad T^{g_2(z)}_{\tilde{g}_1} \circ T^z_{\tilde{g}_2} = T^z_{\tilde{g}_1 \tilde{g}_2} \qquad \text{for all } \tilde{g}_1, \tilde{g}_2 \in \tilde{G},\ z \in \mathfrak{H}_n .$$

Since $T^z_1 = \mathrm{id}$, $T^z_{\tilde{g}}$ is actually an isomorphism of $\mathfrak{F}_z$ onto $\mathfrak{F}_{g(z)}$.

In particular, for $\tilde{u} \in \tilde{V}$, $T^z_{\tilde{u}}$ is a unitary operator on $\mathfrak{F}_z$. Applying the relation (*) to $T^z_{\tilde{u}^{-1}} \phi$ instead of ϕ, one sees at once that $\overline{e^z_{w'}(w)}$ satisfies the conditions (3a), (3b) (for a fixed z). Therefore, by the uniqueness of the kernel function (which can be proved exactly as Lemma 2), one has $\overline{e^z_{w'}(w)} = \gamma\kappa_z(w',w)$ with a non-zero real constant γ. The uniqueness of the kernel function also implies the irreducibility of the unitary representation $(\mathfrak{F}_z, T^z_{\tilde{u}})$ $(\tilde{u} \in \tilde{V})$. This representation is called a *Fock representation* of $\tilde{V}$.

To proceed further, we need the following Lemma, which can be proved by a straightforward calculation (see [6]).

LEMMA 3. *For any* $\tilde{z} = (w, z)$, $\tilde{z}'' = (w'', z'') \in W_C \times \mathfrak{H}_n$ *and* $z' \in \mathfrak{H}_n$, *one has*

$$\int_{W_C} \kappa(\tilde{z}'', \tilde{z}')\kappa(\tilde{z}', \tilde{z})\kappa(\tilde{z}', \tilde{z}')^{-1} d_{z'}w' \tag{8}$$

$$= \det\left(\frac{1}{2i}(z''-\bar{z}')\right)^{1/2} \cdot \det\left(\frac{1}{2i}(z'-\bar{z})\right)^{1/2} \cdot \det\left(\frac{1}{2i}(z''-\bar{z})\right)^{-1/2}$$

$$\times \det(y')^{-1/2} \cdot \kappa(\tilde{z}'', \tilde{z}),$$

where $\tilde{z}' = (w', z')$, $y' = \operatorname{Im}(z')$, *and we choose the branch of the analytic function* $\det(\frac{1}{2i}(z'-\bar{z}))^{1/2}$ $(z, z' \in \mathfrak{H}_n)$ *in such a way that it is* > 0 *for* $z' = z$. *The integral in* (8) *is absolutely convergent; more precisely, one has*

$$\int_{W_C} |\kappa(\tilde{z}'', \tilde{z}')\kappa(\tilde{z}', \tilde{z})|\, \kappa(\tilde{z}', \tilde{z}')^{-1} d_{z'}w' \tag{8'}$$

$$\leq C\, \kappa(\tilde{z}'', \tilde{z}'')^{1/2} \cdot \kappa(\tilde{z}, \tilde{z})^{1/2},$$

where C *is a positive constant depending only on* z, z', z''.

We now define an integral operator $U_{z',z}$ as follows:

$$(U_{z',z}\phi)(w') = \gamma_{z',z} \int_{W_C} \kappa(\tilde{z}', \tilde{z})\kappa(\tilde{z}, \tilde{z})^{-1}\phi(w)\, d_z w, \tag{9}$$

where $\tilde{z} = (w, z)$, $\tilde{z}' = (w', z')$, and

$$\gamma_{z',z} = \det\left(\frac{1}{2i}(z'-\bar{z})\right)^{-1/2} \cdot \det(y)^{1/4} \cdot \det(y')^{1/4}.$$

Since one has

$$\int_{W_C} |\kappa(\tilde{z}', \tilde{z})|^2 \kappa(\tilde{z}, \tilde{z})^{-1} d_z w < \infty$$

by Lemma 3, it follows from Schwarz inequality that the integral $U_{z',z}\phi$ is absolutely convergent, for all $\phi \in \mathfrak{F}_z$, and expresses a holomorphic function on $W_{\mathbf{C}}$. To prove that $U_{z',z}\phi$ actually belongs to $\mathfrak{F}_{z'}$, suppose first that ϕ satisfies the condition

$$|\phi(w)| < C\, e^{-\frac{\pi}{2}((1-\lambda^2)\,y^{-1}[\mathrm{Re}\,(w)]-(1+\lambda^2)\,y^{-1}[\mathrm{Im}\,(w)])}$$

for some $0 < \lambda < 1$ and $C > 0$. Then, by (8′), one sees that the multiple integral expressing $\|U_{z',z}\phi\|^2_{z'}$ is absolutely convergent, so that one has $U_{z',z}\phi \in \mathfrak{F}_{z'}$. Moreover, for ϕ satisfying the above condition, the double integral $U_{z'',z'}(U_{z',z}\phi)$ is also absolutely convergent, and by Fubini's theorem and (8) one has

$$U_{z'',z'}(U_{z',z}\phi) = U_{z'',z}\phi \ .$$

In particular, putting $z = z' = z''$, one has $U^2_{z,z}\,\phi = U_{z,z}\,\phi$. But, since we know that $U_{z,z} = \gamma^{-1}\cdot 1$, one has $\gamma = 1$, i.e., $U_{z,z} = \mathrm{id}$. Furthermore, again by Fubini's theorem, one has

$$\|U_{z',z}\phi\|^2_{z'} = \langle U_{z',z}\phi, U_{z',z}\phi\rangle_{z'}$$

$$= \langle \phi, U_{z,z'}(U_{z',z}\phi)\rangle_z = \|\phi\|^2_z \ .$$

Thus $U_{z',z}$ is an isometry on the subspace of $\mathfrak{F}_z$ consisting of all ϕ satisfying the above condition. Since this subspace is everywhere dense in $\mathfrak{F}_z$, this implies that the operator $U_{z',z}$ is an isometry of $\mathfrak{F}_z$ onto $\mathfrak{F}_{z'}$. Moreover, one has the relation

$$U_{z'',z'} \circ U_{z',z} = U_{z'',z} \quad \text{for all } z, z', z'' \in \mathfrak{H}_n \ . \tag{10}$$

Now, let $z, z' \in \mathfrak{H}_n$ and $\tilde{g} = (*, *, g) \in \tilde{G}$. By a simple calculation from the definitions, one has

$$U_{g(z'),g(z)} \circ T^z_{\tilde{g}} = \varepsilon(g;\, z', z)\,T^{z'}_{\tilde{g}} \circ U_{z',z} \tag{11}$$

with

$$\varepsilon(g; z', z) = \varepsilon(\arg(\det(cz'+d)^{1/2}) - \arg(\det(cz+d)^{1/2})) ,$$

where, for each $g = \begin{pmatrix} a & b \\ c & d \end{pmatrix} \in G$, one fixes once and for all a branch of the analytic function $\det(cz+d)^{1/2}$ $(z \in \mathfrak{H}_n)$, which is possible since $\mathfrak{H}_n$ is simply connected. In particular, one has $\varepsilon(1; z', z) = 1$, which means that $U_{z',z}$ gives a unitary equivalence of the Fock representations $(\mathfrak{F}_z, T^z_{\tilde{u}})$ and $(\mathfrak{F}_{z'}, T^{z'}_{\tilde{u}})$ $(\tilde{u} \in \tilde{V})$. Thus all Fock representations are mutually equivalent; this fact is, of course, a special case of the uniqueness theorem of Stone and von Neumann.

For $\tilde{g} = (*, *, g) \in \tilde{G}$, we put

$$\dot{T}^z_{\tilde{g}} = U_{z,g(z)} \circ T^z_{\tilde{g}} \quad . \tag{12}$$

Then, by (11), one sees that the assignment $\tilde{g} \longmapsto \dot{T}^z_{\tilde{g}}$ is a continuous map from $\tilde{G}$ into the group of unitary operators of $\mathfrak{F}_z$ satisfying the relation

$$\dot{T}^z_{\tilde{g}_1} \circ \dot{T}^z_{\tilde{g}_2} = \alpha_z(g_1, g_2)\, \dot{T}^z_{\tilde{g}_1 \tilde{g}_2} , \tag{13}$$

where $\alpha_z(g_1, g_2) = \varepsilon(g_1; g_2(z), z)$. α_z is a continuous 2-cocycle of G and, if one puts $\varepsilon(g) = \varepsilon(\arg(\det(J(g, z))))$, one has

$$\alpha_z(g_1, g_2)^2 = \varepsilon(g_1)\, \varepsilon(g_1 g_2)^{-1}\, \varepsilon(g_2) \sim 1 .$$

Thus, $\dot{T}^z_{\tilde{g}}$ $(\tilde{g} \in \tilde{G})$ gives rise to a (continuous) projective representation of $\tilde{G}$ with a factor set α_z of order 2, or equivalently, to an ordinary unitary representation of a covering group of $\tilde{G}$ of order 2. The existence of such a representation (in a more general setting) has been proved by Shale [7] and Weil [8.2] using another (equivalent) representation-space $L^2(W)$. *

* A unitary equivalence from $L^2(W)$ onto $\mathfrak{F}_z$ is given by the correspondence

$$L^2(W) \ni \Phi \to \phi(w) = 2^{n/4} \det(z/i)^{-1/2} \det(y)^{1/4} \int_W e^{-\pi i z^{-1}[w-u]}\, \Phi(u)\, du.$$

3. Theta-functions (cf. [8]). Suppose there is given a lattice L in V such that $L = L \cap W + L \cap W^*$ and that $A(\ell, \ell') \in Z$ for all $\ell, \ell' \in L$. We put $L_1 = L \cap W$, $L_2 = L \cap W^*$ and denote by L_1' (resp. L_2') the dual lattice of L_1 (resp. L_2) in W^* (resp. W) with respect to the inner product $<w, w^*> = -A(w, w^*)$ ($w \in W$, $w^* \in W^*$). Then the above condition implies that $L_1 \subset L_2'$, $L_2 \subset L_1'$ and one has $L_2'/L_1 \cong L_1'/L_2$.

A *semi-character* ψ of L is by definition a map $\psi : L \to T$ such that $\psi(\ell) \cdot \varepsilon(\tfrac{1}{2} < \ell_1, \ell_2 >)$ ($\ell \in L$) is a character of L, where one writes $\ell = \ell_1 + \ell_2$ with $\ell_i \in L_i$. For every $r \in V$, $r = r_1 + r_2$, $r_1 \in W$, $r_2 \in W^*$, one can define a semi-character ψ_r by

$$\psi_r(\ell) = \varepsilon(\tfrac{1}{2} <\ell_1, \ell_2> + <\ell_1, r_2> + <r_1, \ell_2>) ,$$

which depends only on r (mod. $L_1' + L_2'$). It is clear that all semi-character of L can be written in this form. When $r \in V_Q = L \otimes Q$, ψ_r is called *rational*.

Now, for $r \in V$, one defines a *theta-function* θ_r as follows:

$$(14) \qquad \theta_r(w, z) = \sum_{\ell_2 \in L_2} \varepsilon(\tfrac{1}{2} z [r_2 + \ell_2] + <w + r_1, r_2 + \ell_2>) .$$

It is well-known that this series is normally convergent and expresses a holomorphic function on $W_C \times \mathfrak{H}_n$. Moreover, it is immediate from the definitions that one has

$$\theta_r(w + \ell_z, z) = \psi_r(\ell) \cdot \eta(\ell, z) \cdot \theta_r(w, z)$$

for all $\ell \in L$, where $\ell_z = \ell_1 - z\ell_2$. We put further

$$\theta_r^{\,z}(w) = \det(y)^{1/4} \cdot \theta_r(w, z)$$

and, for a given semi-character ψ, denote by Θ_ψ^z the linear space formed of all holomorphic functions θ on W_C such that one has $T_{\underline{\ell}}^z \, \theta = \psi(\ell)^{-1}\theta$ for all $\ell \in L$, where one writes $\underline{\ell} = (1, \ell, 1)$ ($\in \tilde{G}$). Then, it is also well-known (and immediate) that $\dim_C \Theta_\psi^z = [L_1' : L_2]$ and that, if $\psi = \psi_{r0}$

with $r^0 \in V$ and if R denotes a complete set of representatives of L_1' modulo L_2, then $\{\theta^z_{r^0+r}\ (r \in R)\}$ forms a basis of Θ^z_ψ over C.

For a given semi-character ψ, let Γ_ψ denote the subgroup of $G = Sp(V, A)$ formed of all $\gamma \in G$ such that $\gamma(L) = L$ and $\psi \circ \gamma = \psi$. [When ψ is rational, Γ_ψ is an arithmetic subgroup of G with respect to its Q-structure defined by L.] Since we have

$$(T^{\gamma z}_{\underline{\gamma \ell}} \circ T^z_\gamma)\theta = (T^z_\gamma \circ T^z_{\underline{\ell}})\theta = \psi(\ell)^{-1}\, T^z_\gamma\, \theta$$

for $\theta \in \Theta^z_\psi$, $\gamma \in \Gamma_\psi$, $\ell \in L$, one has

$$T^z_\gamma\ \Theta^z_\psi = \Theta^{\gamma(z)}_\psi \qquad \text{for } \gamma \in \Gamma_\psi\ . \tag{15}$$

On the other hand, although Θ^z_ψ is not contained in $\mathfrak{F}_z$ itself, the operator $U_{z',z}$ can also be extended to Θ^z_ψ in a natural sense. Namely, if we denote by $(\mathfrak{F}_z)_\infty$ the Gårding space of $\mathfrak{F}_z$, i.e., the subspace of $\mathfrak{F}_z$ formed of all analytic vectors, and by $(\mathfrak{F}_z)_{-\infty}$ the conjugate dual of $(\mathfrak{F}_z)_\infty$, then one has $\mathfrak{F}_z \subset (\mathfrak{F}_z)_{-\infty}$, $\Theta^z_\psi \subset (\mathfrak{F}_z)_{-\infty}$ in a natural sense.* The isomorphism $U_{z',z}$ can therefore be extended canonically to an isomorphism of $(\mathfrak{F}_z)_{-\infty}$ onto $(\mathfrak{F}_{z'})_{-\infty}$. If one puts $f_r(\bar{z}) = \varepsilon(\tfrac{1}{2} z[r_2] + \langle w + r_1, r_2\rangle)$, an easy computation shows that

$$\int_{W_C} \kappa(\bar{z}', \bar{z}) \cdot \kappa(\bar{z}, \bar{z})^{-1} f_r(\bar{z})\, d_z w = \det\left(\frac{1}{2i}(z' - \bar{z})\right)^{1/2} \cdot \det(y)^{-1/2} \cdot f_r(\bar{z}')\ .$$

It follows that one has $U_{z',z}\,\theta^z_r = \theta^{z'}_r$ and hence

* The isomorphism $L^2(W) \cong \mathfrak{F}_z$ given in the second footnote can naturally be extended to an isomorphism from $L^2(W)_{-\infty}$ onto $(\mathfrak{F}_z)_{-\infty}$, where $L^2(W)_{-\infty}$ can be interpreted as the space of tempered distributions on W. Under this isomorphism, the theta-function θ^z_r corresponds to the tempered distribution $S(W) \in \psi \to 2^{-n/4} \Sigma_{\ell_2 \in L_2} \hat{\psi}_{r_1}(r_2 + \ell_2)$, where $S(W) = L^2(W)_\infty$ is the Schwartz space of W and $\hat{\psi}_{r_1}$ denotes the Fourier transform of $\psi_{r_1}(u) = \psi(u - r_1)$.

(16) $$U_{z',z}\Theta^{z}_{\psi} = \Theta^{z'}_{\psi} .$$

Combining the relations (15) and (16), one obtains the relation

(17) $$\dot{T}^{z}_{\gamma}\,\Theta^{z}_{\psi} = \Theta^{\gamma(z)}_{\psi} \qquad \text{for } \gamma \in \Gamma_{\psi} .^{*}$$

This means that, for $\gamma = \begin{pmatrix} a & b \\ c & d \end{pmatrix} \in \Gamma_{\psi}$, one has a transformation formula of theta-functions of the following form:

$$\theta_{r^0 + r}({}^t(cz + d)^{-1} w, \gamma(z))$$

$$= \det(cz + d)^{1/2} \cdot \eta(\gamma, \tilde{z}) \sum_{r' \in R} \lambda^{z}_{r',r}(\gamma)\theta_{r^0 + r'}(w, z)$$

for all $r \in R$. In case $\psi = \psi_{r^0}$ is rational, it is known furthermore that the matrix $(\lambda^{z}_{r',r}(\gamma))$ is independent of z and, when the branch of $\det(cz + d)^{1/2}$ is suitably chosen, gives a representation of Γ_{ψ}, of which the kernel is of finite index in Γ_{ψ}.

University of California, Berkeley

---

* I am indebted to J.-I. Hano for giving me a chance to read a manuscript of his recent paper [9] before its publication, in which he gives a proof of (17) by a somewhat different method.

REFERENCES

[1] Bargmann, V., On a Hilbert space of analytic functions and an associated integral transform, I, *Comm. Pure and Appl. Math.*, 14 (1961), 187-214; II, A family of related function spaces. Application to distribution theory, *ibid.*, 20 (1967), 1-101.

[2] Cartier, P., (1) Quantum mechanical commutation relations and theta functions, Symposium on algebraic groups and discontinuous subgroups, *Proc. Symp. Pure Math., IV*, American Math Soc. 1965, 361-383.

(2) Théorie des groupes, fonctions théta et modules des variétés abéliennes, *Sém. N. Bourbaki, 20e année*, 1967/68, Exp. 338.

[3] Godement, R., Fonctions holomorphes de carré sommable dans le demi-plan de Siegel, *Sém. H. Cartan, E. N. S., 10e année*, 1957/58, Exp. 6.

[4] Harish-Chandra, (1) Representations of semisimple Lie groups, IV, *Amer. J. Math., 77* (1955), 743-777; V, *ibid.*, 78 (1956), 1-41; VI, *ibid.*, 564-628.

(2) Discrete series for semisimple Lie groups, I, *Acta Math.*, 115 (1965), 241-318; II, *ibid.*, 116 (1966), 1-111.

[5] Murakami, S., (1) Cohomology of vector-valued forms on symmetric spaces, Lecture-notes at Univ. of Chicago, Summer 1966.

(2) Facteurs d'automorphie associés a un espace hermitien symétrique, *Geometry of Homogeneous Bounded Domains, Centro Int. Mat. Estivo,* 3° Ciclo, Urbino, 1967.

[6] Satake, I., On unitary representations of a certain group extension (Japanese), Sugaku, Math. Soc., Japan, 21 (1969), 241-253.

[7] Shale, D., Linear symmetries of free boson fields, *Trans. Amer. Math. Soc.*, 103 (1962), 149-167.

[8] Weil, A., (1) *Variétés kählériennes*, Hermann, Paris, 1958.

(2) Sur certains groups d'opérateurs unitaires, *Acta Math.*, 111 (1964), 143-211.

[9] Hano, J. -I., On theta functions and Weil's generalized Poisson summation formula, *Trans. Amer. Soc.*, 141 (1969), 195-210.

'UNIFORMIZATIONS' OF INFINITELY CONNECTED DOMAINS

R. J. Sibner

1. Let $\mathfrak{D}$ be a family of Riemann surfaces and $\mathfrak{S}$ a subfamily. In the theory of conformal mappings, what might be called the "universal uniformization problem," is to show the existence of an element $S \in \mathfrak{S}$ in each isomorphism [i.e. conformal equivalence] class of $\mathfrak{D}$.

For example, if $\mathfrak{D}$ is the family of Riemann surfaces of genus zero and $\mathfrak{S}$ the subfamily of plane domains, the above is Koebe's "general uniformization principle." For $\mathfrak{D}$ the family of simply connected surfaces and $\mathfrak{S}$ the subfamily consisting of the disk, the finite plane and the extended plane, it is the Riemann mapping theorem. Other examples are; for $\mathfrak{D}$ the family of Riemann surfaces of finite type and $\mathfrak{S}$ either the subfamily of surfaces obtained as ramified coverings of the unit disk or else as quotient spaces of the unit disk by finitely generated Fuchsian groups or for $\mathfrak{D}$ the family of symmetric surfaces of finite type and $\mathfrak{S}$ as above except that the group is taken to be a symmetric Fuchsian group.

If $\mathfrak{D}$ is taken to be the family of plane surfaces (i.e. plane domains), $\mathfrak{S}$ is usually described by geometrical properties of the boundary components. Some examples of families which have been investigated are those whose boundary components are:

(1) parallel slits, radial slits, slits along concentric circles etc.

(2) convex curves

(3) circles

---

* Research partially supported by NSF grant GP -8556.

(4) homethetic images of *given* convex curves

(5) analytic Jordan curves

The present state of knowledge on the above problem for these cases is as follows:

(1) Solved for $\mathfrak{D}$ the family of *all* plane domains and $\mathfrak{S}$ as described (as well as many other types of slits). See [9, 14, 16, 22] for references.

(2) If p is the isomorphism of a given domain D onto a horizontal slit domain obtained by variation techniques (on the coefficient of p) and q is the corresponding vertical slit mapping then p + iq maps D onto a domain bounded by convex curves, some of which can reduce to slits unless D is finitely connected. Thus the two problems have been solved: $\mathfrak{D}$ the family of all domains or of finitely connected domains and $\mathfrak{S}$ the respective subfamilies of domains bounded by convex or by strictly convex curves.

(3) The problem in this case is the content of the Koebe conjecture—every plane domain is conformally equivalent to a circle domain [10]. The conjecture was shown to be true for finitely connected domains and for certain symmetric domains by Koebe [11, 12], for domains quasiconformally isomorphic to a circle domain by the author [18] and for domains satisfying various extremal length conditions on the "limit boundary components" by Denneberg [6], Grötzsch [7], and Strebel [20,21]. Other cases have been considered by Sario [15] and Meschkowski [13].

As an application of the methods developed in the present paper we will (in §5) obtain some results on domains admitting symmetries, including a new proof for the domains considered by Koebe.

(4) Courant, Manel and Shiffman [5] treat the case of $\mathfrak{D}$ the family of finitely connected domains, and Strebel [21] has obtained a generalization of domains satisfying conditions on the limit boundary components (as in (3)).

(5) For $\mathfrak{D}$ the family of finitely connected domains, repeated use of the Riemann mapping theorem (or the p + iq maps of case (2) provides a solution. (Of course a solution for any $\mathfrak{D}$ for which a solution is obtained to (3) gives automatically a solution of (5). *One of the main results of this paper*

(§3) *is a solution of* (5) *in the case that* $\mathfrak{D}$ *is the family of infinitely connected domains*. This result has been announced in [19].

2. THE ORDINAL SEQUENCE. With variational techniques, used extensively for slit domains, one does not usually have to look closely at the limit points of the boundary components. For the technique which follows, it is necessary to begin by examining the structure of the limit points.

Let $\Gamma^0 = \Gamma^0(D)$ denote the collection of boundary components of the plane domain D. We form the sequence $\Gamma^\alpha = \Gamma^\alpha(D)$ (cf. [8], also [20])

$$\Gamma^0 \supset \Gamma^1 \supset \cdots \supset \Gamma^\omega \supset \Gamma^{\omega+1} \supset \cdots \supset \Gamma^\alpha \supset \cdots \tag{1}$$

where ω denotes the ordinal type of the natural numbers and α is an arbitrary ordinal number. If α has a predecessor $\alpha - 1$ then Γ^α is defined as the subset of $\Gamma^{\alpha-1}$ of components which contain points which are limit points of the components in $\Gamma^{\alpha-1}$ so that if Γ^0 is given a metric space structure (see [20]) then Γ^α is the *derived* set of $\Gamma^{\alpha-1}$. If α is a limit ordinal, we define $\Gamma^\alpha = \bigcap_{\beta<\alpha} \Gamma^\beta$.

We recall [8] that there exists a (first) ordinal η such that $\Gamma^\eta = \Gamma^{\eta+1}$. Γ^η is called the *perfect* (or *dense in itself*) kernel of Γ^0 and is empty if Γ^0 is countable. If Γ^0 is uncountable then Γ^η is non-empty and every component in Γ^η is a limit boundary component of other components in the perfect kernel. In any case $\Gamma^0 - \Gamma^\eta$ is countable. Note that the "ordinal type" of a boundary component of a domain is a conformal invariant.

Remark. It is easy to show (and well known from general considerations) that if α is a limit ordinal in the above sequence, then there exists an increasing sequence β_j, $j = 1, 2, \ldots$, of ordinals in the sequence such that for any $\beta < \alpha$ there is a β_j with $\beta \le \beta_j < \alpha$. A simple proof can be constructed using the Cantor-Bendixson theorem [8].

Example 1. We construct a closed set of points on the real axis and show that the complementary domain has a non-finite ordinal sequence (1).

We will obtain an ordinal sequence with $\eta = \omega^2$ but it will be clear how the construction can be continued to obtain ordinal sequences of greater "length".

Let Z^+ denote the non-negative integers (including ∞) and let S_0 be the image of Z^+ under a (sense preserving) homeomorphism of (the closed interval) $[0,\infty]$ onto $[0,1]$. Let S_1 denote the closure of the images of S_0 under homeomorphisms of $[0,1]$ onto $[1,1\frac{1}{2}],[1\frac{1}{2},1\frac{3}{4}],\ldots$ so that $S_1 \subset [1,2]$. Let S_2 denote the closure of the images of S_1 under homeomorphisms of $[1,2]$ onto $[2,2\frac{1}{2}],[2\frac{1}{2},2\frac{3}{4}],\ldots$ so that $S_2 \subset [2,3]$. Continuing the construction, writing $S = (\bigcup_{i=0}^{\infty} S_i) \cup \{\infty\}$ and considering $\Gamma^\alpha = \Gamma^\alpha(\hat{C} - S)$ we find that Γ^α is non-empty for $\alpha \leq \omega$. In particular $\{z = n\} \in \Gamma^n$ and $\{\infty\} \in \Gamma^\omega$.

Repeating the entire construction, (starting with T_0 = image of S instead of Z^+), letting T be the resulting point set and considering $\Gamma^\alpha(\hat{C} - T)$ we find that $\{z = n\} \in \Gamma^{n\omega}$ and $\{\infty\} \in \Gamma^{\omega^2}$.

The domains which have been constructed have point boundary components but, of course, we could have considered vertical slits with corresponding x-coordinates.

Example 2. The domain D which is the complement of a Cantor set has non-countably many boundary components. In fact, $\Gamma^0(D)$ is itself the perfect kernel.

3. Domains bounded by analytic Jordan curves. We defer, until §4, a proof of the basic

Lemma 1. *Let* D *be an arbitrary plane domain. Then* D *is isomorphic to the domain whose* isolated *boundary components are analytic Jordan curves.*

and proceed to our main result.

Theorem 1. *Let* D *be a domain with countably many boundary components. Then* D *is isomorphic to a domain bounded by analytic Jordan curves.*

The proof, by transfinite induction on α, follows: (We use the term "analytic Jordan curves" to include points.) Consider the following statement:

S(α): There exist isomorphisms $f(\gamma): D \to D_\gamma$ for all $\gamma \leq \alpha$, and isomor- $h(\gamma,\delta): D_\gamma \to D_\delta$ for $\gamma \leq \delta \leq \alpha$ such that (for $\gamma \leq \delta \leq \epsilon \leq \alpha$)

(i) $h(\delta,\epsilon) \circ h(\gamma,\delta) = h(\gamma,\epsilon)$

(ii) $h(\gamma,\delta)$ is the restriction of an isomorphism $\tilde{h}(\gamma,\delta)$ of the domain bounded by the elements of $\Gamma^\gamma(D_\gamma)$.

(iii) $f(\delta) = h(\gamma,\delta) \circ f(\gamma)$

(iv) The elements of $\Lambda^\gamma(D_\gamma) = \Gamma^0(D_\gamma) - \Gamma^\gamma(D_\gamma)$ are analytic Jordan curves.

S(0) is trivially true for $f(0) = h(0,0) =$ identity. Suppose $S(\beta)$ is true for all $\beta < \alpha$. We show the truth of $S(\alpha)$.

(Case 1) If α is not a limit ordinal and hence has a predecessor $\alpha-1$, then there exists an isomorphism $f(\alpha-1): D \to D_{\alpha-1}$ such that $\Lambda^{\alpha-1}(D_{\alpha-1})$ consists of analytic Jordan curves. Let $\tilde{D}_{\alpha-1} \supseteq D_{\alpha-1}$ be the domain bounded by the elements of $\Gamma^{\alpha-1}(D_{\alpha-1})$. Then $\Gamma^{\alpha-1}(D_{\alpha-1}) - \Gamma^\alpha(D_{\alpha-1})$ are the isolated boundary components of $\tilde{D}_{\alpha-1}$. By Lemma 1 there exists an isomorphism $\tilde{h}(\alpha-1,\alpha)$ of $\tilde{D}_{\alpha-1}$ mapping these boundary components onto analytic Jordan curves. The elements of $\Lambda^{\alpha-1}(D_{\alpha-1})$ are analytic

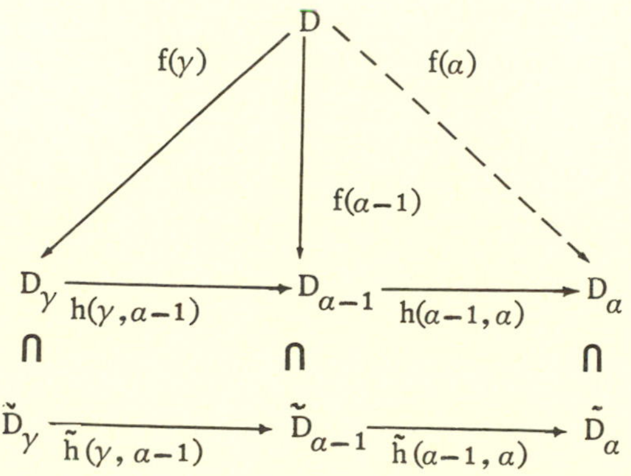

Jordan curves contained in $\tilde{D}_{\alpha-1}$, and hence are mapped by $\tilde{h}$ onto analytic Jordan curves. Let $\tilde{h}(\gamma,\alpha) = \tilde{h}(\alpha-1, \alpha) \circ \tilde{h}(\gamma,\alpha-1)$ and $h(\gamma,\alpha)$ be the restriction of $\tilde{h}(\gamma,\alpha)$ to D_γ. Let $f(\alpha) = h(\gamma,\alpha) \circ f(\gamma)$ which is well defined, independent of γ, by conditions (i) and (iii), and let D_α be the image of D under $f(\alpha)$. Then conditions (i)-(iv) for α can be immediately verified.

(Case 2). If α is a limit ordinal, let $\{\beta_k\}$, $\beta_k < \alpha$, be the sequence of the remark of §2. Then by the induction hypotheses $S(\beta_k)$ holds. For each i, the maps $\tilde{h}(\beta_i, \beta_j): \tilde{D}_{\beta_i} \to \tilde{D}_{\beta_j}$ form a normal family (if they are normalized by requiring, for example, that a fixed element of $\Gamma^{\beta_i}(D_{\beta_i})$ be mapped onto the unit circle.) By the familiar diagonalization procedure we obtain a subsequence j_n. such that, for every i, $\tilde{h}(\beta_i, \beta_{j_n}) \to \tilde{h}(\beta_i, \alpha)$ uniformly on compact subsets of $\tilde{D}_{\beta_i}$. Let $h(\beta_i, \alpha)$ be the restriction of $\tilde{h}(\beta_i, \alpha)$ to D_{β_i}. Clearly, for $k > i$, by condition (i)

$$h(\beta_i, \alpha) = h(\beta_k, \alpha) \circ h(\beta_i, \beta_k) \quad . \tag{2}$$

Since $f(\beta_{j_n}) = h(\beta_i, \beta_{j_n}) \circ f(\beta_i)$ for every i, the $f(\beta_{j_n})$ converge to a to a map $f(\alpha): D \to D_\gamma$ and

$$f(\alpha) = h(\beta_i, \alpha) \circ f(\beta_i) \quad \text{for every } i. \tag{3}$$

Using the remark cf. §2 we can define for arbitrary $\gamma < \alpha$

$$h(\gamma\ \alpha) = h(\beta_\ell, \alpha) \circ h(\gamma, \beta_\ell) \tag{4}$$

where β_ℓ is such that $\gamma \leq \beta_\ell < \alpha$. The map $h(\gamma,\alpha)$ is well defined, independent of β_ℓ.

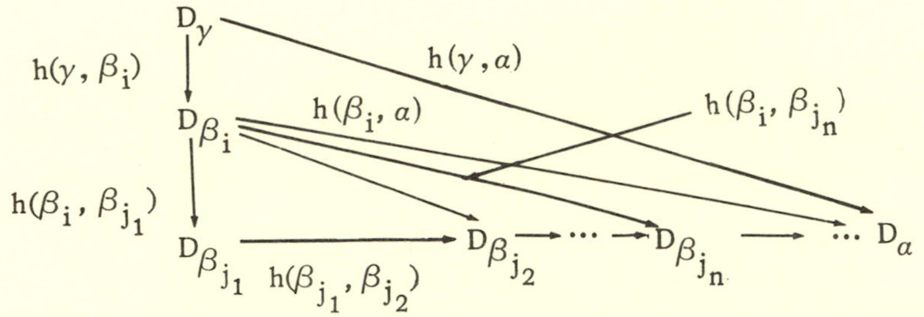

In order to complete the induction step, it suffices to show (for $\gamma \leq \delta \leq \alpha$)

(i′) $h(\delta,\alpha) \circ h(\gamma,\delta) = h(\gamma,\alpha)$

(ii′) $h(\gamma,\alpha)$ is the restriction of an isomorphism of the domain bounded by the elements of $\Gamma^{\gamma}(D_{\gamma})$.

(iii′) $f(\alpha) = h(\gamma,\alpha) \circ f(\gamma)$.

(iv′) The elements of $\Lambda^{\alpha}(D_{\alpha})$ are analytic Jordan curves.

Condition (i′) follows from (4) and (i), condition (ii′) from (4) and condition (iii′) from (4), (iii) and (3).

Since $\Gamma^{\alpha}(D_{\alpha}) = \bigcap_{\beta<\alpha} \Gamma^{\beta}(D_{\alpha})$ we have $\Lambda^{\alpha}(D_{\alpha}) = \bigcup_{\beta<\alpha} \Lambda^{\beta}(D_{\alpha})$ so that if $\gamma \in \Lambda^{\alpha}(D_{\alpha})$ then $\gamma \in \Lambda^{\beta}(D_{\alpha})$ for some $\beta < \alpha$. But $f(\beta)$ maps D onto D_{β} where $\Lambda^{\beta}(D_{\beta})$ are analytic Jordan curves and $h(\beta,\alpha)$ is, by (ii′), the restriction to D_{β} of a map which is conformal in a neighborhood of elements of $\Lambda^{\beta}(D_{\beta})$, Since $f(\alpha) = h(\beta,\alpha) \circ f(\beta)$, the elements of $\Lambda^{\beta}(D_{\alpha})$ are analytic Jordan curves. Hence an arbitrary $\gamma \in \Lambda^{\alpha}(D_{\alpha})$ is an analytic Jordan curve.

This completes the proof of the induction step and by transfinite induction the theorem follows.

From the proof we obtain the following generalization:

COROLLARY. *An arbitrary domain (countably or uncountably many boundary components) is isomorphic to a domain, all of whose boundary components except those in the perfect kernel are analytic Jordan curves.*

4. QUASICONFORMAL MAPPINGS AND THE ISOLATED COMPONENT LEMMA. In this section we recall some facts about quasiconformal mappings and present some technical lemmas to be used in the proof of Lemma 1 and in some further applications of the induction technique.

For a domain N, let B(N) be the open unit ball in $L_{\infty}(N)$. Then for $\mu \in B(N)$ a homeomorphic solution of the Beltrami equation $f_{\bar{z}} = \mu(z) f_z$ (generalized L_2 derivatives understood) is called a quasiconformal or

μ-conformal map of N. We recall [1] that for $\mu \in B(N)$ such a solution always exists and that it is unique up to composition by a conformal map. For For μ in the closure $\overline{B(N)}$ of B(N) the existence statement still holds [3] if $\mu \in B(K)$ for every K with compact closure contained in N.

Let Δ be a disk with boundary c and denote the complex sphere by $\hat{C}$. Then $\mu \in B(\Delta)$ can be extended as an element of $B(\hat{C})$ in such a way that any μ-conformal map of $\hat{C}$ maps c onto a circle. (See [1] and [17] for an explicit extension.) Using this the following is easily obtained.

LEMMA 2. *Let* N^e *be an "exterior neighborhood" of a circle* c *and let* $\mu \in B(N^e)$. *Then there exists an "interior neighborhood"* N^i *of* c *and a* $\mu^* \in B(N)$, $N = N^e \cup c \cup N^i$, *such that* $\mu^* = \mu$ *in* N_e *and any* μ^*-*conformal map of* N *maps* c *onto an analytic Jordan curve. (By an exterior (interior) neighborhood we mean a ring domain whose inner (outer) boundary component is* c.)

By the Riemann mapping theorem we obtain the

COROLLARY. *The above lemma is true for* c *an analytic Jordan curve.*

LEMMA 3. *For* a *and* h *real, let* $E = \{|\mathrm{Re}(z-a)| < h\}$, $\Delta = \{|z-a| > h\}$ *and* U *the upper half plane. Then there exists a quasiconformal map of* $E \cap U$ *onto* $(E-\Delta) \cap U$ *which is the identity on* $\{z : |\mathrm{Re}(z-a)| = h$ *and* $\mathrm{Im}\, z > 0\}$.

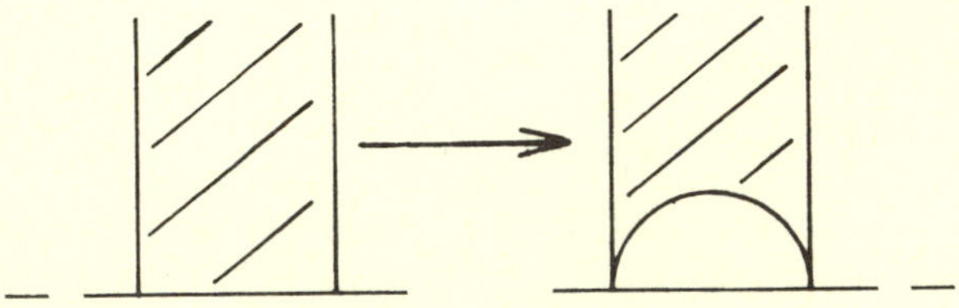

COROLLARY. *If* D *is a domain whose boundary components are slits along the real axis, then there exists a homeomorphism* f *of* D *onto a domain bounded by circles with centers on the real axis, such that* f *is quasiconformal in the exterior (with respect to* D*) of any neighborhood of the limit boundary points of* D. *(Such points occur, of course, as end points of the limit boundary slits.)*

LEMMA 4. *Let* α *be a continua and* β *a circle contained in the interior of an analytic Jordan curve* γ. *Then there exists a quasiconformal map of the ring domain bounded by* α *and* γ *onto the ring domain bounded by* β *and* γ *which leaves* γ *pointwise fixed.*

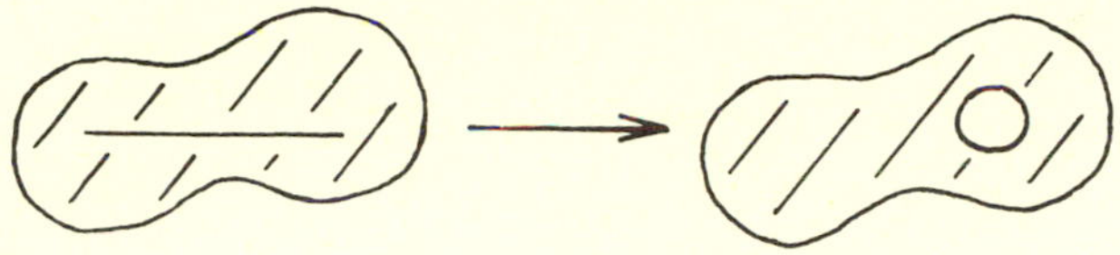

The proof of Lemma 3 is by a trivial construction. A proof of Lemma 4 may be found in [18].

We now prove Lemma 1 of §3; that *every domain* D *is isomorphic to a domain whose isolated boundary components are analytic Jordan curves.*

Proof of Lemma 1. By quasiconformal deformations (Lemma 4) in the neighborhood of each isolated boundary component of D, there exists a domain K whose isolated boundary components are circles $\{c_k\}$, and a homeo- $f: K \to D$ which is quasiconformal in the exterior (with respect to K) of any neighborhood of the limit boundary components of K. Moreover, for each c_k, there exists an exterior domain N_k^e of c_k such that $\mu \in B(N_k^e)$: By Lemma 2 applied to all the c_k there exists a $\mu^* \in \overline{B(K^*)}$, where K^* contains K as well as each c_k, such that if g is a μ^*-conformal map of K^*, $g(c_k)$ is an analytic Jordan curve. Since g and f are both μ-conformal in K, $h = g \circ f^{-1}$ is conformal in D. Then $h(D)$ $(= g(K))$ is the required domain.

Remark. The stronger result that D is isomorphic to a domain whose isolated boundary components are *circles* is given in [19]. The proof there is similar to the one above but makes use of the *full* group of reflections in the isolated (circular) boundary components of K. This result is also implicit in [18, p. 294], [20, p. 20] and [13, p. 395]. See also Note end §5.

5. DOMAINS ADMITTING SYMMETRIES. As a further example of the induction technique introduced in §3 we consider now the "Koebe problem" for symmetric domains. A domain D is said to be symmetric (of degree n) if it admits anti-holomorphic automorphisms $\sigma_1, \ldots, \sigma_n$. Using the terminology of §1 we will show that for various families $\mathfrak{D}$ of symmetric domains the subfamily $\mathfrak{S}$ of circle domains contains an element in each isomorphism class of $\mathfrak{D}$.

THEOREM 2. (Koebe) *Let* D *be a domain with countably many boundary components which admits a symmetry* σ *with* σ^2 = *identity and* D/σ *simply connected. Then* D *is isomorphic to a circle domain.*

Proof: By the Riemann mapping theorem and the reflection principle we can assume that the boundary components of D are slits on the real axis. By the Corollary to Lemma 3 there exists a circle domain E and a homeomorphism $f: E \to D$ which is quasiconformal on the complement E' of any neighborhood of the limit boundary points of E. Then $\mu(z) = f_{\bar{z}}/f_z \in B(E')$ and as is [18] (by a modification of the proof of Theorem 5 in Bers [3]) one obtains a locally quasiconformal homeomorphism $w^\mu: E \to D_0$ and hence an isomorphism $D \to D_0$. A priori, the boundary components of D_0 are as follows:

Each component is symmetric with respect to reflection in the real axis. The isolated components are circles and the limit components consist of two symmetrically situated arcs γ_1 and γ_2 of some circle, together with the "singular" continua α and β as shown. The singular continua correspond, under the isomorphism $D \to D_0$ to the end points of a limit boundary

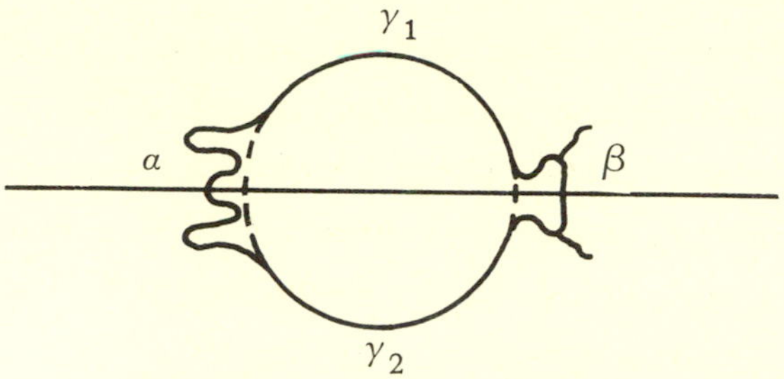

slit of D, and thus consist entirely of limit points of other boundary components of D_0.

To complete the proof we show that all singular continua are, in fact, points. This will be done by the induction method of §3. Here, however, the situation is much simpler.

Consider the following statement:

$S(\alpha)$: The elements of $\Lambda^\alpha = \Lambda^\alpha(D_0)$ are circles.

$S(0)$ is vacously true. Suppose $S(\beta)$ holds for $\beta < \alpha$.

Case 1. (α not a limit ordinal) The singular continua of the elements of Λ^α consists entirely of limit points of the elements of $\Lambda^{\alpha-1}$. But, by the induction hypothesis, these are circles (or points) which are symmetric with respect to reflection in the real axis and hence can accumulate only at *points* on the axis. Since the singular continua of Λ^α are therefore points, the Λ^α are actually circles.

Case 2: (α a limit ordinal) If $\lambda \in \Lambda^\alpha$ then, since $\Gamma^\alpha(D_0) = \bigcap_{\beta<\alpha} \Gamma^\beta(D_0)$ is equivalent to $\Lambda^\alpha(D_0) = \bigcup_{\beta<\alpha} \Lambda^\beta(D_0)$, we have that $\lambda \in \Lambda^\beta(D_0)$ for some $\beta < \alpha$ and hence that λ is a circle.

(Note: Koebe's original proof [12] does not assume that D has countably many boundary components.)

THEOREM 3. *Let* D *be a domain with one limit boundary component* κ *which is either* (i) *the origin or* (ii) *the unit circle, and whose isolated boundary components are slits along the rays* $\arg z = 2\pi k/n$, $k = 1, \ldots, n$. *Suppose that* D *is symmetric with respect to reflection in these rays. Then* D *is isomorphic to a circle domain.*

Proof: By Lemma 4, there exists a circle domain E bounded by κ and circles with centers on the above rays, and a homeomorphism $g: E \to D$ which is symmetrically defined with respect to reflection in the rays and is quasiconformal in the complement of any neighborhood of the limit boundary points (the origin in case (i) and the points $z = e^{2\pi ki/n}$ in the case (ii)). As in Theorem 2 there exists a locally quasiconformal map w^{μ} (where $\mu(z) = g_{\bar{z}}/g_z$) which maps each isolated circular boundary component of E onto a circle. By symmetry, these circles can accumulate only at points so that, again as in Theorem 2, $w^{\mu}(\kappa)$ is (i) a point or (ii) the unit circle. In either case $w^{\mu} \circ g^{-1}$ is a conformal map of D onto a circle domain.

Remark. Recall that the Teichmüller space $T(S_0)$, with respect to a base surface S_0, consists essentially of isomorphism classes of the family of all surfaces quasiconformally equivalent to S_0. (Actually one must consider marked surfaces. See [2] or [4] for details.) In [18] we showed that if a domain D was quasiconformally equivalent to a circle domain D_0 then it was conformally equivalent (i.e., isomorphic) to some circle domain. This clearly can be reformulated: *each element of* $T(D_0)$ *can be represented by a circle domain.* It follows that Theorems 2 and 3 could be generalized to domains admitting "quasisymmetries."

Note. (Added in proof, August 18, 1970). A proof of Lemma 1 can be given using only the Riemann mapping theorem and the standard theory of normal families. This eliminates the dependence upon results in the theory of quasiconformal mappings in the proof of Theorem 1 (§3) and the complete proof can thus be considered as elementary.

REFERENCES

[1] Ahlfors, L. V., and Bers. L., Riemann's mapping theorem for variable metrics, *Ann. of Math.*, 72 (1960), 385-404.

[2] Ahlfors, L. V., *Lectures on Quasiconformal Mappings.* Van Nostrand, Princeton, N.J. (1966), 146 pp.

[3] Bers, L., Uniformization by Beltrami equations, *Comm. Pure Appl. Math.*, 14 (1961), 215-228.

[4] ——— On Moduli of Riemann Surfaces, lecture notes, *Forschungs-institut für Math., ETH,* Zurich (1964).

[5] Courant, R., Manel, B., and Shiffman M., A general theorem on conformal mapping of multiply connected domain, *Proc. Nat. Ac. Sci. U.S.A.*, 26 (1940).

[6] Denneberg, R., Konforme Abbildung einer Klasse unendlich vielfach zusammenhangender schlichter Bereiche auf Kreisbereiche, Diss., *Leipziger Berichte*, 84 (1932) 331-352.

[7] Grötzsch,H., Eine Bemerkung zum Koebeschen Kreisnormierungsprinzip, *Leipziger Berichte*, 87 (1935), 337-324.

[8] Hausdorff, F., Mengenlehre, de Gruyter, Berlin, 1935; English transl., Chelsea, N.Y., 1957. MR 7, 419- MR 19, 111.

[9] Jenkins, H., *Univalent Functions and Conformal Mapping.* Sprınger, Gottingen-Heidelberg: (1958), 169 pp.

[10] Koebe, P., Uber die Uniformisierung beliebiger analytischer Kurven III, *Nachr. Ges. Wiss. Gott.* (1908), 337-358.

[11] ———, Abhandlungen zur Theorie der Konformen Abbildung: VI. Abbildung mehrfach zusammenhängender Bereiche auf Kreisbereiche, etc., *Math. Z.*, 7 (1920), 235-301.

[12] ———, Über die konforme Abbildung endlich-und unendlich-vielfach zusammenhängender symmetrischer Bereiche auf Kreisbereiche, *Acta Math.*, 43 (1922), 263-287.

[13] Meschowski, H., Uber die konforme Abbildung gewisser Bereiche von unendlich hohem Zusammenhang auf Vollkreisbereiche, I, II, *Math. Ann.* 123 (1951), 392-405; 124 (1952), 178-181.

[14] Rodin, B., and Sario, L., *Principal Functions*. Van Nostrand, Princeton, N.J. (1968), 347 pp.

[15] Sario, L., Über Riemannsche Flachen mit habbarem Rand. *Ann. Acad. Sci. Fenn. Ser. A.I.* No. 50 (1948), 79 pp.

[16] Sario, L. and Oikawa, K., *Capacity Functions*, Springer Verlag (1969).

[17] Sibner, R. J., Uniformization of symmetric Riemann surfaces by Schottky groups, *Trans. Amer. Math. Soc.*, 116 (1965), 79-85.

[18] ______, Remarks on the Koebe Kreisnormierungsproblem. *Comment, Math. Helv.* 43 (1968), 289-295. MR 37 #5380.

[19] ______, Domains bounded by analytic Jordan curves, *Bull. A.M.S.* 76 (1970), 61-63.

[20] Strebel, L., Uber das Kreisnormierungsproblem der konformen Abbildung, *Ann. Acad. Sc. Fenn.*, 101 (1951), 1-22.

[21] ______, Uber die konforme Abbildung von Gebieten unendlich hohen Zusammenhangs, *Comment. Math. Helv.*, 27 (1953), 101-127.

[22] Tsuiji, M., *Potential Theory in Modern Function Theory*. Maruzen, Tokyo (1959), 590 pp.

Rutgers University, New Brunswick, N.J.
Institute for Advanced Study, Princeton, N.J.